Robert M. Beckstead

A Survey of Medical Neuroscience

With 255 Illustrations

Springer

Robert M. Beckstead
Department of Physiology
Medical University of South Carolina
Charleston, SC 29425-2258
USA

Library of Congress Cataloging in Publication Data

Beckstead, Robert M.
    A survey of medical neuroscience/by Robert M. Beckstead
      p.   cm.
    Includes bibliographical references and index.
    ISBN 0-387-94488-5 (softcover: alk. paper)
    1. Neurophysiology. I. Title.
    [DNLM: 1. Nervous System—anatomy & histology. 2. Nervous
System—physiology. WL 102 B397s 1995]
    QP361.B36 1995
    616.8—dc20
    DNLM/DLC
    for Library of Congress   95-11272

Printed on acid-free paper.

Production managed by Princeton Editorial Associates and supervised by Karen
Phillips; manufacturing supervised by Rhea Talbert.
Typeset by Princeton Editorial Associates, Princeton, NJ.
Printed and bound by R. R. Donnelley and Sons, Harrisonburg, VA.
Printed in the United States of America.

9 8 7 6 5 4 3 2 1

ISBN 0-387-94488-5 Springer-Verlag New York Berlin Heidelberg

A Survey of Medical Neuroscience

Springer
New York
Berlin
Heidelberg
Barcelona
Budapest
Hong Kong
London
Milan
Paris
Santa Clara
Singapore
Tokyo

Nature, to be commanded,
must be obeyed.
—*Sir Francis Bacon*

For *Vilma, Noah, Jonah,* and *Tyler*

Preface

Neuroscience research is the most rapidly expanding area of study in all of modern science. The ultimate goal of the diverse theoretical and technological disciplines encompassed by neuroscience is to understand the mystery that is the human brain. This noble pursuit is of immeasurable value to mankind and an essential component of the modern biomedical enterprise.

The guarantor of all understanding, the brain is extremely hostile to observation by those who would understand it. If it is to be understood at all, that comprehension will depend on a synthesis of knowledge gained at multiple levels of analysis. Brain function must be studied from a molecular, biochemical, physiological, anatomical, pharmacological, and psychological perspective all at once. This interdisciplinary approach is a modern concept; the antecedents of neuroscience are rooted in the classical disciplines of anatomy, physiology, pharmacology, and psychology. Most neuroscientists active today were originally trained in only one of these disciplines, and, *entre nous,* tend to be a bit distrusting of the value of the others. But the majority of today's younger neuroscientists, and certainly those who will emerge from graduate programs from now on, are a new breed whose academic training is broad and interdisciplinary. Likewise, the medical neuroscience subspecialties of the past—neurology, psychiatry, neurosurgery, ophthalmology—are gradually transforming into *neuropsychiatry, neuroendocrinology,* and so on. Moreover, it is becoming increasingly clear that virtually all other systems of the body are closely regulated in critical ways by the central nervous system—they are, so to speak, supplicants of the brain. This realization has led to a melding of neuroscience with several other medical specialties, giving rise to even more incongruous and unwieldy descriptors (*psychoneuroimmunology*!). This transformation in the practice of medical neuroscience is driven, as it always has been, by the rapid progress achieved at the laboratory bench. And *the research frontier in research frontier in neuroscience moves rapidly.* The well-trained graduate and medical student can no longer afford to settle for what is *known* but must be prepared to deal with the *hypothetical* as well. She must be prepared to assimilate new concepts into a solid framework of theories of brain function. For this reason, it is as important for students to understand *how* new neuroscientific knowledge *is obtained* as it is to know up-to-the-minute *facts* about the brain. Most important of all, today's students must acquire a firm conceptual basis even though today's concepts will surely evolve exuberantly over the duration of their careers.

It is far easier to write, teach, and learn within a single discipline. There are many excellent texts of neuroanatomy or neurophysiology or molecular neurobiology available to the student, and several of them have been instrumental in the writing of this book. However, most medical schools and allied health schools in the United States and elsewhere have recognized (or soon will) the need for a neuroscience course in their first-year curriculum. The texts available for such courses are few, so despite the obvious difficulty of the task, I felt compelled to try, having spent a good deal of the past 20 years teaching the subject to graduate and medical students in large medical schools.

Realistically, a book of this size cannot embrace the full scope of the field. For instance, its treatment of the relationship between the brain and the other organ systems—the immune system or the endocrine system or the digestive system—is more implied than documented. Not described in depth are many important technical details about neurons and their interactions and about the regional anatomy of the brain. Instead, the book portrays the central nervous system. It is a portrait of the brain painted in broad strokes in hopes that the student can, in a single semester, attain an appreciation for the modern concepts that guide its study. Likewise, the brush is dipped in pigments not pure; in most chapters, function is analyzed from the molecular to the circuitry to the behavioral level and from both experimental and clinical perspectives.

In writing the text I have strived for pedestrian simplicity. I wished to write a book that was terse without being turbid, or worse still, shallow; a book that would inspire the imaginations of bright people who were not yet technicians. Having once been a freshman medical and graduate student myself, and having listened to the generously proffered criticisms of several hundred others over the years, I tried to keep ever in mind the peculiar onus of assimilating the diverse series of compressed information packets that are the basic sciences. To this end, I have tried to emphasize concepts over facts, even though most in neuroscience remain to be proven. I have weeded out technical terms and opted for English over the more euphonic Greek and Latin except in those cases where the translation becomes comical or fails entirely (as it sometimes does). I favor intuitive explanations over mathematical ones, and workaday (forgivably prosaic?) examples when warranted. The same philoso-

designed as information-rich resources that can be revisited whenever necessary. Very few of the figures in this book fit that description. Most were designed to convey but a single concept as an icon that is easily apprehended by the mind's eye. Thus, I have not crafted a source of information for reference so much as a primer that, once read, can be discarded without guilt.

In organizing the book, I assumed that students must understand the cell biology of the neurons and glia, the fundamentals of synaptic transmission, and the three-dimensional topographic anatomy of the brain in order to understand the circuits that serve particular functions. Accordingly, the book begins with a treatment of the cellular and molecular biology of neurons and glia and proceeds to an overview of the embryological emergence of central nervous system structure. The largest, central portion of the book is organized around the four functional realms of *behavior* (skeletal and visceral), *emotion, perception,* and *cognition* over which the brain has dominion. In the last chapters, I address the issues of nervous system resiliency.

Thus, although this book is but a modest introduction to the field of neuroscience, it may provide a starting point for some young minds that may go forward to solve one day the mysteries of behavior, perception, emotion, and, ultimately, the cognitive proces itself. I submit the book to students with the hope that it may serve this purpose.

Acknowledgments

I wish I could describe myself as the *author* of this book. In truth, however, I must instead admit that I am but a scribe who has penned the discoveries and ideas of a great many others accumulated over a long period of time. The most personally influential among these has been Walle J. H. Nauta, a man almost hidden in gentleness, of uncommon integrity and generosity, whose genius continues to awe and inspire me, even after his passing, every time I reread one of his many seminal papers or a passage from his eloquent monograph on the anatomy of the human brain. More immediately, each chapter of this book has been critically read by one or more of my friends and colleagues or by anonymous reviewers—an invaluable service for which I am very grateful. Those I know of are, alphabetically, Doctors Narayan Baht, John G. Blackburn, Gregory J. Cole, Robert J. Contreras, E. Earl Kicliter, Nidza Lugo-Garcia, Henry Martin III, Ralph Norgren, Jerome G. Ondo, Philip J. Privitera, Richard A. Schmiedt, Richard L. Segal, and David L Somers. Although not responsible for any of the miscues in the text, they have certainly uncovered many that would have otherwise remained. I am also deeply indebted to the many students who came forward over the years to identify my many foibles and incite me to figure out better ways to put across the concepts. Last, but not least, I must thank William Day and his associates at Springer-Verlag and Peter Strupp and the staff at Princeton Editorial Associates for their efforts to produce the book and their kind tolerance of my frequent evolutions. I am especially grateful to Jonathan Harrington for copyediting the manuscript and to Karen Phillips for her careful processing of the illustrations.

Contents

Preface . . . . *vii*

Acknowledgments . . . . *xi*

1. Neurons and Nervous Systems: First Approximations . . . 1

I. Cell Biology of Neurons and Glia 15

2. The Bioelectrical Activity of the Neuronal Membrane . . 17

3. The Life Cycle of Neurotransmitters 32

4. Synaptic Transmission 45

5. The Response Repertoire of Neurons 60

6. Glial Cells and Their Functions 72

II. Morphogenesis and Topographic Anatomy 81

7. The Basic Plan and Ontogeny of the Spinal Cord 83

8. The Ontogeny of the Brain Stem 95

9. The Ontogeny of the Forebrain 108

III. Skeletomotor Control Systems 121

10. Sensory Transduction and Coding 123

11. Primary Somatosensory Processing 131

12. The Ascending Somatosensory Systems 143

13. Arousal and Sleep 154

14. The Machinery of Movement 163

15. The Motor Nuclei and the Descending Systems 172

16. Reflexes, Rhythms, and Posture 190

17. Gaze 202

18. Volitional Movement 214

19. The Cerebellum and Movement Precision 226

20. The Basal Ganglia System and Movement Sequencing . . 238

IV. Visceral Sensorimotor Control Systems 253

21. Smell, Taste, and Viscerosensation 255

22. Visceromotor Reflexes 265

23. Hypothalamic Regulation of the Internal Environment . . 275

24. Motivation and Emotion 288

V. Special Senses and Higher Brain Functions 299

25. Hearing 301

26. Visual Transduction 314

27. Visual Processing 329

28. Perception, Cognition, and Language 341

29. Memory and Learning 356

VI. Building and Rebuilding the CNS 371

30. The Histogenesis of Nuclei and Cortices 373

31. The Formation and Modification of Connections 385

32. Damage Control and Recovery 397

Epilogue 407

Appendixes 409

A. Blood Supply and Drainage 411

B. Ten Representative Sections 414

Index 421

Neurons and Nervous Systems: First Approximations

1

For an animal species to endure, its members must consume nutrients and avoid being consumed themselves, at least until they can successfully reproduce. To accomplish these tasks, animals have to *sense* conditions in and around themselves and respond accordingly. In all animals above the phylogenetic level of the marine sponge, the capacities to sense and behave are controlled by special cells called **nerve cells** or, more commonly, **neurons,** which assemble themselves into networks called **nervous systems.** Before setting out to study the most advanced of all nervous systems, the human brain, it may be worthwhile to first take a long view of the properties of neurons and nervous systems.

The Phylogeny of Neurons and Nervous Systems

You have heard many times before that a feature common to all living cells is irritability (an inherent property of a plasma membrane riddled with ion channels) and that cells can communicate with one another either by direct physical contact of their membranes or by the release of chemical messengers. In neurons, these common cellular attributes of irritability and signaling are greatly refined, and *neurons are dedicated to the sole purpose of information transfer.*

A single-celled animal that lives independently in nature cannot logically be considered a nerve cell or any other cell "type," since it must carry out many general and specific functions all at once. Thus, it makes good sense to expect neurons and nervous systems to reside only in the multicellular fauna. The simplest of the extant multicellular animals is the sponge. Like other animal forms, it is able to respond (curl up) to an environmental stimulus (the poke of a diver's finger). But is this sensorimotor behavior mediated by neurons? Microscopic examination of

Box 1.1   Neuroanatomical Techniques, in Brief

Before embarking on a discussion of brain systems, it is important to understand how knowledge about pathways is obtained. Postmortem observations on human brains are appallingly crude compared to careful experiments in animals. Therefore, the circuitry described in this book is derived from many observations in a variety of mammalian species. How is information on circuitry obtained in animals? In all of science, discovery must often await technological breakthroughs. So it is with the study of the brain. One of the most significant breakthroughs occurred in the 1870s, when an Italian histologist named Camillo Golgi, like his forebear Columbus who sought Cipangu and instead chanced upon a New World, accidentally invented a neuron-staining technique while attempting to selectively stain the meninges that coat the brain. He applied silver nitrate and potassium dichromate to previously fixed tissue slices and found that the silver chromate would densely infiltrate whole nerve cells. Amazingly, only a few out of every hundred neurons would stain—a circumstance as fortunate as it is enigmatic, since, if all were to stain, the tissue would be black and the technique useless. Since the entire neuron with all its processes is impregnated, the dendritic geometry can be studied as well as the course and branchings of the cell's axon. It was Golgi's technique that carried Cajal on his many journeys along the meandering paths of the vertebrate brain.

In addition to the Golgi stain, early neuroanatomists had available a number of dyes that differentially stained cell bodies and myelinated axons. With such dyes, they studied the nuclear organization and major fiber tracts in the brain. For instance, basic dyes can be used to stain the rough endoplasmic reticulum (Nissl bodies) so abundant in most neuronal cell bodies. The Nissl stain is responsible for many of the named nuclei we identify today.

Working in Europe during World War II and later in the United States, Walle J. H. Nauta developed a revolutionary silver-staining technique in which only degenerating axons are marked. If a group of neurons is destroyed, the axons of those neurons will undergo a progressive degeneration and can be selectively stained by application of the Nauta technique. This method became a powerful tool for the *experimental* study of brain circuitry, and although it was soon supplanted by techniques based on axoplasmic transport (see Chapter 3), it was so expertly applied by Nauta and his many disciples that it set the standard for the modern era of neuroanatomy.

Axonal endings and cell bodies of neurons take up substances from the extracellular fluid by a process of endocytosis. If a visible substance (fluorescent dyes) or a substance that can be rendered visible (some lectins, the enzyme horseradish peroxidase, and viruses that express immunocytochemically detectable antigens) is injected near axonal endings, it will be taken up and transported retrogradely to accumulate in the cell body (see Chapter 5). Thus, one can map the distribution of neurons that send axons to the injection site. Conversely, some substances ($[^3H]$amino acids, lectins, or horseradish peroxidase) are taken up by cell bodies and transported anterogradely along their axons to the terminals. This method, then, allows one to identify the axonal projections of the injected cell group to its targets.

Finally, the behavioral contribution of neuronal cell groups and pathways can be assessed by selectively activating or destroying them and subsequently testing the animal's performance. Nuclei can be destroyed mechanically, thermoelectrically, or by the injection of selective neurotoxins. This approach, begun in earnest over a century ago by Paul Broca, has been valuable in understanding the functional organization of the human brain, where the lesions, of course, are the result of trauma or disease rather than experimental destruction.

sponge tissue reveals that the cells of its outside layer form **gap junctions** (tight membrane appositions at which there is cytoplasmic continuity between the partner cells) with contractile cells that reside in its deeper layers. These gap junctions allow electrical activity to pass rapidly between the coupled cells and represent a simple device that enables the sponge to react as it does to external stimulation. Since the behavior of the sponge is limited largely to shrinking away from forceful contact and its tissues are but a few cells thick, the gap junction is a perfectly adequate device for sensorimotor communication between its cells. However, despite their irritability and capacity for intercellular communication, none of the sponge's cells appear to be specialized exclusively for this purpose, and they fail to qualify as neurons.

The sea anemones are phylogenetically the lowest life form to possess cells that can be unequivocally identified as neurons (as defined by a special staining technique called the Golgi method; Box 1.1). The collection of neurons in the anemone, although hardly a "system," can be considered an approximation of the earliest stage in a hypothetical evolutionary emergence of nervous systems (Fig. 1.1A). The anemone's neurons are interposed between the surface cells and the muscle cells and mediate the conduction of signals from the former to the latter. The first true nervous *system* does not appear until the phylogenetic level of the jellyfishes, where it exists as a netlike array of neurons beneath the surface layer. Moreover, there is a true separation of labor within this nerve net (Fig. 1.1B). Some of the neurons detect external stimuli by way of a filamentous **process** that extends to the outer surface of the animal. Such neurons, which transduce environmental energies into electrical impulses, are called **sensory neurons.** Other neurons trigger movements by way of processes that contact the anemone's muscle cells. Call them **motor neurons** (usually contracted to **motoneurons**). How do these sensory and motor neurons communicate with one another? Instead of gap junctions, the contact between neurons is a unique combination of close membrane apposition (but without actual contact or continuity)

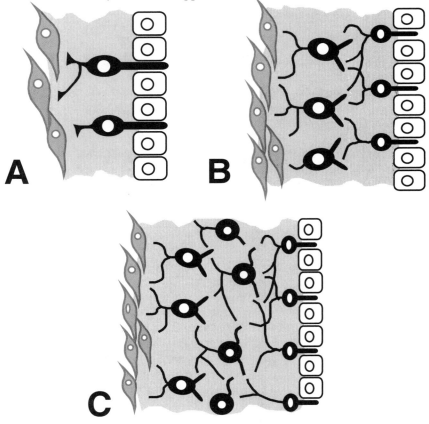

Figure 1.1 The drawings schematize a hypothetical evolution of nervous systems from a primitive one-neuron system (**A**) in which the single class of neuron serves both sensory and motor functions, to a two-neuron system (**B**) consisting of specialized sensory and motor neurons, to a three-neuron system (**C**) with the addition of the interneuronal class. In each sketch, the organism's surface cells are at the right; the deep-lying, red, spindle-shaped cells are the contractile cells; neurons are jet black.

and the secretion of a chemical messenger. We call this contact a chemical **synapse.**

The Chemical Synapse

In the late nineteenth century, the great Spanish neuroanatomist and later Nobel laureate, Santiago Ramón y Cajal, studied the organization of vertebrate nervous systems, relying mainly on a relatively neuron-specific stain discovered by his contemporary Camillo Golgi (who shared the Nobel Prize in Physiology or Medicine with Cajal; see Box 1.1). Cajal's

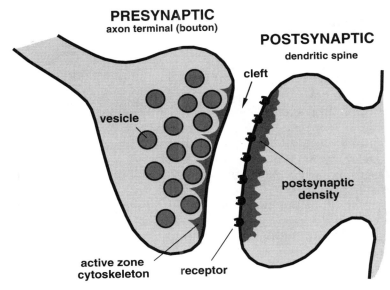

PRESYNAPTIC
axon terminal (bouton)

POSTSYNAPTIC
dendritic spine

cleft

vesicle

postsynaptic
density

active zone
cytoskeleton   receptor

Figure 1.2 The drawing schematizes the elements of a generic chemical synapse.

observations led him to expound a concept that became known as the *neuron doctrine*. It states simply that *the neuron is the anatomical and functional unit of the nervous system*—a notion that eroded the then prevalent dogma that neurons fused to form cytoplasmic continuity with one another. Instead, Cajal surmised, communication between the individual neurons must occur through close membrane appositions of their filamentous processes. Already in the primitive nervous system of the jellyfish, the connection from the sensory neurons to the motor neurons takes the form imagined by Cajal: a special junction that has been called a synapse since the term was coined by the founder of modern neurophysiology, Charles Sherrington.

The synapse is not as tight as the gap junction, and there is no cytoplasmic continuity between the partner cells (Fig. 1.2). Communication across the synapse is mediated by a chemical messenger called a **neurotransmitter** that is released by one neuron and triggers a response in another. This has two important consequences for signaling. First, whereas the gap junction can conduct bioelectrical activity equally well in either direction, *the synapse is polarized.* Information travels from the **presynaptic** neuron, which releases the neurotransmitter, to the **postsynaptic** neuron, which receives it. Second, whereas the gap junction is an all-or-none device—any activity in one cell is transmitted full-blown to its partner—*the synaptic response is graded* in proportion to the amount of chemical messenger released and the responsiveness of the postsynaptic membrane. A final important feature of chemical synaptic transmission is that the neurotransmitter can, in many cases, *inhibit* the activity of the postsynaptic membrane rather than *excite* it.

The Archetypal Neuron

Thus, through evolution, Mother Nature has designed a type of animal cell that is specialized for *rapid, long-distance, and selective intercellular signaling*—the neuron. Now that you know the biological role of neurons, you are ready for a description of an actual neuron. Alas, here we confront a serious problem: neurons are the most diverse cells in biology. They come in a wide variety of shapes and sizes and exhibit a plethora of biochemical traits. Fortunately, however, most of them conform to a *basic plan*. At this point, it serves us best to describe a generic brand of neuron as schematized in Figure 1.3, leaving the many variations on this basic plan to be elaborated later.

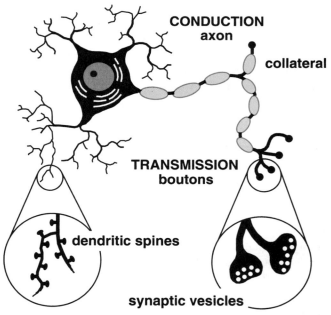

Figure 1.3 Most central neurons conform to the basic plan shown in this schematic. The lower-case labels indicate the structural elements of the neuron, and the upper-case labels indicate their functions. The red wrappings around the axon and its collateral represent the discontinuous myelin sheath.

For all neurons, one can identify a region called the **cell body** (or soma) that contains the nucleus and usually an abundance of **rough endoplasmic reticulum** (or Nissl substance). Extending away from the cell body are one or more (usually more) filamentous processes. The processes are of two types called **dendrites** and **axons.** Most neurons give rise to at least a few dendrites, which are usually shorter than axons. In contrast, only one axon emerges from the cell body, usually from a conical eminence of the soma called the **axon hillock.** The axon can extend from a few micrometers up to a meter or more in length, whereas dendrites rarely exceed a millimeter in length. Dendrites taper with distance from the cell body, whereas axons tend to retain a relatively constant diameter throughout their length. Functionally, dendrites are a receptive part of the neuron upon which synapses are formed by the axons of other neurons. The synapses often occur on special excrescences called dendritic **spines.** In contrast, the axon is specialized to conduct bioelectrical impulses called **action potentials** (or **spikes**) to distant synaptic sites. This important functional difference between dendrites and axons is a product of their distinct membrane properties. Conduction of membrane potentials by the dendrites is said to be *decremental:* when electrical activity is established at some point in the dendritic membrane, it will spread in all directions and will diminish in magnitude with distance from the site of origin and over time. This is not true for axons. If an impulse is triggered at the so-called **initial segment** of an axon (near its origin at the cell body), it will continue unabated away from the cell body along the axonal membrane to its end. Because of this self-propagating quality of the action potential, it is described as *nondecremental* conduction.

Another conspicuous difference between many axons and dendrites is that axons often have a special wrapping called a **myelin sheath** that is formed by cells that are themselves not neurons. They are a type of cell called **glia** (Chapter 6). The myelin sheath serves to increase the speed of impulse conduction along the axon and to insulate the electrical activity of the axon (Chapter 2). Dendrites never have this wrapping.

Incidentally, both dendrites and axons depend on the cell body for the replenishment of the many proteins that turn over during synaptic transmission. For this reason, there is usually a well-developed protein synthetic apparatus, typified by an abundance of rough endoplasmic reticulum (Nissl substance), in the cell body and a well-organized cytoskeleton that extends throughout the neuronal processes. The delivery of raw materials to the distant ends of axons and dendrites is a microtubule-dependent process called **axoplasmic transport.** Axons that are cut off from their parent cell bodies do not survive for very long.

The classic route of signal flow in the neuron is from the dendrites and cell body to the axon initial segment, where an impulse is established and **conducted** along the axon to the presynaptic ending. The arriving impulse causes the release of a chemical messenger, the neurotransmitter, from the axon terminal. This release is an active process of exocytosis—the Ca^{2+}-dependent fusion of neurotransmitter-containing **vesicles** with the presynaptic membrane (Fig. 1.2). The interior of the presynaptic membrane is associated with a cytoskeletal specialization that is important for vesicle release. This region of vesicle exocytosis is called the **active zone** of the axon terminal. The newly released neurotransmitter diffuses through the extracellular fluid of the narrow synaptic **cleft** to the surface of the postsynaptic neuron, where it interacts with **receptor** proteins embedded in the postsynaptic membrane. Depending on the nature of the receptor, the postsynaptic response can be either an increase (**excitation**) or decrease (**inhibition**) in the membrane's excitability state. This process of neurotransmitter release, diffusion, and activation of a postsynaptic membrane is called **synaptic transmission.** The binding of neurotransmitter molecules to their postsynaptic receptors is transient; the influence of the neurotransmitter on the postsynaptic membrane is halted when the neurotransmitter diffuses out of the cleft, is enzymatically degraded, or is taken up either back into the presynaptic terminal or by a type of glial cell called an **astrocyte** (Chapter 6).

The significance of this phenomenon of synaptic transmission cannot be overstated. When we learn and remember, when we forget, and when we experience derailments of brain function, it is because of changed synaptic transmission. The great frontier in brain science lies at the molecular biology of synaptic transmission. The ultimate question is how molecular changes in synaptic signaling give rise to altered information transfer and behavior.

As stated above, there is no shortage of variation on the basic plan of the neuron. The cell bodies of neurons can range in size from a few cubic micrometers up to volumes that are discernible by the unaided eye. Some neurons have no dendrites at all, whereas others have enormously branched dendritic trees. The axons of some neurons extend no more than a few micrometers from the cell body, whereas the axons of others reach lengths that exceed a meter. Although axons most often synapse either on dendrites or on cell bodies, it is not uncommon to find synaptic contacts that occur from axon to axon or even from dendrite to dendrite.

Rudiments of Synaptic Connections

Before going on, let's consider some important consequences of a nervous system with two classes of neuron, sensory and motor, which communicate by way of chemical synapses. In such a system, there can be some degree of **convergence** (several sensory neurons form synaptic contacts on a single motoneuron) or **divergence** (one sensory neuron contacts several motoneurons) (Fig. 1.4). The synapses between the sensory and motor neurons need not all be excitatory. In some cases, they can inhibit the activity of the motor neurons. As a result, there occurs an *algebraic summing* of excitation and inhibition by the motoneurons in the system. Finally, not all synapses are created equal; they can be more or

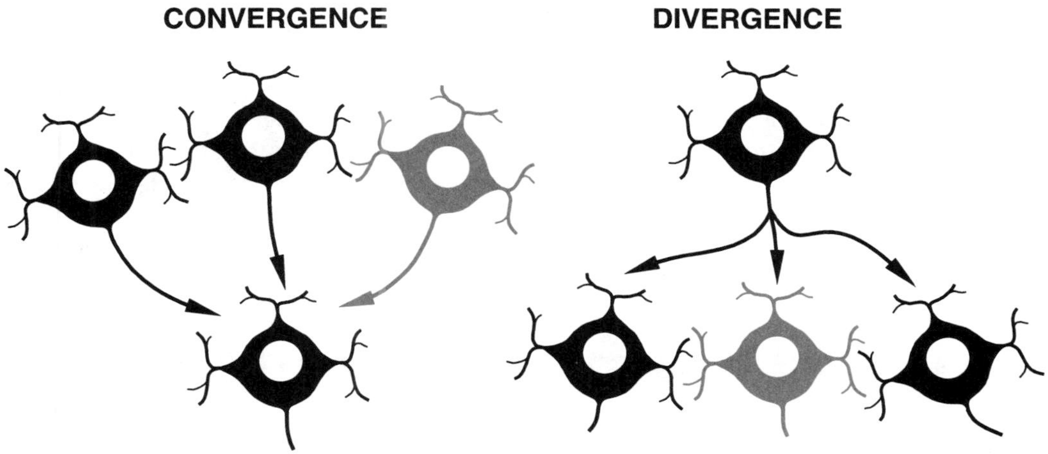

CONVERGENCE   DIVERGENCE

Figure 1.4 Convergence and divergence are two fundamental principles of synaptic connections and signal integration. Although only a handful of neurons are depicted, in reality single neurons of the human CNS can receive synaptic input from well in excess of a thousand other neurons. As a rule, both excitatory (black) and inhibitory (red) inputs converge on individual neurons.

less **potent** in eliciting a bioelectric response. All of these features lead to a degree of *variability* in the response of the system to a given sensory input.

Suppose now that a third class of neuron is added to the system, one that is neither strictly sensory nor motor, but instead is interposed between the two. This important event in the evolution of nervous systems appears first in the more highly evolved jellyfishes and in the mollusks where, interposed between the sensory and motor neurons, sit the intermediate neurons or **interneurons** (Fig. 1.1C). In this system, the signaling from sensory to motor neurons is indirect, mediated by the network of interneurons. In a sense, the addition of this third class of neuron is the most important step in the evolution of nervous systems. From here on, Mother Nature increases the potential for interaction and response variability by simply increasing the number of interneurons. The increasing numbers of interneurons are assembled into increasingly specialized circuits and subcircuits in which the fundamental rules of convergence–divergence, excitation–inhibition, and differential synaptic potency remain in effect. In addition, as their numbers increase, the neurons become increasingly selective about which other neurons they contact. This, in turn, leads to an ever-increasing degree of spatial organization within the nervous system.

The remaining evolutionary alterations that occur are mainly for economy of packaging increasingly large and complex nervous systems. Beginning with the flatworms, higher organisms are polarized—they have head and tail ends. In such creatures, the interneurons tend to cluster first into aggregates we call **ganglia** and later into a centralized nervous system. As we ascend the phylogenetic ladder, the leading end of the central nervous system expands rapidly in size to accommodate the increasing numbers of interneurons, a process called **encephalization.** And so it goes until we reach the pinnacle of nervous system evolution where we find the massive and elaborate human brain. Obviously, with the increasing body dimensions of higher animals, neurons are compelled to make synaptic contacts at great distances. They accomplish this by extending long axons that conduct their bioelectrical activity to the next neuron in line.

The Human Central Nervous System: Elementary Observations

Let us take a first look at the formidable human central nervous system (CNS) to establish some elementary organizational features. In the human, the CNS consists of a large **brain** that sits at the upper end of a long, narrow **spinal cord.** As in the lowly mollusk, however, there are just three broad categories of neuron. As schematized in Figure 1.5, the sensory neurons carry information from the periphery to the CNS. *The cell bodies of the sensory neurons* (with a few exceptions) *reside outside the CNS in clusters called* **ganglia.** *The axons of the sensory neurons enter the CNS to synapse on central neurons. The motoneuron cell bodies reside within the CNS and send their axons out to the periphery,* where they innervate muscles or synapse in ganglia with other neurons that in turn innervate muscles or secretory glands. After tallying the sensory and motor neurons, we are left with the interneurons, which make up the rest of the CNS.

Now we confront a challenge to comprehension: a realistic estimate of the total number of neurons in the human CNS is 10^{12}. To make matters

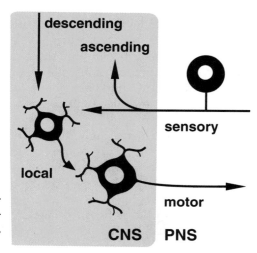

Figure 1.5 General descriptors are used frequently in reference to axonal projections. Abbreviations: CNS, central nervous system; PNS, peripheral nervous system.

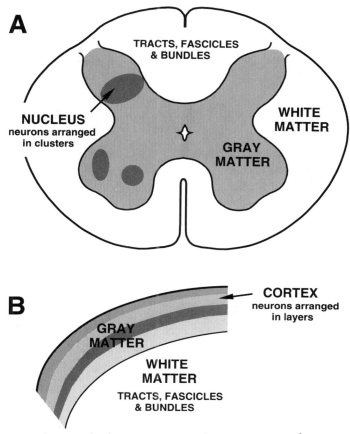

A TRACTS, FASCICLES & BUNDLES

NUCLEUS
neurons arranged
in clusters

WHITE MATTER

GRAY MATTER

B **CORTEX**
neurons arranged
in layers

GRAY MATTER

WHITE MATTER

TRACTS, FASCICLES
& BUNDLES

Figure 1.6 Gray and white matter constitute two general compartments found in the mammalian CNS. Sketch **A** shows a cross section through the human spinal cord. Sketch **B** represents a small slice of the layered cerebral cortex cut through its depth.

still more difficult, only about two to three million of these are motoneurons; *more than 99.99% of the neurons in the human CNS are interneurons.* Although the number of synaptic inputs varies widely from one neuron to the next, an average of about one thousand per neuron is a reasonable estimate. This places the total number of synaptic contacts between neurons of the human CNS at 10^{15}. The only other scientific discipline that has to deal with so many zeros is astral physics. It remains to be seen how much we must know about each of these synaptic connections in order to comprehend the workings of the human brain.

The least intimidating place to begin an exploration of the CNS is where it appears at its simplest: the spinal cord. If you examine a cross section through the spinal cord and the tissue is fresh, you can readily distinguish two regions. As drawn in Figure 1.6, you will see a central, butterfly-shaped zone of grayish material embedded in a surrounding

FIBER
CAPSULE

Figure 1.7 Nuclei can be identified on the basis of the size, shape, and packing density of their resident neurons. To the left, in this drawing of neuronal cell bodies from a region of the brain, is a nucleus of small, densely packed cells that is separated by a fiber capsule from a more diffuse aggregate of larger, spindle-shaped neurons. To the far right is another group of neurons of varied shapes and sizes.

matrix of glistening white material. The butterfly has a high density of cell bodies that give it that gray tinge. We call it **gray matter.** There are some motoneurons here, but the number of interneurons is already legion. The surround appears white because it contains an abundance of lipid-rich myelin ensheathing tens of thousands of axons. Call it **white matter.**

The cell bodies of neurons in the CNS are often collected into groups that we call **nuclei.** There are several nuclei here in the gray matter of the spinal cord, though you cannot see them without special staining (Fig. 1.6). Nuclei are rather like the ethnic enclaves typical of large American cities. That is, the neurons within a nucleus tend to have a lot in common with one another and differ from the neurons in neighboring nuclei.

Today, nuclei are defined by a multiplicity of morphological, functional, and biochemical criteria. The oldest and simplest means of defining nuclei are based on architectural considerations (Fig. 1.7). Thus, a nucleus is a cluster of one or a few types of neuron with a similar size and shape. Such clusters are often more or less distinguishable from the surrounding tissue because of the distinct morphology of their constituent neurons and (or) a difference in their packing density. Sometimes nuclei are set off clearly from the surrounding tissue by a fiber bundle or capsule. Many studies have shown these morphologically identifiable nuclei also have unique sets of **afferent** (incoming) and **efferent** (outgoing) connections associated with particular functional domains (e.g., visual sensation, jaw movement, vasodilation). The neurons within a nucleus also have a common and sometimes quite distinctive neurotransmitter complement and typically express the same set of receptors for neurotransmitters. Thus, *nuclei are phenotypic and functional enclaves within the CNS.*

Neurons can also be arranged into a layered sheet called a **cortex** (Fig. 1.6). No such sheets are present in the spinal cord, but in the brain, cortices are a common form of gray matter (e.g., the cerebral cortex).

In the small sampling of white matter that appears in a spinal cord cross section, a few thousand of the axons, the sensory ones, have entered the CNS from the periphery, while still fewer, the motoneuron axons, are on their way out of the CNS. Most of the axons in the white matter arise from cell bodies in widespread nuclei of the CNS and extend to other equally disparate parts. Some **ascend** to make synaptic contacts on neurons at higher levels of the spinal cord and in the brain, while others **descend** to synapse on cells below (Fig. 1.5). Axons of a common origin and destination often travel together in **bundles** (or **fascicles**) we call **tracts.**

In the chapters ahead you will read much about the anatomical and functional organization of the various cell groups and axonal pathways of the brain and spinal cord and how they relate to sensory, motor, and higher mental functions. The discourse will begin with a detailed analysis of the cell biology of neurons and glia so that you understand the principles of their workings. After you gain familiarity with these cellular elements, you will be introduced to the general topographic anatomy of the human CNS. Since understanding the structure of the mature brain is often facilitated by an examination of its morphogenesis, this portrait of the brain will be painted in the medium of embryology. Next, you will embark on an integrative, multilevel analysis of the functional systems of the brain that enable sensation and perception, movement, emotion, and the higher functions such as language, cognition, and learning.

Detailed Reviews

Bullock TH, Horridge GA. 1965. *Structure and Function of the Nervous System of Invertebrates.* WH Freeman, New York.
Stevens CF. 1979. The Neuron. *Sci Am* 241: 148.

CELL BIOLOGY OF NEURONS AND GLIA

I

If a nervous system is to serve its owner effectively, it must fulfill three criteria. First, many events in an animal's environment, such as the approach of a predator or prey, demand a rapid response. For a nervous system to mediate such responses, the information transfer between its component neurons must necessarily be rapid, on a time scale of milliseconds. A second challenge confronting nervous systems, especially in larger animals, is the necessity to signal over distances that are many hundreds or even thousands of times greater than the dimensions of a typical eukaryotic cell. Importantly, this capacity for long-distance signaling cannot come at the expense of speed. Last, most animals that inhabit complex environments exhibit discrete, patterned responses of individuated effectors (limbs and fingers) in three-dimensional space. This finer behavioral control is achieved by a high degree of spatiotemporal order in the signaling between the neurons within the nervous system. Thus, the biological role of a nervous system is *to process and transfer information with spatiotemporal specificity over relatively long distances at high speed.* Neurons are designed precisely to fulfil this role.

In the chapters that follow, we will examine this design and how it enables neurons to function appropriately as purveyors of information. In higher nervous systems, such as the human brain, the neurons are closely assisted by the second category of cells called glia. We'll examine how glial cells facilitate the data-processing functions of neurons.

The Bioelectrical Activity of the Neuronal Membrane

2

In all tissues of the human body, the cytoplasm of the constituent cells has an ionic composition very different from that of the extracellular fluid. The plasma membranes of the cells are adorned with an assortment of regulated ion channels and pumps that control this differential distribution of ions. As a result of these unbalanced ion distributions, the plasma membranes of all living cells exhibit a bioelectric charge that represents a form of potential energy that the cells can exploit for various purposes. Neurons have greatly refined this universal feature of plasma membranes and bent it to the task of information processing.

The Resting Membrane Potential

If two recording microelectrodes (see Box 2.1) are placed into the extracellular fluid near a live but quiescent neuron and connected to an amplifier and oscilloscope, the potential difference between the two electrodes will be manifest on the oscilloscope screen as a flat line at zero. If one of the electrodes is now gingerly advanced through the membrane of the neuron so that its tip rests in the cell's cytoplasm, the oscilloscope will immediately register a steady-state potential difference between the two electrodes; the cytoplasmic electrode will be some 70 millivolts *negative* with respect to the extracellular electrode. This potential difference across the neuronal membrane is called the **resting membrane potential.** *Much of this potential difference exists because the proportions of charged molecules (ions) on the two sides of the membrane are different* (Table 2.1). The inside of the neuron is chock-full of large, negatively charged biological molecules held captive by the membrane. Very few large anionic molecules are present in the extracellular fluid. In addition, the concentration of Na^+ and Cl^- is about tenfold higher in the extracellular fluid than in the cytoplasm, whereas the concentration of

17

Box 2.1 Recording Techniques, in Brief

Knowledge about the brain must come by proxy, from instruments of special capability. One such instrument is the **microelectrode.** Information on the functional nature of neurons and their interconnections can be obtained *electrophysiologically* by recording the electrical correlates of their behavior with glass micropipettes that are heated and pulled to finely tapered tips and filled with a conductive buffer solution. By connecting such microelectrodes to an amplifier and oscilloscope, the electrical activity of neurons can be recorded while peripheral sensory systems or central nuclei are being stimulated, while movements are being executed, or while drugs are being applied. Electrophysiology has its roots in the work of Charles Sherrington in the early part of the twentieth century. Today it has evolved to such a high level of sophistication that we can monitor the ionic current flow through a single channel in the plasma membrane of a neuron!

Microelectrodes can be placed in the vicinity of the neurons to be studied, or they can be inserted through the membrane of single neurons or even axon terminals if the cellular elements are large enough to accommodate them. To prevent nonlinear membrane responses such as action potentials from occurring, the membrane potential is often **voltage clamped** by connecting the electrode to a feedback current source. The experimenter sets the voltage level at the desired value, and the feedback current source injects enough current to balance any spontaneous or induced currents that occur due to channel openings. The injected current becomes a direct, albeit inverse, measure of naturally occurring or experimentally induced membrane current at any given point in time. This voltage clamp method allows the experimenter, for instance, to determine whether an applied stimulation causes current to flow into (excitatory) or out of (inhibitory) a neuron.

In the most recent refinement of microelectrode recording, small areas of membrane are electrically isolated from the rest of the neuron by using a technique called a **patch clamp.** Special micropipettes with tiny (0.5 micrometer) fire-polished tips are placed against the membrane, and gentle suction is used to form a tight, gigohm seal. The patch can even be pulled free of the cell so that the solutions on the two sides of the membrane can be precisely controlled. Such a patch clamp, in conjunction with voltage clamping and statistical noise analysis, enables experimenters to examine the behavior of single channels in response to, for example, the application of a neurotransmitter or drug.

K^+ is higher inside the neuron than outside. Although they are of a far lower concentration than the monovalent ions, several divalent cations, like Ca^{2+}, Mg^{2+}, and Zn^{2+}, are present in the extracellular compartment. *These differential ion concentrations are maintained because the resting membrane is both stubbornly resistant to the passage of these ions and equipped with a large number of energy-dependent protein pumps—the Na^+-K^+ ATPases—that*

Table 2.1 The approximate concentrations (in millimoles) and equilibrium potentials (in millivolts) of the major ionic species on the two sides of the neuronal membrane in mammalian brain

Ion	Cytosol	Extracellular Fluid	E_{ion}
Na^+	20	180	+65
K^+	420	25	−85
Cl^-	15	150	−70
An^-	370	—	—

constantly pump Na^+ out of and K^+ into the cell with a stoichiometry of three Na^+ ions to two K^+ ions.

As a result of the separation of charge across the membrane, an *electromagnetic driving force* exists that would propel cations into the cell if free passage through the membrane were allowed. Superimposed on this force is a *chemical driving force* created by the asymmetric distribution of major ion constituents on the two sides of the membrane. Thus, the membrane can be viewed as a closed thermodynamic system, which, like all such systems, is subject to entropy. If tiny holes were opened in the membrane so that the small monovalent ions had free passage, they would move across the membrane. *Whether and in which direction a species of ion would move across the membrane is a function of the combined electromagnetic and chemical driving forces.* An ion species will move one way or the other until it reaches *electrochemical equilibrium,* the point at which the electromagnetic and chemical driving forces are precisely balanced.

The electrochemical equilibrium potential for each of the three major ion species separated by the neuronal membrane can be calculated using the Nernst equation (Box 2.2). This equation specifies the electrical potential across the membrane that would exactly balance the driving force due to the concentration gradient. As it happens, the calculated equilibrium potentials for K^+ (about −85 millivolts) and Cl^- (about −70 millivolts) are near or at our measured resting potential of −70 millivolts. Thus, if pores were opened through the membrane that selectively allowed K^+ or Cl^- to pass, only a minor flux, if any, of these ions would occur. In contrast, Na^+ is far from its electrochemical equilibrium of about +65 millivolts; both the chemical and the electrical driving forces would send Na^+ ions rushing into the cell if selective pores were suddenly made available. Unchecked, this Na^+ flux would continue until the membrane potential reached some +65 millivolts, where the electrical and chemical driving forces would achieve a balance.

Now it so happens that although quiescent neuronal membranes are impermeable to Cl^- and allow only a negligible amount of Na^+ to move across, they are more leaky to K^+ ions. *This higher permeability to K^+ has a*

Box 2.2   Useful Biophysical Equations

Much of what we know about membrane potentials is based on formulas derived about mid-century by several biophysicists using mainly squid giant axon and frog muscle preparations. The term *membrane potential* refers to the voltage (V) across the membrane. Its instantaneous level is controlled by the resistance (R) of the membrane to ionic current flow (I) of the three monovalent ion species unequally distributed on the two sides of the membrane. Therefore, the membrane potential must obey Ohm's law: $V = IR$. Since it is sometimes preferable to think about membranes in terms of conductance (g), which is equal to $1/R$, Ohm's law can be rewritten as $V = I/g$.

The Nernst equation allows us to calculate the potential (E_{ion}) at which a given ion species is at equilibrium. This equilibrium potential can be compared to the resting membrane potential to see how much of an electrochemical driving force is able to influence the ion species.

$$E_{ion} = (RT/ZF) \times \ln ([ION]_o / [ION]_i)$$

The gas constant (R), temperature (T), and faraday (F) are relatively stable in the living mammalian CNS. The valence (Z) is +1 for Na^+ and K^+, and $_-1$ for Cl^-. Thus, the important variable in the equation is the ion *concentration* on the outside (o) and inside (i) of the neuron.

The Goldman–Hodgkin–Katz equation predicts the membrane potential (E_m) that would result from a specific set of *permeabilities* (P) of the membrane for the three major ion species.

$$E_m = (RT/F) \times \ln \frac{P_K[K^+]_o + P_{Na}[Na^+]_o + P_{Cl}[Cl^-]_i}{P_K[K^+]_i + P_{Na}[Na^+]_i + P_{Cl}[Cl^-]_o}$$

The measured time constant (τ) for a potential change across a membrane is defined as the time it takes for the potential to fall to 37% ($1/e$) of its resting value. It is computed by the formula $\tau = RC$, where R is membrane resistance and C is membrane capacitance.

The exponential decay of the membrane potential with distance is computed by the formula $\lambda = R_m/R_a$ (the membrane resistance over the axoplasmic resistance).

significant influence on the resting membrane potential. The Goldman–Hodgkin–Katz equation (Box 2.2), which predicts the potential difference across the membrane that results from different permeabilities to the three ion species, allows us to estimate the resting membrane potential. Because the resting membrane is somewhat permeable to K^+ and almost impermeable to Na^+ and Cl^-, the resting membrane potential tends to hover near the Nernst equilibrium potential of K^+. It is not precisely at the K^+ equilibrium potential, because the K^+ permeability is not absolute. Moreover, the membrane is not absolutely impermeable to Na^+ at rest; the slight permeability to Na^+ keeps the resting potential a bit less polarized than it would otherwise be. Since the Na^+-K^+ ATPase pumps three cations out for every two it pumps into the cell, it also contributes slightly (about 1 or 2 millivolts) to the resting potential. Thus, it should be evident that *the resting membrane potential of a neuron is caused mainly by the uneven distribution of charged molecules on the two sides of the plasma membrane and the relatively higher permeability of the membrane to K^+ ions.*

Neuronal Membranes Are Restless

The potential across neuronal membranes is rarely at rest. It bounces up and down in predictable ways according to incoming synaptic activity. Let's analyze the ionic basis for such fluctuations in the membrane potential. Using the GHK equation, we can see that if the membrane permeability to K^+ were to increase, the membrane potential would become slightly more negative inside (closer to the Nernst equilibrium of –85 millivolts) due to the increased net movement of K^+ ions out of the cell. We call this condition a **hyperpolarization** of the neuronal membrane. Since the equilibrium potential for K^+ is only slightly more negative than the resting potential, hyperpolarizations are always of modest amplitude. If, on the other hand, the membrane were to become more permeable to Na^+, the membrane potential would move toward 0 millivolts or **depolarize.** The magnitude of the depolarization is proportional to the amount of Na^+ influx. Obviously, since the equilibrium potential for Na^+ is around +65 millivolts, a Na^+-mediated potential change can be large, even to the point of repolarizing the membrane in the opposite direction. Notably, if the membrane were to become more permeable to Cl^-, no potential change would occur at all, since Cl^- is at equilibrium at our resting potential of –70 millivolts. As we shall see in Chapter 4, however, increased Cl^- permeability has a great dampening effect on membrane *excitability* since it can, up to a point, compensate for any concurrent increases in Na^+ permeability.

Membrane Permeability Is Mediated by Ion Channel Proteins

Neuronal membranes are studded with pores called **ion channels** that are relatively selective for the passage of different species of ion. These pores are made of proteins that span the lipid bilayer of the membrane and provide an aqueous lumen through which the ions can flow. The

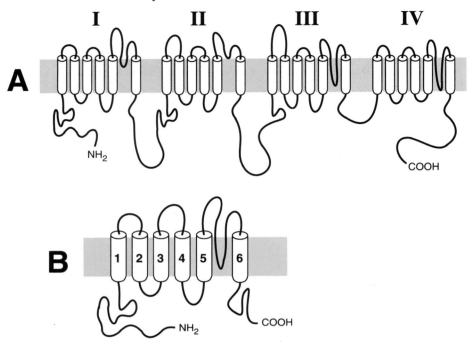

Figure 2.1   The structure of the voltage-gated Na^+ channel is diagrammed in sketch **A.** It is a single polypeptide with four membrane-related domains, each of which contains six membrane-spanning sequences (cylinders). One of the polypeptides that combine together to form the voltage-sensitive K^+ channel is diagrammed in sketch **B.** A functional K^+ channel is made up of four to six such subunits.

shape of the channel protein (called its **conformation**) can change so that the lumen (called the **ionopore**) is either open or closed. For many types of ion channel, *the conformation it assumes at a given instant in time is highly regulated by the membrane potential.* Such channels are said to be **voltage-gated** because depolarization of the membrane from the resting level increases the probability that they will assume the open conformation. Newly acquired molecular data about the structure of the Na^+ and K^+ channel proteins suggest a model of how they are able to conduct ions selectively in response to voltage shifts.

The voltage-gated Na^+ channel is a single large polypeptide with four internally homologous domains (Fig. 2.1). Each domain contains six segments that extend through the thickness of the cell's plasma membrane. These membrane-spanning segments are connected by extracellular and cytoplasmic loops of the polypeptide. The first four membrane-spanning segments have an α-helical structure (something akin to a coil spring), and all but the fourth are hydrophobic (or lipophilic) so that they meld comfortably with the lipid bilayer. The fourth segment has an amphipathic structure (having both hydrophilic and lipophilic properties) in which every third amino acid is a positively charged arginine or lysine. This region is thought to function as the voltage sensor. A plausible model of the channel has the four major domains arranged in the membrane such that the nonhelical fifth and sixth transmembrane regions of each form the walls of the ionopore. A depolarization of the membrane affects the charge relationships in the fourth membrane-spanning segments of each domain in such a way that each undergoes a slight change in shape—a little twist, perhaps. The shape change is transmitted to the membrane-spanning segments that line the aqueous pore so that the ionopore will now allow Na^+ ions to pass. Usually, there are two smaller polypeptides (called β1 and β2 subunits) associated with the Na^+ channel polypeptide (itself referred to as the α subunit to distinguish it from its associates). The function of these associated proteins is unknown.

Although we do not know exactly how the parts of the polypeptide interact with each other and with the local membrane milieu, we do know from electrophysiological recording experiments (Box 2.1) that the Na^+ channel can assume different conformations that correlate with three distinct functional states: (1) closed and voltage-sensitive, (2) opened, or (3) closed and inactivated. Each Na^+ channel randomly fluctuates *in sequence* through these states. *A membrane depolarization increases the probability of Na^+ channel opening so that more channels are likely to be open at the same time. Having opened, the Na^+ channel is committed to cycle through the closed-inactivated state, during which it will not respond to a voltage shift, before it can return to a closed, voltage-sensitive state.* As we shall soon see, these functional states of the Na^+ channel are essential to the proper responses of the neuronal membrane.

There are several types of voltage-gated K^+ channels, with different kinetic properties. However, they are all homo- or heteromultimeric proteins in which each subunit exhibits six transmembrane domains (Fig. 2.1). Like the transmembrane domains of the Na^+ channel, the fourth α-helix is amphipathic and may function as a voltage sensor, whereas the fifth and sixth nonhelical domains form the ionopore. Importantly, *K^+ channels respond more slowly to voltage changes than do Na^+ channels and, once activated, remain open longer.* Moreover, *they do not experience an inactivated state.* It is important to keep in mind these differences in the behavior of voltage-gated Na^+ and K^+ channels; they figure prominently in the initiation and conduction of action potentials, which will be discussed below.

Voltage-Gated Channels Are Selective for Particular Ions

We know that voltage-gated Na^+ and K^+ channels are highly selective for the particular cation they pass. How does a channel act as a selectivity filter? To answer this question, we must appreciate certain physical features of ions in an aqueous solution. With regard to cations, the smaller the ion, the stronger its hydration energy. This is presumably because the negative polar part of the water molecules can get closer to the positively charged center of the ion. As a consequence, it is more difficult for Na^+ to shed its associated water molecules than it is for K^+. Even though a potassium atom is larger than a sodium atom, the hydration *shell* of Na^+ makes its *ionic size* greater than that of K^+. We might reasonably surmise that a channel pore that is relatively large and lined with positive charges will allow Na^+ to enter and shed its associated water molecules. A channel pore that is smaller and minimally charged will not favor Na^+ passage, but will allow K^+ with its smaller ionic size and weaker hydration to pass. Although hypothetical, these physical properties of K^+ and Na^+ channels are consistent with biophysical data on their behavior and offer a plausible explanation of channel selectivity.

Current Flow and the Spread of Depolarization

Now let's consider a hypothetical patch of membrane, like the one depicted in Figure 2.2, of, say, a few square millimeters in area, that separates an ionically normal cytoplasmic compartment from a normal extracellular fluid compartment. Voltage-gated channels are sparse in this patch of membrane, except for a few square micrometers near its center, where a cluster of voltage-gated Na^+ channels resides. If we briefly depolarize the membrane by injecting a small amount of current through a microelectrode (in other words, mimic the current injection that occurs physiologically during a synaptic transmission event), the Na^+ channels will be induced to open for a few microseconds and a small net movement of ions into the cell will occur—a positive inward current—that depolarizes the membrane in this central region. Let's analyze this ion-based depolarization (Fig. 2.3A). The lipid bilayer itself is a thin, poorly conducting membrane that lies between two highly conductive compartments. This property defines the membrane in electrical terms as a charged capacitor. The opening of ion channels converts the membrane into a better conductor, and like all conductors, it displays some finite resistance to current flow. Thus, the membrane can be viewed in terms of electrical equivalence as a resistor in parallel with a capacitor. The opening of Na^+ channels constitutes a decrease in the resistance of the membrane, and a *resistive current* (of ions) flows across it from the extracellular to the cytoplasmic compartment. The resulting depolarization of the membrane causes the capacitor to discharge, resulting in a *capacitative current* that is also carried by the movement of charged ions toward and away from the two sides of the membrane. Since the charge of the capacitor (membrane) is equal to its capacitance (a constant) times

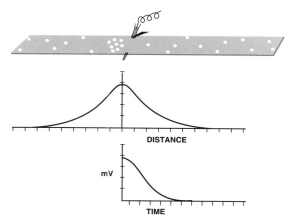

DISTANCE

mV

TIME

Figure 2.2 A recording electrode is placed in a hypothetical sheet of membrane near a patch of voltage-gated Na^+ channels (white holes). The passive spread of depolarization along the membrane (length constant) and its duration (time constant) at the patch are charted after being induced by a brief, artificial, inward current near the patch.

the voltage (membrane potential), a change in the charge across the capacitor leads to a change in voltage across the membrane—in this case of inward current, a depolarization. From knowledge of the amount of capacitor discharge needed to produce a given change in voltage and the number of ions required to produce a discharge, one can calculate that *only a small fraction* (in the range of 10^{-18} moles) *of the external Na^+ ions need enter the cell to produce a significant* (say, tens of millivolts) *shift in membrane potential due to the capacitative current.* Thus, the neuronal membrane is exquisitely sensitive to small ion fluxes. Moreover, the ionic concentration on the two sides of the membrane can remain relatively stable despite frequent depolarizations and repolarizations of the membrane potential.

At this juncture, we can ask two relevant questions about our hypothetical patch of membrane. *How long will a depolarization last?* and *How far along the membrane will it spread?* To answer these questions, we must consider the passive electrical properties of the membrane. Like any capacitor, the membrane exhibits time and length constants (Box 2.2) for current conduction such that the amplitude of the capacitative current will decay exponentially over time as the capacitor approaches its new equilibrium charge, and over distance away from the site at which the current is initiated (Fig. 2.2). If our cluster of Na^+ channels were induced to open a second time before the first capacitative current had diminished to zero, the newly generated current would algebraically sum with the residual current, leading to an increase in membrane depolarization. Since the neuronal membrane has a uniform thickness and lipid composition, it will have consistent capacitative properties everywhere that it

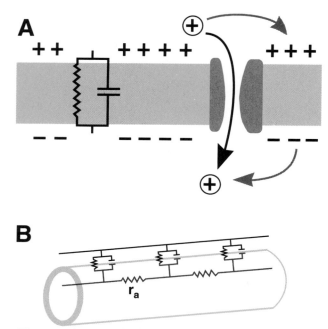

Figure 2.3 The membrane (shaded red) behaves as if it were a resistor and capacitor wired in parallel (**A**). When an ion channel opens (e.g., to Na^+), a *resistive* (ionic) current will flow (black arrow). This will discharge the capacitor, causing a *capacitative* current to flow (gray arrow). The cytoplasm also behaves like a weak resistor (r_a) to ionic current flow (**B**). As a result, ionic current will travel through the cytoplasm, decaying with distance.

separates cytoplasm from extracellular fluid, and there will be little variability in the time constant from one patch of membrane to the next. *The length constant, however, is a function of cytoplasmic and extracellular resistance as well as membrane resistance and will vary dramatically from one part of the neuron to the next, depending on the local geometry.* To illustrate this point, let's roll up our patch of membrane into a tube filled with cytoplasm and bathed in extracellular fluid (Fig 2.3B). If we now open our cluster of Na^+ channels briefly and depolarize the membrane locally, the resulting current flow will meet with high resistance in adjacent (relatively channel-free) parts of the membrane, but very low resistance in the cytoplasm and extracellular fluid. Applying the formula for the determination of length constants (Box 2.2), it is apparent that *high* membrane resistance coupled with *low* resistance in the separated fluids will lead to a substantial, but finite, length constant. If other voltage-gated Na^+ channels were present in the membrane so that they fell within the domain of depolarization dictated by the length constant (a distance

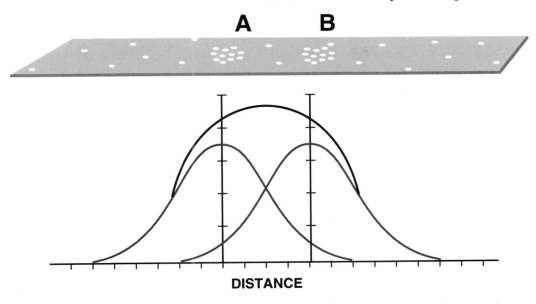

Figure 2.4 Algebraic summation over space (black line) occurs when the membrane is depolarized (gray lines) at two sites (**A** and **B**) simultaneously.

one might measure in micrometers), they also might be induced to open momentarily and add to the initial depolarization (Fig. 2.4). Thus, it can be appreciated that the denser the distribution of Na^+ channels in the membrane leading away from any site of initial depolarization, the greater will be the spread of depolarization along the membrane.

If the cluster of channels in our hypothetical patch of membrane were voltage-gated K^+ channels, the response to a brief depolarization would be very different. Transient opening of the K^+ channels would hyperpolarize the membrane. Like the Na^+-mediated depolarization, however, the hyperpolarization would exhibit time and length constants such that the membrane potential would exponentially recover to the resting level over distance and time. Thus, the Na^+ and K^+ channels mediate opposing changes in the membrane potential. Unlike our fictitious patch of membrane, Na^+ and K^+ channels occur together. But because the Na^+ channels respond more rapidly to voltage changes, the neutralizing effect of the K^+ channels will be delayed and transient depolarization will usually occur. *The conductivity of the neuronal membrane, then, will depend to a large extent on the relative densities of voltage-gated Na^+ and K^+ channels that are present.*

What would be the effect of changing the diameter of the cylinder? Increasing the cross-sectional area of the cylinder of cytoplasm decreases its resistance to current flow and consequently lengthens the spread of current through the cytoplasm. Conversely, decreasing the diameter of the cylinder increases its resistance and attenuates the current flow. As you might suspect, the tubular shape was chosen to illustrate this point about current spread because both the dendritic and axonal elements of the neuron are roughly cylindrical but of varied diameter.

The Action Potential

Keeping in mind the dynamic properties of the voltage-gated Na^+ and K^+ channels and the electrical properties of the neuronal membrane, let us now consider a real bioelectrical event that is critical to neuronal function: the action potential, or *spike*, as it is often called. If we were to insert the tip of our recording electrode into the cytoplasm of a real axon while the other electrode rested in the extracellular fluid, we would intermittently record a massive and transient shift in membrane potential. This shift is plotted with respect to time (or distance along an axonal membrane) in Figure 2.5. First, we would witness an abrupt depolarization and an opposite repolarization of the membrane, called the *rising phase*, followed by an equally steep return to a hyperpolarized level, the *falling phase*. The **afterhyperpolarization** gradually returns to the resting potential.

Generally speaking, every spike that we record from the impaled axon will have the same amplitude and duration. That is, *a spike will always achieve maximum amplitude*. If we were to choose a different axon from which to record, the amplitude of its spikes might be somewhat larger

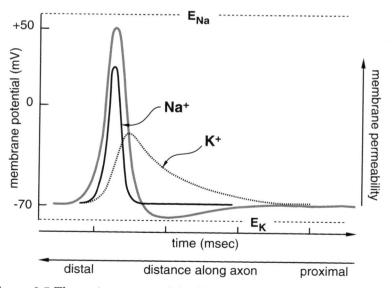

Figure 2.5 The action potential (red line) and the conductances of Na^+ (black line) and K^+ (dotted line) are plotted as a function of time or distance along the axon.

or smaller than those of the first axon, but every spike would be equal in amplitude to every other spike as long as we kept our electrode within the same axon. Thus, one is led to infer that since the spike amplitude is invariant, *individual neurons are left to encode information only in terms of spike frequency.* Although this inference is accurate as a first approximation, we will see in the discussion of synaptic transmission (Chapter 4) that synaptic mechanisms exist that can modify the peak and duration of the action potential as it invades the axon terminal.

Having described the action potential as an electrical phenomenon, let's now explain it in terms of what we know about ion channel behavior and equilibrium potentials. The shallow depolarization at the outset of the rising phase of the spike is due to the passive capacitative current flowing along the axon membrane, as the spike approaches our recording site. The steeply rising phase is due to the rapid and transient opening of a large number of voltage-gated Na^+ channels present in the axonal membrane, which drives the membrane potential toward the Na^+ equilibrium potential. It usually falls short of the Na^+ equilibrium potential, because the Na^+ channels rapidly reclose and the more slowly activated K^+ channels, also abundant in the axonal membrane, begin to open up. The falling phase is due to opening of voltage-gated K^+ channels; the resultant K^+ efflux drives the membrane potential toward the K^+ equilibrium potential. The afterhyperpolarization is a consequence of the lingering open state of the K^+ channels. As the K^+ channels gradually close up, the membrane potential recovers to the resting level.

During the falling phase of the action potential and the afterhyperpolarization, the inactivated state of the Na^+ channels and the open state of the K^+ channels impede renewal of the spike. This **refractory period** is absolute during the falling phase: *no amount of current can elicit a depolarization because the Na^+ channels have cycled into their closed and unresponsive state.* During the afterhyperpolarization, the refraction is at first strong and gradually wanes as the ion channels recover. Relatively greater currents are necessary to trigger a spike during this relative refractory period because of the lingering open state of the K^+ channels, which allows a compensatory efflux of positive K^+ current in response to any Na^+ influx. Obviously, the durations of these refractory periods have an impact on the maximum firing frequency of axons. Axons range in maximum firing frequency from around 100 spikes per second to as high as one thousand per second.

Getting the Spike Started

Synaptic potentials elicited in the dendrites and cell body of a neuron are conducted according to the passive properties and channel composition of the neuronal membrane. Generally speaking, the dendritic and somal regions of the neuronal membrane contain a much lower density of voltage-gated Na^+ and K^+ channels than does the axon. The local potentials that occur across dendritic and cell body membranes will be discussed more fully in Chapter 4. For now, suffice it to say that they sum algebraically with one another in accordance with the local time and space constants. If a sufficient magnitude of depolarization arrives at the initial segment of the axon, it will trigger an action potential. *The spike can be triggered only at the initial segment, because the membrane there is special: it is loaded with voltage-gated Na^+ and K^+ channels.* As a depolarization invades the initial segment, it triggers the opening of a few of the Na^+ channels. At the same time, the depolarization tends to drive K^+ ions out of the axon through channels that are much more leaky to begin with and are induced to open a little later than the Na^+ channels. If the net K^+ efflux can balance the net Na^+ influx, the spike will fail. However, if the initial depolarization is large enough, so many Na^+ channels will be induced to open that the K^+ efflux will be unable to compensate for the Na^+ influx. At this point, a positive feedback phenomenon ensues, in which the incoming Na^+ ions further depolarize the membrane and trigger the opening of still more Na^+ channels, leading to the explosive generation of the spike. The point at which K^+ flux can no longer compensate for the Na^+-induced depolarization is called the **threshold** level. Generally, thresholds for the initial segments of neurons lie at about –55 to –60 millivolts—about 10 to 15 millivolts of depolarization. Since depolarizations that do not reach threshold level do not trigger the spike, but those that do give rise to a full-blown impulse, the action potential is said to be *all-or-nothing.* Because of this positive feedback system, the spike is self-sustaining and propagates the length of the axon, usually invading any collateral branches as well. *It does not propagate back into the cell body or reverse itself along the axon at any point, because the Na^+ channels behind the spike are obliged to pass through the closed-deactivated state before they can respond anew.*

Conduction is Faster in Large-Caliber Myelinated Axons

If we now returned to the axon in which we first recorded an action potential and inserted a second recording electrode exactly one millimeter distal to the first, we could determine the velocity at which the spike traveled down the axon. Something around one hundred meters per second could reasonably be expected. However, a spike might be conducted much more rapidly or more slowly in a different axon, depending on its diameter and whether or not it was myelinated. As explained above, the larger the axon caliber, the less its resistance to axoplasmic current flow. For this reason, *larger axons conduct faster than smaller ones.* Increased diameter also increases the length constant, allowing for greater internodal distance in myelinated axons (below).

Myelinated axons conduct faster because the increased resistance and negligible capacitance introduced by the layers of myelin membrane shunt all the current through the axoplasm rather than across the axonal membrane. In such axons, the voltage-gated Na^+ and K^+ channels are concentrated only at the nodes of Ranvier, where the impulse is regenerated full strength. Needless to say, the internodal distance falls well within the axon's length constant, for, if it did not, the spike would fail. Although ionic current flows through the internodal axoplasm, the membrane cannot depolarize there. Instead, the spike appears to jump from node to node in a manner we call **saltatory conduction.** In addition to increased speed of conduction, this situation reduces the energy demand on the axonal membrane, since the Na^+-K^+ ATPase need perform its strenuous duty only at the nodes.

As a result of some autoimmune reactions or exposure to certain toxins present in the environment, the myelin coating around axons can degenerate. When an axon is demyelinated, its rate of conduction is at least slowed down. More often, however, conduction fails entirely. This spike failure happens because the internodal segments of myelinated axons are deficient in the voltage-gated ion channels needed to renew the membrane depolarization and because the now-present capacitative current flow across the membrane reduces the axoplasmic current proportionally.

Now that you understand how the impulse is generated and conducted by neurons, you no doubt are eager to know what happens when the spike enters the presynaptic ending of the axon. The short answer is that it triggers the release of a neurotransmitter and thereby initiates synaptic transmission. After learning some basics about neurotransmitters themselves in Chapter 3, we will analyze synaptic transmission in detail (Chapter 4).

Detailed Reviews

Catterall WA. 1989. Structure and function of voltage-sensitive ion channels. *Science* 242: 50.

Hille B. 1992. *Ionic Channels in Excitable Membranes* (2nd ed). Sinauer Associates, Sunderland, MA.

Stuhmer W, Conti F, Suzuki H, Wang XD, Noda M, Yahagi N, Kubo H, Numa S. 1989. Structural parts involved in activation and inactivation of the sodium channel. *Nature* 339: 597.

3 The Life Cycle of Neurotransmitters

Because neurotransmitters are the very currency of brain commerce, disturbances of their metabolism can devastate normal synaptic signaling. Since it is possible to intervene at one or more stages in the life cycle of neurotransmitters for therapeutic purposes—a fact that fuels the activity of the global pharmaceutical industry—understanding neurotransmitter biochemistry is of paramount importance in medical neuroscience.

How Do Chemicals Qualify As Neurotransmitters?

Chemical synaptic transmission requires first and foremost that a neurotransmitter substance be *produced* by a neuron and *transported to the axon terminal* from which it can be *released in a spike-dependent manner*. The neurotransmitter must be capable of producing a *response in the postsynaptic neuron*—a condition that, as we shall see in the next two chapters, depends on the presence of **receptor** proteins that specifically recognize and bind the neurotransmitter molecules. Finally, there must exist one or more mechanisms for *terminating the postsynaptic action* of the neurotransmitter. These phases in the life cycle of a neurotransmitter are used to test for the neurotransmitter status of any candidate substance (Fig. 3.1). Let's examine some of the substances so far identified as neurotransmitters.

Neurotransmitter Classification

Neurotransmitters are classified by their biochemical or functional characteristics. Biochemically most neurotransmitters fall into three categories: *amino acids*, *biogenic amines*, and *peptides*. Acetylcholine (ACh) and

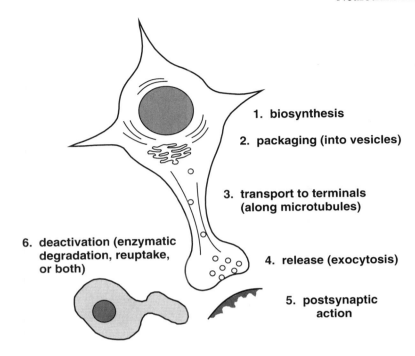

1. biosynthesis

2. packaging (into vesicles)

3. transport to terminals
   (along microtubules)

6. deactivation (enzymatic
   degradation, reuptake,
   or both)

4. release (exocytosis)

5. postsynaptic
   action

Figure 3.1 The phases in the life cycle of a neurotransmitter substance also constitute the criteria for neurotransmitter status. The small, gray cell is a glial cell called an astrocyte.

the nucleoside **adenosine** can be considered special cases, since they do not fit neatly into any of these categories. The most prominent of the amino acid neurotransmitters are **glutamate,** aspartate, and homocysteate (together called *excitatory amino acids*), γ-**aminobutyrate** (GABA), and **glycine.** The biogenic amines include **dopamine, norepinephrin,** and, to a more limited extent, **epinephrin** (together called *catecholamines*), and the *indolamine* **serotonin** (5-hydroxytryptamine; 5HT). The imidazole **histamine** is another biogenic amine that fulfills some of the criteria for transmitter status.

There are now several neuroactive peptides (usually from five to thirty or more amino acids in length) that have been shown to fulfill several of the criteria for neurotransmitter status (Table 3.1). However, since they are commonly found in terminals that also contain either a biogenic amine or an amino acid neurotransmitter, we often call them **cotransmitters.**

An important phenotypic trait of neurons is the combination of neurotransmitters they express. Although some exceptions appear to exist, a population of neurons will usually express only one of the biogenic amine or amino acid neurotransmitters, together with one or more peptide cotransmitters. *Whatever combination a given neuron expresses, however, it will use that same combination in all of its synaptic endings.* This principle of

Table 3.1 Common peptide cotransmitters and the conventional neurotransmitters with which they have been found to colocalize in axon terminals based on immunocytochemical observations

Cotransmitter	Neurotransmitter
Cholecystokinin	Dopamine, GABA, 5HT
Dynorphin	GABA
Leucine-enkephalin	GABA
Methionine-enkephalin	GABA, norepinephrin, 5HT
Neurokinin	GABA
Neuropeptide tyrosine	GABA, norepinephrin
Neurotensin	Dopamine
Somatostatin	ACh, GABA, norepinephrin
Substance P	ACh, GABA, glutamate, 5HT
Vasopressin	Norepinephrin
Vasoactive intestinal polypeptide	ACh, GABA

neurotransmitter phenotypy has been exploited to identify nuclei and axon systems in the CNS (Box 3.1).

Functionally, neurotransmitters can be described as *fast-acting* or *slow-acting*. If a neurotransmitter binds to a receptor protein that is also an ion channel, causing the channel to open and thereby inducing a rapid change in membrane potential (see Chapter 4), it is said to be a fast transmitter. Examples of fast neurotransmitters commonly found in the CNS are glutamate, GABA, and glycine. A slow transmitter, on the other hand, is one that binds to a receptor protein that is *not* an ion channel itself, but that activates ion channels through intermediary proteins (see Chapter 5); because of the additional metabolic steps, the response of the postsynaptic membrane will be slightly delayed and usually longer-lasting. The biogenic amines and peptides work this way. Because their effects are longer-lasting and sustain an altered state of responsiveness to fast transmitters, they are also called *neuromodulators*. In reality, these functional distinctions are more properly ascribed to the *receptors* than to the neurotransmitters themselves. We will learn in the following chapters that *the same neurotransmitter can act as a fast or slow transmitter, depending on which of its receptor types it activates.*

Neurotransmitters Are Produced and Replenished in Different Ways

Acetylcholine is synthesized in a one-step reaction from the two substrates *acetyl coenzyme A* and *choline* by the enzyme **choline acetyltrans-**

Box 3.1 Neurons Can Be Identified by Biochemical Phenotype

There are techniques that exploit the presence of characteristic substances endogenous to certain neuronal populations. Much of the time, the substances are neurotransmission-related, such as peptide cotransmitters, enzymes dedicated to the biosynthesis of particular neurotransmitters, receptor proteins, and signal transduction molecules. For example, over 30 years ago, autofluorescence of monoaminergic neurons (catecholamines exhibit a green fluorescence, whereas serotonin fluoresces yellow when thin tissue sections are exposed to formaldehyde vapors) was used to map the dopamine, norepinephrin, and serotonin systems of the brain. Some populations of neurons express marker enzymes, such as acetylcholinesterase, NADPH diaphorase, or cytochrome oxidase, that yield a visible reaction product when treated histochemically with the proper reagents.

Immunocytochemical identification of neurons with unique antigens to which antibodies have been raised has proven to be among the most powerful methods for deciphering phenotypic cell groups and axons. In this method, the purified antigen is used to inoculate an animal (often, a rabbit or goat), which then produces an antiserum that can be used to detect the antigen in brain tissue sections. Highly specific and pure antibodies called monoclonal antibodies can be mass-produced using **hybridoma cells**—fusions of mouse antibody-producing cells with immortal cancer cells. The antibodies themselves can be labeled with a visible dye or, as is more often done, revealed by a second antibody that carries the label. Peptide cotransmitters, in particular, lend themselves to this approach; the locations of several peptidergic nuclei and their terminal distributions have been immunocytochemically mapped in recent years in postmortem human as well as animal brains. Antibodies to glutamate decarboxylase, choline acetyltransferase, and tyrosine hydroxylase have been used to map, respectively, the GABAergic, cholinergic, and catecholaminergic systems of the brain.

Finally, neurons that express particular genes can be revealed by **in situ hybridization,** a technique in which a labeled antisense base sequence anneals to a specific mRNA in the cytoplasm of the cell body. Once a gene has been cloned and sequenced, **oligodeoxynucleotide probes** can be synthesized that are specific for short segments of the transcript. Both in situ hybridization and immunocytochemistry are amenable to quantification and therefore can be used to examine changes in the biosynthesis of neurotransmission-related molecules.

ferase (ChAT; Fig. 3.2). ChAT is uniquely expressed by cholinergic neurons of the CNS and is, therefore, a marker for such neurons. Ordinarily, the enzyme is rate-limiting in the production of ACh, but since choline is not synthesized by neurons, the availability of precursor choline can be rate-limiting if it is deficient in the diet.

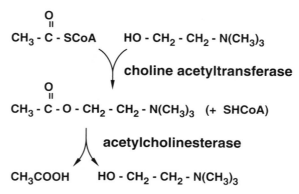

Figure 3.2   ACh is synthesized from choline and acetyl coenzyme A and degraded to choline and acetate.

The catecholamines dopamine, norepinephrin, and epinephrin are important neurotransmitters in the CNS that are synthesized from the amino acid tyrosine by four enzymes (Fig. 3.3). The first, **tyrosine hydroxylase,** is rate-limiting and a marker for catecholaminergic neurons. It converts tyrosine to L-dihydroxyphenylalanine (L-DOPA), which is immediately decarboxylated to form dopamine. In neurons that express tyrosine hydroxylase, but not **dopamine β-hydroxylase,** dopamine is the neurotransmitter end product. In neurons that express dopamine β-hydroxylase, dopamine is an intermediary metabolite that is immediately converted to norepinephrin. Dopamine β-hydroxylase is a marker enzyme for noradrenergic neurons. Some central neurons, such as the chromaffin cells of the adrenal medulla that secrete hormonal epinephrin, express the enzyme **phenylethanolamine-*N*-methyltransferase** (PNMT), which uses S-adenosylmethionine as a methyl donor to methylate norepinephrin to epinephrin.

The biosynthesis of catecholamines is regulated in a rather unique way: the activity of tyrosine hydroxylase is inhibited by its own catecholamine end products, presumably because they interact competitively for a pteridine cofactor-binding site on the enzyme. By way of this *negative feedback inhibition, the catecholamines are able to regulate their own synthesis according to their rate of activity-induced turnover.*

Serotonin, another important CNS neurotransmitter, is synthesized from tryptophan by two enzymes (Fig. 3.4). **Tryptophan hydroxylase** is rate-limiting and produces the intermediary metabolite 5-hydroxytryptophan, which is converted to 5HT by L-aromatic amino acid decarboxylase. Serotonin does not appear to inhibit tryptophan hydroxylase. The first important step that affects serotonin biosynthesis is the uptake of precursor tryptophan from the systemic circulation. Plasma tryptophan is obtained mainly from food, and *a reduction of dietary tryptophan can lower the brain levels of serotonin.*

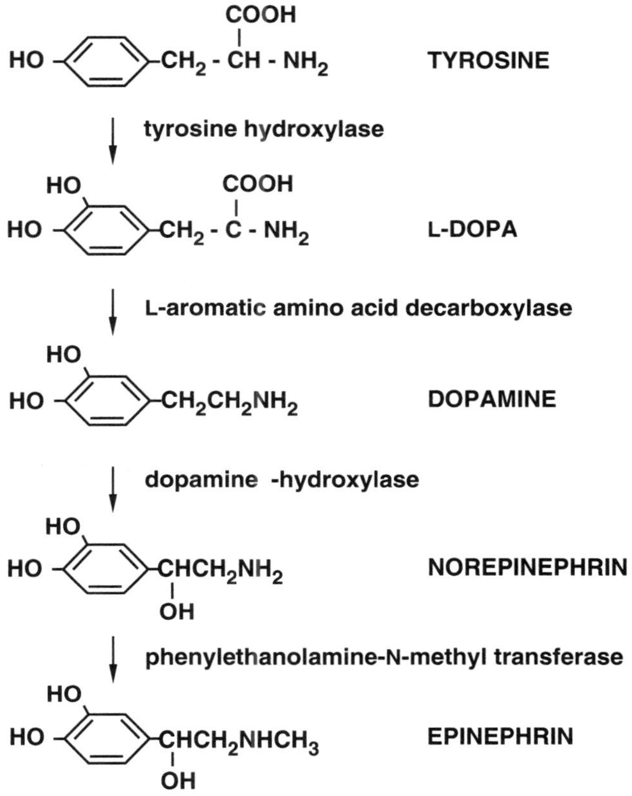

Figure 3.3 Catecholamines are synthesized from tyrosine. The end product catecholamine that is produced depends on the presence of the appropriate enzymes. That is to say, dopaminergic neurons do not express dopamine β-hydroxylase and noradrenergic neurons do not express phenylethanolamine-*N*-methyltransferase.

Glutamate, the major excitatory neurotransmitter of the CNS, is synthesized from glutamine by the neuron-specific enzyme **glutaminase** (Fig. 3.5). After release from the presynaptic ending, most glutamate is taken up by local astrocytes via a specific glutamate transporter protein (below) and converted to glutamine by the astrocyte-specific enzyme glutamine synthase. The glutamine is then released into the extracellular fluid and taken up by the terminals of glutamatergic neurons for conversion back to neurotransmitter glutamate by glutaminase. This recycling of used glutamate is referred to as the *glutamate–glutamine cycle*. Drugs that inhibit glutamine synthase cause a diminution in the availability of glutamine for conversion to neurotransmitter GLU and result in decreased synaptic levels of glutamate.

GABA and glycine are the two major inhibitory neurotransmitters of the CNS; GABA seems to be the inhibitory transmitter of choice in the

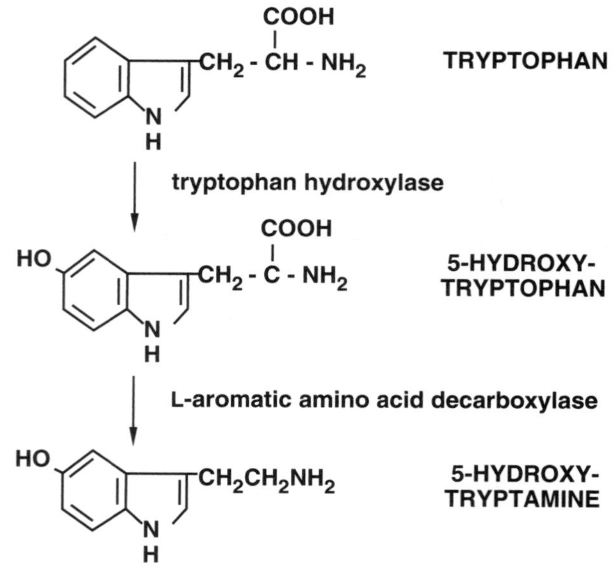

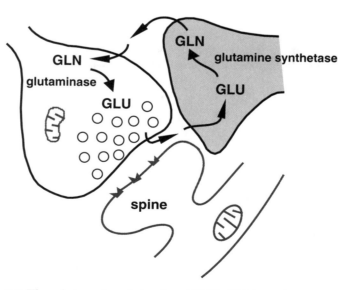

COOH
CH2 - CH - NH2
N
H
TRYPTOPHAN
tryptophan hydroxylase
COOH
HO
CH2 - C - NH2
N
H
5-HYDROXY-
TRYPTOPHAN
L-aromatic amino acid decarboxylase
HO
CH2CH2NH2
N
H
5-HYDROXY-
TRYPTAMINE

Figure 3.4 Serotonin is synthesized from the essential amino acid tryptophan.

GLN
glutamine synthetase
GLU
GLN
glutaminase
GLU
spine

Figure 3.5 The glutamate–glutamine (GLU–GLN) cycle involves glial cells called astrocytes that extend processes (gray) near the synapse.

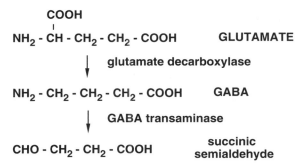

$$\underset{\text{GLUTAMATE}}{NH_2 - \underset{\underset{COOH}{|}}{CH} - CH_2 - CH_2 - COOH}$$

↓ glutamate decarboxylase

$$NH_2 - CH_2 - CH_2 - CH_2 - COOH \qquad \text{GABA}$$

↓ GABA transaminase

$$CHO - CH_2 - CH_2 - COOH \qquad \text{succinic semialdehyde}$$

Figure 3.6 The flow diagram shows the synthesis and catabolism of GABA.

brain, whereas glycine is preferred in the spinal cord. The former is produced by the decarboxylation of glutamate by the enzyme **glutamic acid decarboxylase** (Fig. 3.6). Glutamic acid decarboxylase is a marker for GABAergic neurons. Less is known about the maintenance of neurotransmitter glycine. It may be synthesized from serine by the enzyme *trans*-hydroxymethylase.

Since the enzymes necessary for the synthesis of biogenic amines and amino acid transmitters exist in the vesicles of the axon terminal, transmitters of this type can be synthesized and, thus, replenished locally. Peptide cotransmitters, however, depend on transcription and translation in the cell body for their replenishment (Fig. 3.7). The peptide cotransmitters are often segments of larger precursor polypeptides called **propeptides** that are translated from large mRNA molecules.

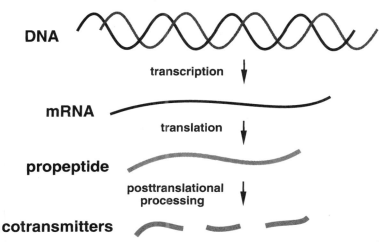

Figure 3.7 Peptide cotransmitters are synthesized in the nucleus and cell body.

These propeptides are cleaved at the appropriate places by peptidases to produce the active cotransmitters.

The Packaging and Transport of Neurotransmitters

Neurotransmitters and associated molecules such as their synthetic enzymes, cofactors, and ATP are packaged together in vesicles by the Golgi apparatus. The mechanisms that specify the contents of particular vesicles and the molecular details of the vesicles themselves are not fully known at present.

In contrast, recent research has revealed considerable detail about the axoplasmic transport of neurotransmitter vesicles to the axon terminals. Such transport occurs along the length of the axon in association with **microtubules** and is driven by unidirectional motors in a process that depends on the hydrolysis of ATP (Fig. 3.8). Axoplasmic transport occurs in both *anterograde* (toward the terminal) and *retrograde* (toward the cell body) directions. Neurotransmitter-containing vesicles are carried in the anterograde direction. Although retrograde transport is always relatively fast (around 3 millimeters per hour), anterograde transport can be either fast or slow (several millimeters per day). Synaptic vesicles travel in the fast lane.

Recent cell biological research using the squid giant axon has shown that the motors that drive transport in the two directions along the axon are different. **Kinesin** is an ATPase that binds to microtubules and vesicles and drives anterograde axonal transport; if antibodies to kinesin are infused into the axoplasm, anterograde but not retrograde transport is halted. **Cytoplasmic dynein,** a soluble form of the same protein that moves the microtubules found in motile flagella, is also an ATPase that binds microtubules. It will transport carboxylated latex microspheres toward the minus end of microtubules and is thus the best candidate to

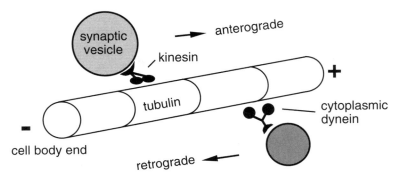

Figure 3.8 The fast, bidirectional transport of materials through the axoplasm occurs along microtubules.

date for a retrograde motor. Other cytoskeletal proteins are undoubtedly involved in the complete transport mechanisms, but their interactions are poorly characterized at present. How the two motors achieve their directional specificity remains a mystery as well.

Transport systems can be poisoned by compounds that depolymerize microtubules. This has proven to be a useful experimental tool in the study of axonal transport. Axoplasmic transport can also be exploited to study neuronal interconnections by injecting visually detectable substances that are taken up by neurons and transported within the axon (see Box 1.1).

Neurotransmitter Release

Neurotransmitters are stored in vesicles that are concentrated in the presynaptic ending of the axon (Fig. 3.9). They are released into the synaptic cleft by a processes of **exocytosis** that is triggered by the high concentration of cytosolic Ca^{2+} that occurs when a spike invades the axon terminal; the spike triggers the opening of voltage-gated Ca^{2+} channels that are present in the membrane of the active zone. The mechanisms of neurotransmitter release will be discussed more fully in the next chapter.

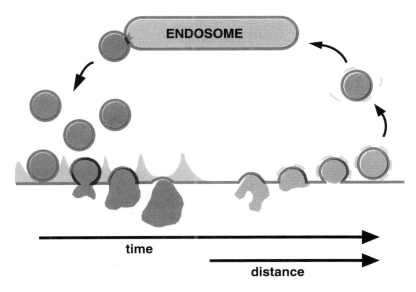

Figure 3.9 Neurotransmitters are released from vesicles at the active zone (red peaks on the inside of the plasma membrane) by exocytosis and can be taken back up into clathrin-coated vesicles by endocytosis. The clathrin coat is removed and the endocytosis vesicle is transported to and fuses with the endosomal compartment.

Neurotransmitter Deactivation

Released neurotransmitter molecules diffuse across the synaptic cleft, and some of them bind transiently to receptor proteins in the postsynaptic membrane. When unbound from their receptors, the transmitter molecules join the majority of their fellows who never met a receptor at all and are taken up either by the presynaptic terminal that released them or by astroglial cells whose processes surround the periphery of the synaptic cleft. Dopamine, norepinephrin, 5HT, GABA, glutamate, and glycine are all taken up by high-affinity **transporter** (or carrier) proteins. Such transporters are selective for the particular neurotransmitter and are phenotypically expressed by the neurons that secrete the neurotransmitter. Most cotransport Na^+ ions with the neurotransmitter molecules. Glial cells have been shown to express transporters for glutamate, GABA, and 5HT. Since such transporters can take up 50% or more of the released neurotransmitter, *a drug that blocks a transporter will cause a prolonged action of the neurotransmitter that the transporter would ordinarily clear from the synaptic cleft.* For example, the beneficial effects of the tricyclic antidepressants are correlated with their ability to block the 5HT and norepinephrin transporters. Some of the neurotransmitter that is recovered by uptake is used to replenish the releasable stores, while some is degraded by enzymes that reside within the cell (below). The fate of neurotransmitter molecules that are not taken up is usually degradation by enzymes that reside in the extracellular fluid.

Acetylcholine is deactivated by **acetylcholinesterase** (Fig. 3.10). This enzyme is present at high concentration in the basal lamina at the neuromuscular junction and in the synaptic cleft of cholinergic synapses in the CNS. The enzyme hydrolyzes ACh to choline and acetate. Choline is transported back into the nerve terminal cytoplasm by Na^+-dependent transporter proteins and used to synthesize more ACh. Newly synthesized ACh is rapidly sequestered in synaptic vesicles by specific transporters associated with the vesicle membrane. The action of ACh at the neuromuscular junction can be augmented by the addition of inhibitors of acetylcholinesterase such as *neostigmine.* Acetylcholinesterase inhibitors can be used to treat *myesthenia gravis,* a heterogeneous disease that is often caused by a reduced number of ACh receptors at the neuromuscular synapse or, in other forms, by a reduced release of ACh from motoneuron terminals.

The enzyme GABA transaminase (GABA-T) catabolizes GABA to form the intermediary succinic semialdehyde (Fig. 3.6). The latter is rapidly transformed to succinic acid by succinic semialdehyde dehydrogenase. GABA-T is believed to reside predominantly in the synaptic cleft and in association with the mitochondria of *post*synaptic neurons.

The preferred route of catecholamine catabolism involves two major enzymes: catechol-*O*-methyl transferase (COMT) and monoamine oxidase (MAO; Fig. 3.11). MAO is associated with the outer membrane of

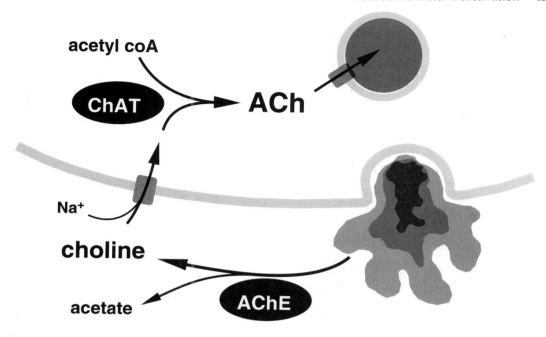

Figure 3.10 Acetylcholine (ACh) can be replenished locally at the synapse. The extracellular choline produced by the degradative enzyme acetylcholinesterase (AChE) is taken up by specific transporter proteins of the presynaptic membrane. The cytoplasmic choline is converted by choline acetyltransferase (ChAT) to produce ACh. The newly synthesized ACh is transported into synaptic vesicles by specific transporter proteins of the vesicle membrane.

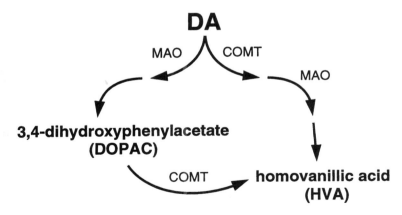

Figure 3.11 The catabolism of dopamine by monoamine oxidase (MAO) and catechol-*O*-methyl transferase (COMT) leads to two major metabolic products. These metabolites can be measured in the cerebrospinal fluid as an index of the activity of dopaminergic neurons.

mitochondria. Although conventional wisdom has it that COMT acts extraneuronally, there is little evidence to support such a notion. Two of the major metabolites generated, homovanillic acid and 3,4-dihydroxy-phenylacetate (DOPAC), can be assayed in samplings of the cerebro-spinal fluid and used as indicators of the activity level of dopaminergic neurons in the brain. Despite the intracellular location of MAO, it appears that drugs that inhibit MAO can prolong the action of the catecho-lamine neurotransmitters, an effect that can be therapeutic in some cases of clinical depression. Exactly how MAO inhibitors enhance catecho-lamine activity is unclear. A major disadvantage of therapies that affect the active levels of neurotransmitters is their lack of spatiotemporal specificity, which usually causes unwanted side effects. As we shall see later, therapies with drugs that selectively activate or antagonize specific *receptor* subtypes suffer considerably less from this problem.

Detailed Reviews

Cooper JR, Bloom FE, Roth RH. 1991. *The Biochemical Basis of Neuropharmacology* (5th ed). Oxford University Press, New York.

Koob GF, Sandman CA, Strand FL. 1990. A decade of neuropeptides: past, present and future. *Ann NY Acad Sci* 579: 1.

Sossin WS, Fisher JM, Scheller RH. 1989. Cellular and molecular biology of neuropeptide processing and packaging. *Neuron* 2:1407.

Sudhoff TC, Jahn R. 1991. Proteins of synaptic vesicles involved in exocytosis and membrane recycling. *Neuron* 6: 1.

Synaptic Transmission 4

Invertebrates employ both the chemical synapse and the electrotonic contact for interneuronal signaling. Although still present to a minor extent in vertebrate nervous systems, electrotonic coupling has been largely abandoned in favor of the more versatile chemical synapse. Because of the complexity of the synapses in the human brain, much remains to be discovered about synaptic transmission. It is likely that many idiopathic brain disorders that frequently confront neurologists and psychiatrists result from defects of synaptic transmission. Since there is still much to be learned about the complexities of synaptic transmission, we are ignorant about diseases that affect it. In this chapter and the next, let's begin to examine the complexities and the versatility of synaptic transmission, beginning with the sequence of events that ensues when a spike invades a presynaptic ending. Synaptic transmission has two phases: a presynaptic phase of neurotransmitter release and a postsynaptic phase in which the neurotransmitter binds to receptors to induce changes in the bioelectrical activity of the postsynaptic membrane.

The Presynaptic Phase

Back in the 1950s, Bernard Katz and his colleagues used the neuromuscular junction to show that the release of the neurotransmitter ACh occurs in finite increments they called *quanta*. Soon after, it was determined that *a quantum of neurotransmitter corresponds to the amount contained in a single synaptic vesicle.* At the neuromuscular junction, an action potential causes the release of some 250 quanta, indicating the exocytosis of some 250 vesicles at the active zone. In the CNS, the mode of release is essentially the same as at the neuromuscular junction, although the specific neurotransmitter and quantal size vary from neuron to neuron.

Today, it is clear that *most, if not all, neurotransmitter release occurs by way of vesicle fusion with the membrane at the active zone.*

Calcium-Dependent Vesicle Exocytosis

It is commonplace nowadays to grow mammalian brain cells in a culture dish. The neurons will mature, grow processes, and make synaptic contacts with one another. Using such a system, the experimenter can readily record the activity of the neurons and manipulate the medium that bathes them. If the growth medium in which the neurons are raised is replaced with a medium that does not contain Ca^{2+}, there will be no synaptic transmission in the dish. One can still record spikes in individual neurons; the absence of Ca^{2+} has no effect on the neurons' ability to conduct action potentials. Likewise, if a known neurotransmitter substance is applied through a micropipette to the surface of a neuron, the neuron will respond with the predicted change in membrane potential (below). If Ca^{2+} is not necessary for the spike or for the postsynaptic response, how does its absence block synaptic transmission? It does so by preventing the release of neurotransmitter from the presynaptic ending. In fact, transmitter release can be induced in the absence of a spike by artificially increasing the free Ca^{2+} concentration in the presynaptic cytosol. Thus, *Ca^{2+} is both necessary and sufficient for neurotransmitter release.*

Neurotransmitter release requires a 10^4- to 10^5-fold increase in Ca^{2+} concentration. This massive local increase is made possible by a *high density of voltage-gated Ca^{2+} channels in the membrane of the active zone* (Fig. 4.1). Voltage-gated Ca^{2+} channels that have been cloned so far are all structurally similar to one another and to the voltage-gated Na^+ channel discussed in Chapter 2 (see Fig. 2.1). But there are multiple forms that differ from one another in their kinetic properties. At least three types are present in nerve terminals (called N, P, and Q types) that exhibit large conductances, slow activation at high depolarizations, and slow inactivation. The proximity of these Ca^{2+} channels to the vesicles in the active zone ensures that the vesicles will be rapidly exposed to high local Ca^{2+} concentration increases when the channels open. This requirement for a high Ca^{2+} concentration also enables the necessarily rapid termination of neurotransmitter release, since the Ca^{2+} quickly diffuses and is diluted in the cytosol. It is subsequently removed by a Na^+-Ca^{2+} exchanger protein in the plasma membrane and in certain organelle membranes that, respectively, pump it out of the neuron and sequester it in noncytosolic compartments.

Thanks to the power of molecular technology, a vesicle-associated protein called **synaptotagmin** has now been identified that binds Ca^{2+} and perhaps triggers the fusion of the vesicle membrane with the plasma membrane (Fig. 4.1). Synaptotagmin binds Ca^{2+} with a low affinity and associates closely with the cytoplasmic domain of the Ca^{2+} channel

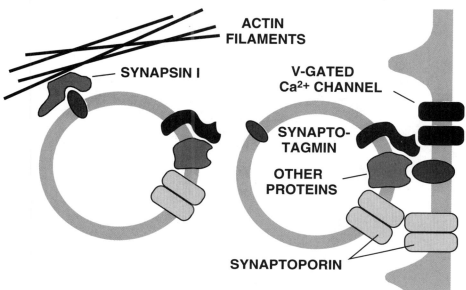

ACTIN FILAMENTS

SYNAPSIN I

V-GATED Ca^{2+} CHANNEL

SYNAPTO-TAGMIN

OTHER PROTEINS

SYNAPTOPORIN

Figure 4.1 The vesicle to the right is ready for release at the active zone membrane. A few of the proteins presumed to be involved in docking and exocytosis have recently been identified. The vesicle to the left is held away from the active zone by synapsin I-mediated attachment to actin filaments. The identification of such proteins proceeds apace, and our knowledge of the exocytosis process will soon be more nearly complete.

proteins in the active zone. In mutant fruit fly (*Drosophila melanogaster*) larvas that are synaptotagmin-deficient, neurotransmitter release is greatly reduced at neuromuscular junctions, and muscular contraction is slow and weak. Such flies do not survive beyond the larval stage.

Another protein called **synaptoporin,** which is found in both the vesicle membrane and the plasma membrane of the active zone and which engages in homophilic binding, appears to be essential in opening the initial pore for neurotransmitter extrusion (Fig. 4.1). Still other proteins are involved in the fusion of the vesicle membrane with the plasma membrane, in forcible extrusion of the vesicle contents, and in reformation of the vesicles (endocytosis). Exactly which proteins are involved and how they function is still under investigation.

Cotransmitter Release Occurs by a Different Mechanism

Larger vesicles that contain peptide cotransmitters are present in many terminals in the CNS (Fig. 4.2). Such vesicles remain distant from the active zone and presumably fuse with a part of the membrane remote

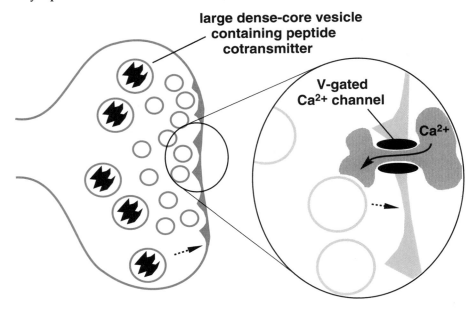

Figure 4.2 Neurotransmitter release from a presynaptic ending occurs by exocytosis at the active zone where voltage-gated Ca^{2+} channels abound. Peptide cotransmitter release from larger, dense-core vesicles may occur by exocytosis at a site distant from the active zone.

from the synaptic cleft. The Ca^{2+}-binding proteins associated with peptide-containing vesicles may have different kinetic properties as well. According to one hypothesis, they have a higher affinity for Ca^{2+}, thus reducing the required concentration for exocytosis. Because they are farther from the Ca^{2+} channels at the active zone, a Ca^{2+} concentration high enough to produce fusion of such vesicles with the membrane may be attained only when the terminal is invaded by a high frequency of spikes.

Preparation of Vesicles for Release

When a spike invades the ending, only a small fraction of the total number of vesicles disgorge their contents. The ones that do, therefore, must be *prepared* in some way that the others are not. Just as nuclear missiles are armed immediately before being loaded into the launcher, the synaptic vesicle must be "cocked" and "docked" at the active zone (Fig. 4.2). Likewise, after a round of vesicles is spent, they must be replaced at the active zone by a new set. Thanks to molecular techniques, rapid progress is being made in identifying the relevant proteins and their interactions.

One protein has so far been identified as a potentially important player in the preparation of synaptic vesicles for exocytosis. **Synapsin I** is a cytoskeletal protein that binds to vesicles and to actin filaments near the active zone (Fig. 4.1). Synapsin I is phosphorylated by a Ca^{2+}-calmodulin-dependent kinase, which itself is activated by the same Ca^{2+} that enters the terminal to trigger vesicle fusion (see Chapter 5). The binding capacity of synapsin I is greatly diminished when the protein is phosphorylated, and it may, therefore, unleash vesicles so that they are able to dock in the active zone.

Neurotransmitter Receptors

When neurotransmitter is released into the synaptic cleft, it diffuses to the postsynaptic membrane, where it encounters a dense concentration of receptor proteins usually specific for the neurotransmitter that is released. *The response of the postsynaptic cell depends on the nature of these receptors.* Incidentally, how the transmitter phenotype of the presynaptic neuron and the receptor phenotype of the postsynaptic neuron are matched during development remains a mystery. Pharmacological and molecular biological techniques have revealed much in recent years about neurotransmitter receptors (Box 4.1).

Neurotransmitter receptors can be categorized into two general classes. One class consists of proteins that are themselves ion channels. These ion channels are not regulated by voltage, as were the channels we encountered in Chapter 2. Instead, they are regulated by the binding of neurotransmitter molecules. We call them **ionotropic** receptors or **ligand-gated** ion channels. *Such receptors mediate fast and transient electrical responses of the neuronal membrane.* The other class of receptors is made up of proteins that are *not* ion channels themselves but instead regulate separate ion channel proteins via one or another of a variety of *signal transduction* mechanisms. Since the signal transduction mechanisms almost always involve enzyme action, we call such receptors **metabotropic.** The responses mediated by metabotropic receptors are typically slower and more complex. They will be considered in Chapter 5. In the present chapter, we will deal with ionotropic receptor-mediated postsynaptic responses. The most studied example of an ionotropic receptor is the nicotinic ACh receptor (nACh-R). Others are the type A γ-aminobutyrate receptor (GABA$_A$-R), the glycine receptor (GLY-R), and several subtypes of excitatory amino acid receptors.

The nACh-R at the Neuromuscular Junction

This nACh-R is an assembly of five independent glycoprotein subunits (Fig. 4.3). Two α subunits combine with one β, one γ, and one δ subunit. The subunits bear high homology to one another. Each exhibits four

Box 4.1 Receptor Pharmacology

Working at the beginning of the twentieth century, Henry Dale found that the plant alkaloid *nicotine* could duplicate the effect of ACh when it was applied at neuromuscular junctions, but had no effect on autonomic responses that were known also to be elicited by ACh. He found further that *muscarine*, another alkaloid, had no effect on muscle, but mimicked the action of ACh on the autonomic nervous system. He correctly surmised that the differential effects of the two alkaloids resulted from the existence of two different types of cholinergic receptor. Today, of course, we call them nicotinic and muscarinic ACh receptors according to Dale's discovery. Since Dale's pioneering work, receptors for many hormones and neurotransmitters have been identified. Electrophysiological responses (see Box 2.1) can be used to assess the effect of specific neurotransmitters or analog drugs applied to neuronal membranes suspected of containing a particular receptor. A broad diversity of neurotransmitter receptors, and even multiple subtypes of individual receptors, have been discovered by using purified toxins (from plants, spiders, mollusks, and snakes) and synthetic compounds as **ligands** to examine their ability to activate (**agonists**) or oppose the activation (**antagonists**) of receptors.

Receptor ligands have been used extensively to bind receptors and estimate their numbers and affinities for neurotransmitters and drugs as well as their precise locations within districts and cell populations of the CNS. Such radioligand-binding experiments are done by incubating brain tissue homogenates or thin sections in a buffer that contains the radiolabeled ligand of interest. Once a highly specific ligand with strong affinity for the receptor is in hand, other suspected agonists and antagonists can be placed into the buffer at increasing concentrations to see if they compete for binding with the known ligand. In this way, potential new drugs can be screened for their pharmacodynamic qualities.

The molecular cloning of receptors allows the study of their molecular structure, which in turn yields insights into the way they interact with the neuronal membrane, with ligands, and with intracellular signal transduction proteins. Moreover, cloned receptors can often be expressed (as can ion channels) in *Xenopus* oocytes or other expression cells in order to better control the experimental conditions under which they are studied.

membrane-spanning domains in the form of an α-helix and has both its N- and C-terminal domains in an extracellular position. The subunits assemble symmetrically around a central pore with their second transmembrane domains lining the pore. The ionopore has a diameter of about 10 Å that will pass Na^+, K^+, and Ca^{2+} when open. The N-terminal domain of the α subunit contains the binding site for ACh. Accordingly, two ACh molecules must bind, one to each α subunit, to fully activate the ionopore.

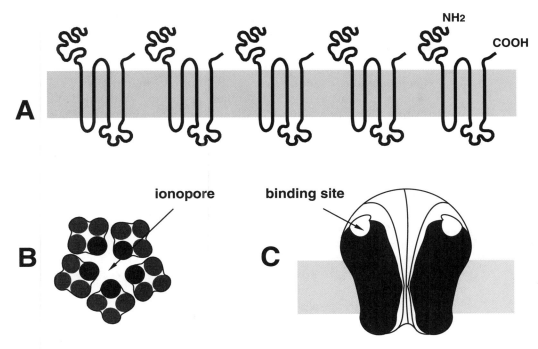

Figure 4.3 The drawings depict a popular structural model of the nACh-R. Sketch **A** shows the five subunits that compose the receptor, each of which contains four transmembrane segments. Sketch **B** is an end-on view to show the presumed symmetrical arrangement of the subunits around a central pore lined by the second transmembrane segment of each subunit. Sketch **C** shows a side view of the receptor in which the α subunits have been cut through and the γ subunit has been removed so that the ionopore is visible.

When bound to ACh, the receptor begins to alternate rapidly between the open and closed conformations in bursts (of several microseconds duration) that persist as long as the ACh is bound (several milliseconds). The influx of Na^+ ions through the channel predominates because of the greater driving force on Na^+. This results in depolarization of the muscle membrane and production of an action potential. Incidentally, many naturally occurring toxins affect the muscle cell nACh-R. For instance, if the receptor is exposed to ACh for seconds to minutes, as might occur from an insect venom with anticholinesterase activity, the receptor ceases to open. This **desensitization** involves a Ca^{2+}-dependent phosphorylation of the receptor's cytoplasmic domains. Some snake venoms (α-bungarotoxin) and black widow spider venom bind competitively at the ACh binding site but do not open the channel. We call such compounds **antagonists,** because they block the action of the endogenous ligand.

The Subunit Composition of Ionotropic Receptors

Heteropentameric ionotropic receptors from the mammalian CNS have been isolated and analyzed. All of them appear to be patterned structurally after the muscle cell nACh-R, but they exhibit important functional differences. For instance, the $GABA_A$-R and GLY-R are pentamers that pass Cl^- rather than cations through their ionopores. Moreover, there is a rather broad diversity within the individual subunits that compose central ionotropic receptors. The neuronal nACh-R pentamer is made up of only two different subunits (α and β) rather than the four found in the muscle version. Furthermore, five different variants of the α subunit of the central nACh-R have been cloned. Importantly, *different subunit variants are combined in different ways that imbue the receptors with different pharmacokinetic properties.* Different neuronal populations appear to express versions of receptors with distinct subunit compositions, suggesting important functional differences.

Preliminary experiments indicate that the expression of receptor protein subunits may be systematically regulated in different subsets of central neurons and even in individual neurons as a function of experience. Consider, for example, the family of ionotropic glutamate receptors (GLU-R). Like the nACh-R, each is a multimer of varying subunit composition. Some fourteen distinct subunits have been cloned so far. When mRNAs that encode the subunit proteins are selectively injected into *Xenopus* oocytes in different combinations, the pharmacokinetic and channel properties of the resulting GLU-Rs differ according to the subunit combination that was injected. Recent evidence suggests that the subunits are also differentially expressed in nature by subsets of brain neurons. And within individual neurons, the exact combination of subunits may be carefully regulated. The potential points at which regulation can occur in the life cycle of receptor proteins (and, perhaps, voltage-gated ion channels as well) are summarized in Table 4.1. Such regulation begins already at the level of gene transcription and extends to local events at the synaptic sites that establish the receptor as a functional molecule in the neuronal membrane. At the genetic level of regulation, the transcription rate of the subunit genes and the selection of particular splice variants in a particular neuron can be modulated according to the nature of signals that the neuron receives at any point in time (see Chapter 5). At the next level, translation, there is evidence that the mRNAs that encode some of the GLU-R subunits are subject to RNA *editing*—removal or replacement of bases. At the third level, the subunit proteins themselves may be posttranslationally modified in ways that are important for their assembly into functional multimers. At the final level, there is evidence that receptors can be retained in a cytoplasmic reservoir or added to the membrane under different signal-

Table 4.1 The ultimate functional character of ligand-gated ion channels (ionotropic receptors) and multimeric voltage-gated ion (K^+ and Ca^{2+}) channels can be modified by regulation at one or more specific points in each of the four stages of their life cycles

Stage	Point
Genetic	Transcription frequency
	Differential splicing
mRNA	Editing
	Translation rate
	Stability
	Transport
Protein	Modification
	Stability
	Phosphorylation state
Receptor/channel	Subunit composition
	Transport
	Stability
	Phosphorylation state
	Sequestration

ing conditions. Thus, the numbers and types of GLU-Rs that ultimately appear in the membrane of a neuron can be regulated at different points in each stage of the life cycle of the receptor. It is tempting to surmise that such fine regulation could be used to adjust the responsiveness of neurons to glutamate transmission and hence change their computational states. The variations and subtle regulation of different subtypes of ionotropic receptors should make it obvious that viewing neurons as static on-off switches similar to those found in a computer circuit is inexcusably naive. Instead, we must acknowledge that *neurons are exquisitely complex and dynamic units that can vary their computational status over time*. We'll expand on this concept in the next chapter, which deals with long-term postsynaptic effects mediated by metabotropic receptors.

Postsynaptic Potentials

Postsynaptic potentials in the CNS are excitatory or inhibitory. Most excitation in the CNS is mediated by neurotransmitter glutamate (and other excitatory amino acids such as aspartate and homocysteate), which, upon release from a presynaptic ending, can interact with some six distinct receptor types that have so far been identified. The fast potential changes in the postsynaptic membrane with which we are concerned in this chapter are mediated by two ionotropic types that both

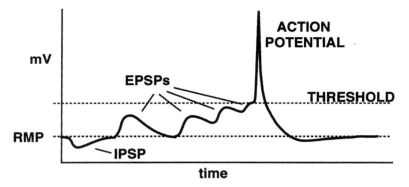

Figure 4.4 The graph shows a plot of postsynaptic potentials that one might record with a microelectrode near the initial segment of a neuron. The temporal summation of subthreshold EPSPs can lead to threshold depolarization and trigger a spike. Abbreviations: EPSP, excitatory post-synaptic potential; IPSP, inhibitory postsynaptic potential; RMP, resting membrane potential.

pass Na^+ and K^+. The two types are named for the agonist drugs that are used to distinguish them. One is activated by the glutamate analog kainic acid and the other by α-amino-3-hydroxy-5-methylisoxazole-4-propionic acid (AMPA). When glutamate binds to either kainic acid or AMPA receptors, a channel is opened to monovalent cations, allowing an influx of Na^+ that exceeds the K^+ outflow (remember that, at the resting membrane potential, the driving forces are far stronger for Na^+ than K^+). As a result of this net inward current, a depolarization of the postsynaptic membrane occurs (Fig. 4.4). Since the neurotransmitter release is quantal, *the magnitude of the postsynaptic depolarization is in general proportional to* the amount of neurotransmitter released from the presynaptic ending. We call this local, transmission-induced depolarization an **excitatory post**synaptic potential (**EPSP**).

Inhibitory neurotransmission is mediated mainly by the neurotrans-mitters glycine in the spinal cord and GABA in the brain. The ionopores of the receptors for these two neurotransmitters pass Cl^- but not cations. The increased membrane conductance to Cl^- will tend to oppose any depolarizations that might occur due to inward Na^+ current. This elec-trical phenomenon is called *shunting* by the electrophysiologists. In real molecular terms, the increased permeability to Cl^- tends to negate any inward Na^+ current, because Cl^- ions are free to enter the cytoplasm with Na^+, thereby neutralizing part or all of the inward positive current.

Inhibition of postsynaptic membrane activity can be induced also by increasing the membrane permeability to K^+. In this case, excitability is decreased in two ways. First, the membrane potential is hyperpolarized (remember the equilibrium potential for K^+ is around –85 millivolts), making it that much more difficult to attain the threshold level of depolarization. Second, the facilitated outflow of K^+ can readily oppose a depolarizing Na^+ influx—the shunting phenomenon. The inhibition mediated by K^+ channels is produced by *metabotropic* receptors (Chapter 5) and is generally slower than that mediated by ligand-gated Cl^- channels. In either case, if a hyperpolarization of the postsynaptic membrane is manifest, we call it an **inhibitory postsynaptic potential (IPSP)**. Obviously, *any deflection of the membrane potential away from threshold will decrease the probability that cotemporal EPSPs will reach threshold and trigger a spike.* Although IPSPs can be readily recorded in the large motoneurons of the spinal cord, they are less evident in other neurons. This is not at all because other central neurons are devoid of inhibitory synaptic input; few, if any, escape significant inhibition. It is simply because Cl^- (and, for that matter, K^+) is so near its equilibrium potential that little, if any, deviation from the resting potential can be seen with the recording apparatus. The postsynaptic neuron is nonetheless inhibited, in the sense that it is *less excitable* because of current shunting.

Postsynaptic Potentials Sum Algebraically Over Time and Space

If an excitatory synaptic ending were to release a second bolus of glutamate before an immediately preceding EPSP decayed to the resting membrane potential level, it would trigger a second EPSP that would add to the residuum of the first. We call this phenomenon **temporal summation.** Likewise, if a different axonal ending were to simultaneously release glutamate onto the same neuron at a nearby synapse, the two postsynaptic depolarizations would add together according to their spread along the intervening membrane. Call this **spatial summation.** In like fashion, cotemporal and (or) nearby inhibitory transmissions subtract from the sum of postsynaptic excitation. Whether or not a neuron fires an action potential depends on the tally of excitation and inhibition arriving at the initial segment of its axon.

Not All Synapses Are Created Equal

Voltage-gated ion channels and dendritic structure influence the potency of postsynaptic potentials. You'll recall from Chapter 2 that voltage-gated Na^+ and K^+ channels are activated by depolarizations of the membrane. Therefore, the greater the density of Na^+ channels in the vicinity of the postsynaptic membrane, the more effective will be the postsynaptic excitation evoked by an excitatory transmission event. An abundance of K^+ channels will tend to dampen membrane excitability. Moreover, since the extent of current spread over the postsynaptic membrane is a function of the number of Na^+ channels (resistive membrane current) as well as the cytoplasmic resistance and membrane capacitance, *the spread of depolarization will depend on the relative densities of Na^+ and K^+ channels in the membrane leading toward the initial segment.* These densities vary from one synaptic site to the next and in different parts of the dendritic and somal membrane, and thereby influence synaptic potency. *The diameter of the dendrite also affects the flow of current and the spread of depolarization* (see Chapter 2). Recalling that dendrites are essentially cylinders whose diameter progressively increases with proximity to the cell body (and initial segment), it can be appreciated that *synapses closer to the cell body are generally more potent than those on more distal segments of the dendrites.*

Inhibitory Inputs Can Impose Patterns on Spike Activity

Inhibitory synaptic transmissions are very important and can be extremely powerful. They often serve to modulate the firing frequency or spike patterns of tonically active neurons in the CNS (Fig. 4.5A). Tonic inhibition of otherwise active neurons is also frequently observed (Fig. 4.5B). In such cases, the inhibitory neurons themselves are phasically inhibited so as to "release" their target neurons from tonic inhibition. This sort of tonic inhibition and phasic **disinhibition** appears to be a common mode of circuit activation in the motor control systems of the brain (dealt with in later chapters). Finally, an inhibitory synapse near the initial segment of an axon is an especially potent inhibitor of the neuron's ability to fire spikes. We will see examples of such synapses in action in the context of circuit function described in later chapters.

The Special Case of Presynaptic Inhibition and Excitation

As mentioned in Chapter 1, axons frequently end on other axons, especially near their terminals. Obviously, such an arrangement enables a significant degree of control over the transmission event. Presynaptic inhibition is often mediated by receptors associated with the opening of K^+ channels (Fig. 4.6). It usually does not cause spike failure at the terminal, but because K^+ efflux is enhanced, *the absolute peak and (or) duration of the spike is reduced.* This in turn *reduces the amount of Ca^{2+} entry through the voltage-sensitive channels in the active zone and, correspondingly, the amount of transmitter released.* Since the potentials elicited in a postsynaptic membrane are proportional to the amount of transmitter released, the effect of presynaptic inhibition is to reduce the impact of the synaptic transmissions at the postsynaptic element.

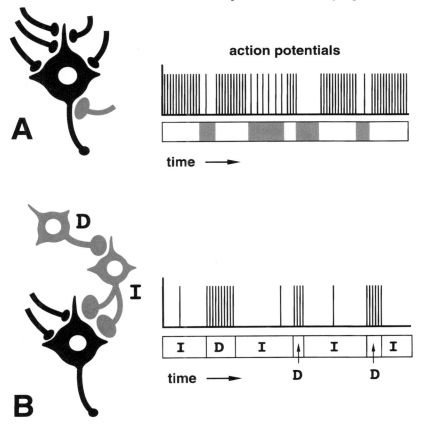

Figure 4.5 Two examples of common integrations performed by central neurons are schematized. The discharge of the neuron in **A** is usually driven at a high frequency by dominant excitatory inputs (jet black terminals) on its dendrites. This *tonic excitation* is *phasically modulated* (decreased) or even interrupted completely by periodically active inhibitory inputs (red terminal), which are often located near the initial segment. The periods during which the inhibitor input is active are shown by the red bars along the abscissa in the activity plot to the right. The opposite situation holds in **B;** the black neuron is *tonically inhibited* by forceful inputs from neuron I. Occasionally, the discharge of neuron I is itself decreased by inhibitory terminals it receives from neuron D. When this happens, the black neuron is *released* from the inhibition by neuron I and can fire bursts of spikes in response to activation of its excitatory inputs (black terminals). The activity recorded from the black neuron is shown to the right. The periods along the abscissa show the phases during which the inhibitor (I) and the disinhibitor (D) neuron are active.

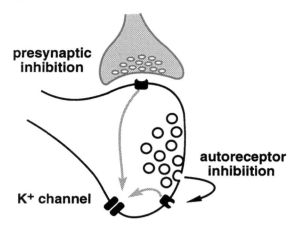

Figure 4.6 Two mechanisms of presynaptic inhibition—an axoaxonic synapse and presynaptic autoreceptors—are shown. By opening K^+ channels, either mechanism can reduce the absolute amplitude and duration of membrane depolarization caused by the spike as it invades the bouton. This reduces the influx of Ca^{2+} through voltage-gated channels at the active zone and, hence, the amount of neurotransmitter released.

Axoaxonic synapses can also augment the coupling of the action potential to neurotransmitter release. In this case of presynaptic excitation, the axoaxonic transmission event leads to the temporary closure of certain types of K^+ channels present in the axon terminal. Action potentials that invade the terminal during this period of reduced K^+ conductivity will have a longer duration, which in turn causes more neurotransmitter release and an enhanced postsynaptic response.

Receptors specific for neurotransmitters released by an axon are often present in the *pre*synaptic membrane itself (Fig. 4.6). These **autoreceptors** are of the metabolic variety and usually mediate an inhibition of the presynaptic ending. *By binding to autoreceptors, a neurotransmitter can modulate its own subsequent release.* Due to this negative feedback, the more active the ending, the more it will dampen its release of neurotransmitter.

Special Functions of the Dendritic Spine

Dendritic spines are a common, though not ubiquitous, element on many dendrites. When spines were originally noticed, many investigators speculated that they were Mother Nature's way of increasing the surface area of dendrites to accommodate more synaptic contacts. It soon became evident that this is not so; many dendrites have contacts on their spines while their shafts remain synapse-free. So why do many neurons seem to prefer this mode of synaptic contact? Our first clue comes from the typical shape of the spines; they often, though not always, have a bulbous head that is connected to the dendritic shaft by a narrow neck. Using our knowledge of the passive electrical properties of the neuron, we know that the cytoplasmic resistance to current flow will be much greater in the constriction of the neck than it is in the head of the spine. Such a geometry would restrict the spread of depolarization so that no single synapse could rule the activity of the postsynaptic neuron. Instead, a "consensus" in the form of several simultaneously active synapses would be required to drive the cell.

Spines also serve to isolate a metabolic *compartment* of the neuron. As mentioned above, some receptors activate biochemical messengers within the cytoplasm that are multifunctional and must, therefore, be restricted in range, lest they produce unwanted effects. You will better appreciate the significance of this compartmentalization after reading Chapter 5. Finally, spines (and axon terminals) appear to be more labile than other elements of the neuron. Evidence has begun to accumulate that spines undergo considerable remodeling as a consequence of differential synaptic activity (see Chapter 29).

Detailed Reviews

Bennett MK, Scheller RH. 1993. The molecular machinery for secretion is conserved from yeast to neurons. *Proc Natl Acad Sci (USA)* 90: 2559.

Betz H. 1990. Ligand-gated ion channels in the brain: the amino acid superfamily. *Neuron* 5: 383.

Burgoyne RD, Morgan A. 1995. Ca^{2+} and secretory vesicle dynamics. *Trends Neurosci* 18: 191.

Claudio T. 1989. Molecular genetics of acetylcholine receptor-channels. In Glover DM, Hames BD (eds), *Frontiers in Molecular Biology: Molecular Neurobiology*. IRL Press, New York.

Unwin N. 1989. The structure of ion channels in membranes of excitable cells. *Neuron* 3: 665.

5

The Response Repertoire of Neurons

Neurotransmitters produce their effects in postsynaptic neurons by coupling selectively and transiently with specific receptor proteins present in the postsynaptic membrane. The response of the postsynaptic cell depends on the nature of these receptors. In the preceding chapter, you learned that some receptors are ion channels that mediate fast and transient electrical responses of the neuronal membrane. Many neurotransmitter receptors, however, are not ion channels themselves, but regulate separate ion channel proteins via one or another of a variety of *signal transduction* mechanism. In this chapter, we will briefly consider the signal transduction mechanisms that link these metabotropic receptors to the ion channel responses. We will see also that the signal transduction mechanisms allow the postsynaptic cells to respond in ways that go far beyond fleeting changes in ion flux across the cell membrane.

Metabotropic Receptors Associate with Guanine Nucleotide-Binding Proteins

The metabotropic receptors form a superfamily of evolutionarily related proteins. Some examples of metabotropic receptors are the muscarinic ACh receptor (mACh-R), most of the monoamine receptors, the GABA$_B$ receptor, the receptors for many of the peptide cotransmitters, the opsins of photoreceptor cells, and odorant receptors. All of them function as monomers that conform to a certain structural motif that contains exactly **seven membrane-spanning** (hydrophobic) **segments** in the form of α-helices (Fig. 5.1). *Portions of the transmembrane segments cooperate to form a single ligand-binding site.* In addition, *the third cytoplasmic loop* and perhaps the carboxyl terminus *form the site of coupling to a* **guanine nucleotide-binding protein** or **G protein,** found in the cytoplasmic

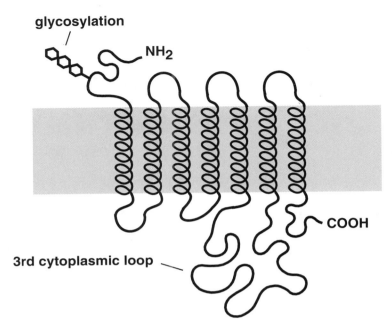

glycosylation

NH₂

COOH

3rd cytoplasmic loop

Figure 5.1 All G protein-coupled receptors conform to a motif that contains seven membrane-spanning segments. The transmembrane segments take the form of α-helices.

leaflet of the plasma membrane (Fig. 5.2). The G protein is a heterotrimeric protein, with subunits designated α, β, γ, that binds GDP, when inactive, and GTP upon activation.

G Proteins Mediate Membrane Excitability Changes

One attractive model for the mechanism of action of G protein-coupled receptors works on a principle called *collision-coupling* (Fig. 5.2). The β-adrenergic receptor serves as a lucid example. Three membrane proteins play primary roles in the signal transduction response to the neurotransmitter norepinephrin: the β-adrenergic receptor itself, the G protein, and the enzyme **adenylyl cyclase.** All three of these proteins move about laterally in the plasma membrane and frequently bump into one another. Most of the time they ignore such random encounters, however, unless norepinephrin is bound to the receptor. When it is, the conformation of the third cytoplasmic loop of the receptor is altered in such a way that it now binds to the G protein upon collision. This causes the α subunit of the G protein to exchange its GDP molecule for a GTP,

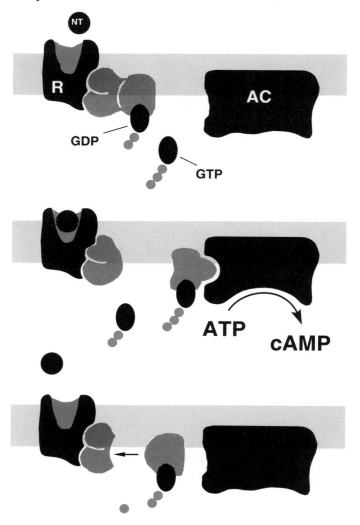

Figure 5.2 One way that metabotropic receptors may mediate signal transduction follows a collision coupling model. Abbreviations: AC, adenylyl cyclase; NT, neurotransmitter; R, metabotropic receptor protein.

an event that alters the conformation of the α subunit. As a result, it separates from the βγ dimer and, upon colliding with adenylyl cyclase, activates the enzyme to catalyze the transformation of ATP to cAMP.

The cAMP floats off into the cytoplasm to activate a protein kinase (**cAMP-dependent protein kinase** or PKA), which in turn can phosphorylate any of a large number of substrate proteins (Fig. 5.3). When bound to adenylyl cyclase, the α subunit acts as a GTPase; it dephosphorylates its bound GTP to GDP, dissociates from the cyclase, and recombines with

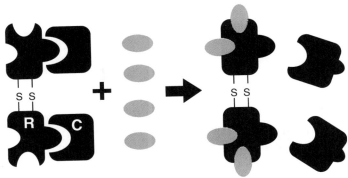

Figure 5.3 cAMP-dependent protein kinase (PKA) requires the binding of four cAMP molecules (red) to cause separation of the catalytic subunits (C) from the regulatory subunits (R). Among the many substrates phosphorylated by PKA are voltage-gated K^+ and Ca^{2+} channels.

an available βγ dimer, thus resetting the system. Incidentally, this system is directly affected by the toxin of the cholera-causing bacterium *Vibrio cholerae*. Cholera toxin causes the transfer of an ADP-ribose to the α subunit, which prevents it from hydrolyzing its bound GTP. The α-GTP is rendered persistently active at the adenylyl cyclase, resulting in an overproduction of cAMP.

Some of the substrates of PKA are ion channel proteins that, when phosphorylated, undergo conductance changes (Fig. 5.4). In some cases ion conduc-

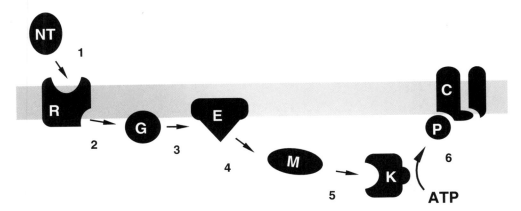

Figure 5.4 The fundamental steps in channel phosphorylation by activated kinases occur in sequence and take more time than the fast channel responses mediated by ionotropic receptors. Abbreviations: C, ion channel; E, effector enzyme; G, GTP-binding protein; K, kinase; M, intermediary protein; NT, neurotransmitter; P, phosphate; R, receptor.

tion through the ionopore is increased, in others it is decreased. How might phosphorylation bias a channel toward a particular conduction state? According to one plausible hypothesis, channel states are controlled by electrochemical interactions between nearby polar residues of the channel protein that result in conformational changes that cause the ionopore to exist in an open or closed state. When one of the interactive groups is phosphorylated, the charges are shifted to an unfavorable relationship so that electrochemical interaction cannot occur; the channel becomes locked in either the closed or the open conformation. Since the conductance change that occurs will influence membrane excitability, it is evident that G protein-mediated neurotransmission can directly influence the activity of postsynaptic neurons.

The apparent benefits of these rather complicated metabotropic receptor-induced, G protein-mediated excitability changes are at least two-fold. First, there is a considerable *amplification of effect*, since the binding of one neurotransmitter molecule to a single receptor can trigger the production of many cAMP molecules. Second, the excitability effect, *although still temporally modulated* (switched on and off by neurotransmission), *is longer-lasting than fast neurotransmission effects*; ion channel phosphorylations last from seconds to minutes. The **phosphatases** that abound in the cytoplasm see to the restoration of the channel to the dephosphorylated state.

There are several variations on the β-adrenergic receptor theme. One important variation has been discovered in which activated α *subunits bind directly to ion channel proteins*, thus bypassing the enzymatic intermediary steps (Fig. 5.5). Many metabotropic receptors that produce IPSPs use this sort of mechanism. In this instance, the activated α-GTP subunit binds to and opens a type of K^+ channel. Thus facilitated, K^+ outflow hyperpolarizes the membrane and can neutralize inward Na^+ currents that would otherwise excite the neuronal membrane. The mACh-R present in heart muscle works this way, enabling ACh to reduce cardiac output.

Signal Transduction Pathways Are Multiple and Interactive

Heterotrimeric proteins that bind GTP come in multiple forms. More than a dozen homologous genes that encode functionally different α subunits have been cloned so far. Although the βγ dimer seems somewhat less variable, there are different forms of its subunits as well that no doubt contribute to the specificity of the G protein interactions with different receptors and types of α subunit. We have seen one form of the G protein whose α subunit stimulates cAMP production (designated Gs). Another α subunit has been identified that inhibits adenylyl cyclase (Gi) and is sensitive to *Bordetella pertussis* toxin; the toxin blocks the ability of

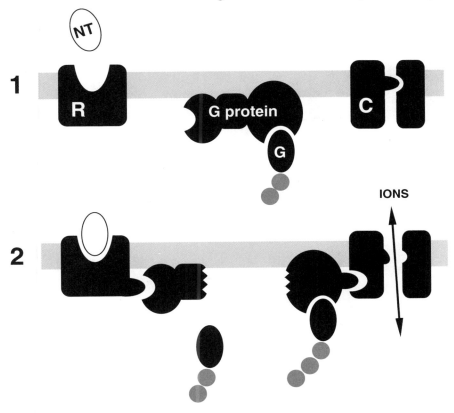

Figure 5.5 Direct interaction of the α subunit of the G protein with a channel protein can sometimes occur. Abbreviations as in Figure 5.4.

Gi to bind GTP. Still other forms of the α subunit have been isolated that do not interact with adenylyl cyclase, but that interact with other effector enzymes involved in signal transduction processes, such as cGMP phosphodiesterase and **phospholipase C.** The latter enzyme cleaves the membrane phospholipid **phosphatidylinositol bisphosphate** (PIP$_2$) into the second messengers **inositol triphosphate** (IP$_3$) and **diacylglycerol.** IP$_3$ in turn binds to a ligand-gated ion channel in the membrane of the smooth endoplasmic reticulum that allows Ca^{2+} (sequestered in this organelle by the action of Ca^{2+}-ATPase pumps) to be released into the cytoplasm. Once cytosolic, Ca^{2+} combines with a calcium-binding protein called **calmodulin** or with diacylglycerol, both of which can then activate protein kinases. The kinases, in turn, can affect ion channel behavior.

Thus, *different neurotransmitter–receptor interactions can activate different signal transduction pathways.* Moreover, *the ensuing second messenger cascades can interact with one another* (Fig. 5.6). For instance, the Ca^{2+} mobi-

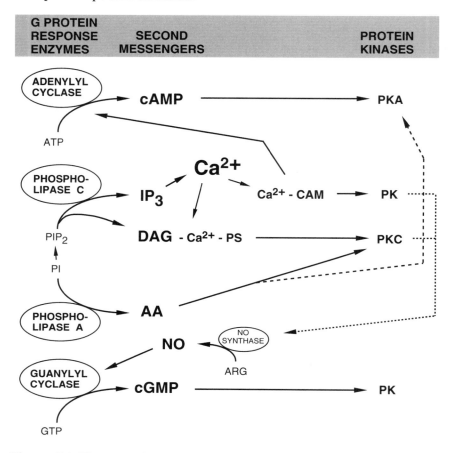

Figure 5.6 The so-called *second messengers* are intermediaries in signal transduction pathways that are highly interactive. Abbreviations: AA, arachidonic acid; ARG, arginine; CAM, calmodulin; DAG, diacylglycerol; IP$_3$, inositol triphosphate; NO, nitric oxide; PI, phosphatidyl inositol; PIP$_2$, phosphatidylinositol bisphosphate; PK, protein kinase; PKA, cAMP-dependent protein kinase; PKC, protein kinase C; PS, phosphatidyl serine.

lized by IP$_3$ activates not only diacylglycerol; when combined with calmodulin, it also activates adenylyl cyclase. To complicate matters still further, in many cases *there are multiple receptor* **subtypes** *for the same neurotransmitter* (for instance, there are three known subtypes of the β-adrenergic receptor) *that are differentially coupled to the signal transduction systems.* Finally, *receptors that bind different neurotransmitters can interact with the same form of G protein, leading to an agonistic convergence of effect.* Since individual neurons often express a half dozen or more receptor subtypes, the possibilities for subtle modulation of the postsynaptic response are legion.

Neurotransmission Can Have Far-Ranging Effects on Cell Function

In a single sensory neuron removed from the sea snail *Aplysia*, seventeen proteins are phosphorylated when cAMP is elevated by a single application of the neurotransmitter serotonin. This is not surprising, since, as we have just learned, the activation of metabotropic receptors can set into motion a cascade of phosphorylation events within the neuron. But what are these many phosphoproteins that are influenced by serotonin transmission? Some are, no doubt, ion channel proteins. But at least some of the others have been identified as metabolic enzymes and cytoskeletal components. While the answer is still incomplete, it is clear that G protein-mediated neurotransmission has significant consequences that cannot be described simply as excitation or inhibition of the excitability of the postsynaptic membrane. It also modulates structural and biosynthetic activities that may alter the computational states of the neuron, presumably to serve a physiological agenda of the nervous system yet to be discovered.

Synaptic Transmission Can Induce Genes

Another recently discovered effect of neurotransmitter–receptor interactions is the induction of gene transcription. If a gene responds *rapidly* (within minutes) by increasing its transcription rate, we call it an **immediate early gene** (IEG). Many IEGs correspond to the so-called **proto-oncogenes** that encode **nuclear regulatory proteins** (Fig. 5.7). If so, the IEG products that are translated in the cytoplasm move rapidly back into the nucleus where they bind to particular nucleotide sequences to enhance or suppress the transcription of downstream target genes. The response of IEGs is transient as well as rapid: their products remain active for minutes to hours after the genes are first induced, a time frame that would allow for regulation by neurotransmission events.

Different Receptors Can Mediate Synergistic Effects

In addition to up-regulation of IEG products, neurotransmission can activate (through phosphorylation) nuclear regulatory proteins extant in the cell (Fig. 5.8). For instance, the transcription rate of certain genes is indirectly increased when cAMP is elevated. Such genes have been shown to possess a special base sequence upstream of the coding sequence that is necessary for this response. This site in the gene is accordingly called a **cAMP response element** (CRE) to distinguish it from

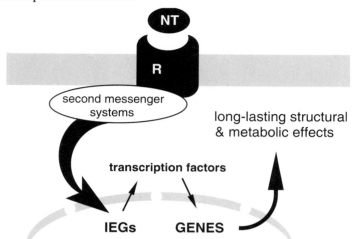

Figure 5.7 Genes can be induced and regulated by neurotransmission. Abbreviations: IEGs, immediate early genes; NT, neurotransmitter; R, receptor.

similar elements that have been identified, which respond to, for instance, glucocorticoid hormones (GRE) or phorbol ester-induced protein kinase C (PRE). The CRE binds a protein aptly named the **CRE-binding protein** (CREB). The affinity of CREB for CRE is greatly enhanced when CREB forms a homodimer by way of a *leucine zipper* motif in its peptide sequence. Although CREB homodimers are constitutively bound to the CRE, they do not engage the protein complex that makes up the transcription machinery unless properly phosphorylated. It now appears that CREB is a substrate for PKA, which is activated by cAMP; protein kinase C (PKC), which is activated by Ca^{2+}-diacylglycerol; and calcium calmodulin-activated kinase type II (CAM kinase II). Each kinase phosphorylates a different set of serine, threonine, or tyrosine residues in CREB. Phosphorylation by CAM kinase II, for instance, inhibits the activity of CREB. In contrast, phosphorylation by PKA or PKC increases CREB-mediated transcription. Theoretically, the dimerization of CREB and its activity as a nuclear regulatory protein can be modulated according to which combination of kinases is activated. Since a given neurotransmitter–receptor interaction will usually activate only one signal transduction pathway, it follows that two or more transmitter–receptor events involving different neurotransmitters that occur in close temporal proximity could subtly modulate CREB activity. Thus, in the case of CRE-containing genes, *different neurotransmitters could act via different signal transduction mechanisms in a synergistic manner to regulate the transcription rate of CRE-containing genes*. The extent to which this principle applies to other regulatory proteins, such as the many IEG products (which are, by the way, phosphoproteins that engage in homo- or het-

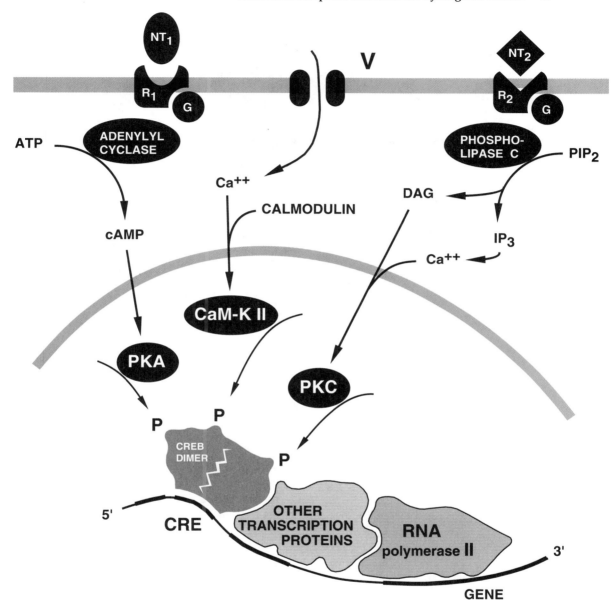

Figure 5.8 The diagram schematizes a model for neurotransmitter regulation of the cAMP response element-binding protein (CREB) by differential phosphorylation. Presumably, the differential phosphorylation triggered by different signal transduction pathways can increase or decrease the efficiency of CREB interaction with the many proteins that regulate gene transcription rate. Abbreviations: CaM-K II, calcium-calmodulin-dependent protein kinase type II; CRE, cAMP response element; DAG, diacylglycerol; G, GTP-binding protein; IP3, inositol triphosphate; NT1, NT2, neurotransmitters; P, phosphate; PIP2, phosphatidylinositol bisphosphate; PKA, cAMP-dependent protein kinase; PKC, protein kinase C; R1, R2, receptors.

erodimerization), remains to be established. But the notion that different combinations of second messengers can lead to differential gene expression provides insight into the question of why so many different neurotransmitters and receptors exist. It could be that, once again, Mother Nature has devised an exquisitely sensitive means for the fine regulation of important postsynaptic responses to synaptic transmission.

Some Regulated Genes Encode Neurotransmission-Related Products

Some of the genes influenced by synaptic transmission encode the very currency of synaptic transmission itself—peptide cotransmitters, transmitter-related enzymes, receptors, and ion channels. This can be demonstrated in studies in which transmitter–receptor interactions are manipulated pharmacologically and measurements are made of the transcription rate and the levels of mRNAs and translation products. The most clear-cut example comes from studies of the biosynthesis of peptide cotransmitters.

In Chapter 3, we learned that the presynaptic level of all neurotransmitters must be regulated so as to maintain a steady state in the face of activity-induced turnover. Recall that the biosynthesis of the amino acid and biogenic amine classes of neurotransmitter is regulated mainly locally in the axon terminal by direct actions on the relevant catalytic enzymes (e.g., end-product inhibition of tyrosine hydroxylase by catecholamines). In contrast, the several neuroactive peptides, which have emerged more recently as a third major class of synaptic signal molecule, are regulated neither locally nor directly. Instead, they are largely dependent on transcription and translation events at the nucleus and cell body for maintenance of their levels at distant synaptic sites.

Over the past decade, neuroscientists have discovered that the neurotransmitters released by axons that synapse on peptidergic neurons to control their membrane activity patterns (EPSPs and IPSPs) also serve as signals that specifically and differentially influence the biosynthesis of the peptide cotransmitters used by the peptidergic neurons. This in turn can affect the amount of peptide cotransmitter available for release at synaptic endings. For example, chronic attenuation of dopamine transmission (in a particular brain system called the nigrostriatal pathway) decreases the biosynthesis of one known peptide cotransmitter, substance P, and increases that of another, enkephalin. These alterations are preceded by and proportional to changes in the levels of the mRNAs that encode the respective precursor peptides, suggesting that the regulatory point influenced lies at the transcriptional level. Likewise, changes in serotonin, GABA, and glutamate neurotransmission have been shown to influence the genes encoding a variety of neuroactive peptides used as synaptic transmitters in the brain.

Implications for Nervous System Function

To summarize, we now know that when a neurotransmitter binds to its postsynaptic receptor, it can engage a broad response repertoire of intracellular events that have several consequences for neuronal function. In addition to the immediate and relatively brief fluctuations in membrane excitability mediated by changes in ion channel conductance, a host of changes in enzyme activity can occur. Like some forms of hormonal signaling, neurotransmission can cause a modification of target neuron function through effects on the transcription of genes that encode molecules relevant to the process of synaptic transmission itself. These neurotransmitter-induced alterations of metabolism and gene expression hold powerful implications for understanding the molecular mechanisms of learning and memory and will be resurrected in Chapter 29.

Many of the neurological diseases encountered in the clinic result from abnormalities of synaptic transmission that are often associated with a primary loss of a particular neurotransmitter. The loss of a neurotransmitter at a relatively circumscribed set of synaptic contacts can have far-reaching effects on the neurotransmitter content of otherwise intact downstream neurons. For example, alterations in peptide cotransmitters following disrupted dopamine transmission (as in Parkinson's disease) or glutamate transmission (as in Alzheimer's dementia) imply that the motor and cognitive deficits in these pathological states may be due in part to functional changes induced in the peptidergic neurons deprived of input mediated by the deficient neurotransmitters. As we learn more about such downstream effects, they become increasingly accessible to selective pharmacotherapeutic intervention.

Detailed Reviews

Berridge MJ, Irvine RF. 1989. Inositol phosphates and cell signalling. *Nature* 341: 197.

Bourne H, Sanders D, McCormick F. 1991. The GTPase superfamily: conserved structure and molecular mechanism. *Nature* 349: 117.

Edelman A, Blumenthal D, Krebs E. 1987. Protein serine/threonine kinases. *Annu Rev Biochem* 56: 567.

Nicoll RA. 1988. The coupling of neurotransmitter receptors to ion channels in the brain. *Science* 241: 545.

6

Glial Cells and Their Functions

Despite the central role of neurons for information processing, there are about tenfold more glial cells in the CNS than there are neurons. Glial cells (or just **glia**) are close cousins to neurons; both arise from a common ancestral population of **progenitor** (or **stem**) **cells** at the earliest stages of CNS development. Stem cells give rise to **neuroblasts** and **glioblasts** that are *committed* by genetic programming to become, respectively, neurons and glia. The glia of the CNS arise from stem cells in the ventricular zone of the neural tube, and the glia of the PNS arise from progenitors in the neural crest (see Chapter 7). How do glia differ from neurons? Structurally, glia have smaller cell bodies (5 to 10 micrometers in diameter) than most neurons, and they extend only one kind of process rather than two. We can also draw three functional distinctions. First, *glia retain the ability to proliferate under certain conditions, whereas neurons are postmitotic.* The second and third distinctions are closely related: *glia do not conduct action potentials* and *do not form chemical synapses with other cells.* In other words, the glia do not directly participate in the conduction or transmission of information. However, the glia assist the signaling capacity of neurons in many essential ways.

The Types of Glia and Their Selective Markers

Within the CNS, we can identify two general kinds of glial cell: **oligoden-drocytes** ("cells of few branches") and **astrocytes** ("star-shaped cells"), which are morphologically and functionally distinct (Fig. 6.1). The PNS counterpart of the oligodendrocyte is the **Schwann cell** found in the many nerves that course throughout the body and in the sensory and autonomic ganglia, where they are called **satellite cells.** There are no astrocytelike cells in the PNS. *The oligodendroglia and Schwann cells are*

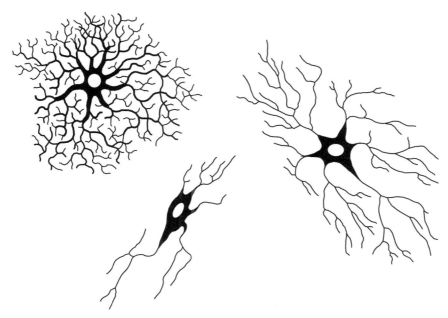

Figure 6.1 The different types of glia found in the CNS are protoplasmic astrocytes (upper left), fibrous astrocytes (right), and oligodendrocytes (lower left).

specialized to form the myelin that ensheaths axons and enables saltatory conduction of action potentials. Many oligodendroglia of the gray matter and the satellite cells of the peripheral ganglia do not elaborate myelin, but lie in close apposition to neuronal cell bodies; their function is unknown.

Astrocytes are found in all parts of the CNS intermingled with neurons. Astrocytes come in two main types, called **protoplasmic** (type 1) because of their generally short and pudgy processes and **fibrous** (type 2) because of their long, slender processes chock-full of intermediate filaments (Fig. 6.1). Within these two main types, there are subtypes of astrocytes that are highly specialized in shape, function, and location. One example consists of the various forms of **radial glia** that function as scaffolds for neuronal migration during development (see Chapter 30). Another is the monolayer of **ependymal cells** that line the fluid-filled cavities (ventricles) at the core of the CNS.

Mature protoplasmic and fibrous astrocytes extend multiple processes that insinuate between the dendrites and axons of the neurons in their vicinity (Fig. 6.2). As a result, *the cellular elements of the CNS are tightly packed together and the volume of fluid-filled extracellular space is minimized.* This abundance of astrocytes and their proximity to neuronal elements should not be construed as evidence that glia exist as "packing material"

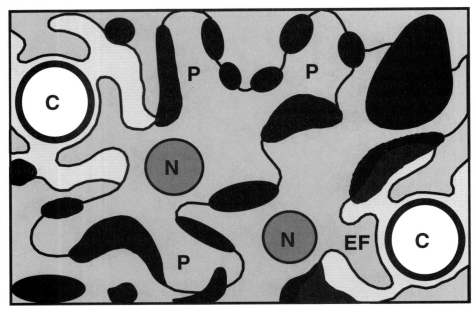

Figure 6.2 This drawing of a high-power micrograph shows the tightly packed tissue elements in the CNS. The astrocyte processes (shaded red) surround neuronal elements (jet black) and extend end-feet near capillary walls (black rings). Abbreviations: C, capillary; EF, astrocyte end-foot; N, astrocyte nucleus; P, astrocyte processes.

for the brain. Much to the contrary, the glia serve a multitude of functions. To list but a few, they isolate the electrical fields of neuronal membranes from one another, they maintain local ionic conditions in the face of activity-induced changes, and they scavenge excess neurotransmitter from the synaptic cleft, sometimes to perform essential recycling.

Astrocytic processes also lie in close apposition to the endothelial cells of brain capillaries and to the pial surface of the CNS, where they form bulbous endings called **end-feet** (Fig. 6.3). Though once considered as a physical manifestation of the blood–brain barrier, the gap junctions that loosely connect the astrocytic end-feet are unable to block the passage of even large molecules such as globular proteins. In contrast, *the endothelial cells, which form tight junctions with one another, do occlude the passage of fluid-borne molecules.* Although the function of the end-feet remains uncertain, it may involve the regulation of water balance and ionic composition in the extracellular fluid compartment of the CNS (see below).

Because they all exhibit processes, astrocytes of different types often resemble one another, oligodendrocytes that have not elaborated myelin, and even small multipolar neurons. A convenient method used these days to distinguish glia from neurons, and one type of glial cell from

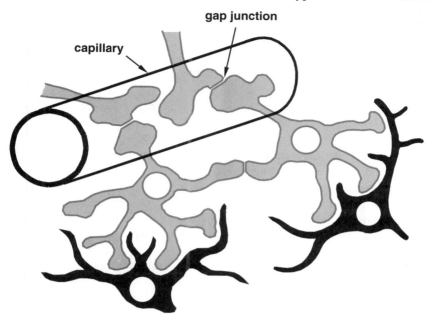

Figure 6.3 Astrocyte processes (shaded red) lie close by all neural elements (jet black), form gap junctions with one another, and extend end-feet near capillary walls.

another, exploits particular antigenic molecules that are specifically expressed by each. Thus, *oligodendrocytes express a glycolipid called* **galactocerebroside,** *whereas astrocytes are galactocerebroside-negative. Astrocytes, but not oligodendrocytes, express a protein called* **glial fibrillary acidic protein** (GFAP). Neurons express neither of these "marker" molecules. Thus, antibodies that specifically recognize one or the other of the two molecules can be applied in an immunocytochemical procedure to tell an oligodendrocyte from an astrocyte. More recently, antigenic cell surface proteins have been discovered that are specific to each of the two types of astrocyte so that they, too, can now be distinguished. Using such antigen-based indentifications, it has been possible to show that although astrocytes are ubiquitous in the CNS, the *protoplasmic tend to keep more to the gray matter and the fibrous seem partial to the white.* The ability to clearly identify glial subtypes is clinically useful in the diagnosis of tumors of glial origin. Since glia remain mitotically competent throughout life and neurons do not, brain tumors in adults are almost always of glial origin. The same immunocytochemical labeling that enables neuroscientists to study the distributions and functions of the different classes of glia also allows pathologists to distinguish between glioblastomas, astrocytomas, ependymomas, and so forth. One day, the distinct phenotypic traits may provide a means for addressing therapeutic drugs to the offending cell type.

Myelin

All axons of the CNS and those in peripheral nerves are surrounded and isolated from one another by, respectively, oligodendrocytes and Schwann cells. If the axons are of sufficiently large diameter, these glial cells elaborate an extensive sheet of plasma membrane that wraps spirally around a length of axon cylinder and compacts to form tight concentric layers enclosing little, if any, cytoplasm. In general, the number of wrappings (up to a hundred) and the length of axon covered are proportional to the diameter of the axon. *These layers of glial cell membrane constitute the myelin sheath* (Fig. 6.4). Each length of myelin, called an **internode,** is provided by a single glial cell and is separated from the next internode by a short stretch of nonmyelinated axon called a **node** (of Ranvier). The first internode occurs immediately distal to the initial segment of the axon, after which they occur in a series separated by nodes until a point is reached near the terminal of the axon, which is also nonmyelinated. By extending processes, single oligodendrocytes are able to form an internode around several nearby axons and a few adjacent internodes on the same axon, whereas Schwann cells dedicate themselves to the formation of a single internode on one axon.

In the process of myelin formation, the myelinating cell membrane first surrounds the axon (Fig. 6.5). When the faces of membrane come into contact, one tucks under the other and grows along the surface of the axon membrane. As the growing edge of membrane completes each cycle around the cylinder of axon, it displaces the previous membrane layers outward from the axon surface. While the external surfaces of the myelin membrane remain in contact throughout the process, the subsequent compaction (squeezing out of cytoplasm) brings the inner (cytoplasmic) surfaces of the myelin membrane into close contact as well. The close apposition of these inner faces of membrane creates a line, which, because of the abundance of integral membrane protein, appears thicker and denser in the electron microscope than does the line created by the external surface appositions. As a result, the concentric spirals of myelin are described as alternating **major dense lines** (inner membrane appositions) and **intraperiod lines** (external membrane appositions).

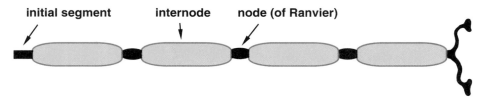

Figure 6.4 Axons are often myelinated like the one drawn here. The myelin is shaded red and the axon jet black.

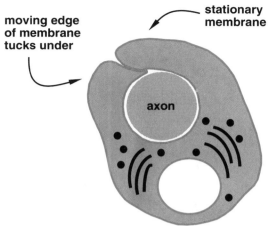

moving edge of membrane tucks under

stationary membrane

axon

Figure 6.5 The first step in myelination of an axon by an oligodendrocyte or a Schwann cell (red) requires one edge of its plasma membrane to tuck under the other.

Although central and peripheral myelin differ, both exhibit a molecular composition of phospholipids, proteolipids, and glycolipids that is unique among eukaryotic membranes. Molecular cloning research has revealed three major genes expressed by myelinating cells that code for **protein zero** (P_0), **proteolipid protein** (PLP), and **myelin basic protein** (MBP). P_0 is an integral membrane glycoprotein that is the major structural protein in Schwann cell myelin; oligodendrocytes do not express it. Conversely, PLP is the major integral membrane protein of central myelin. Both central and peripheral myelinating cells express MBP. P_0 is a homophilic cell adhesion molecule that promotes the adhesion of the extracellular surfaces during myelination in the CNS. A mutant mouse strain that lacks the PLP gene (called Jimpy) has poorly myelinated axons and associated motor disturbances. Studies in Jimpy mice indicate that PLP may perform a cell adhesion role for central myelination similar to that of P_0 in the PNS. MBP is not an integral membrane protein, but microscopic observations in a mouse mutant lacking the MBP gene (this strain is called Shiverer) indicate that it is essential for the compaction step in myelin formation.

The role of myelin as an insulator of axonal membranes and an augmentor of conduction velocity was introduced in Chapter 2. The degeneration of myelin can result from an autoimmune reaction, as in *multiple sclerosis*, or from exposure to certain toxins present in the environment. Since the breakdown of myelin causes deficits in axon conduction, the symptoms will depend on the parts of the CNS or PNS that are affected and can range from simple sensory and motor dysfunctions to cognitive impairments.

In addition to this role in conduction, oligodendrocytes and Schwann cells may take part in the guidance of axons to their synaptic targets during development and during the regrowth of axons after injury to the nervous system (see Chapters 31 and 32).

Astrocyte Functions

You learned in Chapter 2 that the electrical activity changes of neurons are mediated by ion fluxes across their membranes. Whenever neurons are depolarized (excited) or hyperpolarized (inhibited), a small amount of K^+ escapes into the extracellular fluid. Since, as noted above, the volume of the extracellular fluid compartment is very low compared to the volume of the cytoplasmic compartment, the concentration of K^+ would soon build to an unhealthy level were it not for the tireless pumping of the Na^+-K^+ ATPases. Much of the time, however, neurons are so active that they need help in maintaining the correct ionic compositions on the two sides of their membranes. The omnipresent astrocytes provide this help; their membranes are studded with Na^+-K^+ exchangers that rapidly rid the extracellular fluid of excess K^+. In fact, there is so much more astrocytic than neuronal membrane per unit volume of CNS tissue that the astrocytes take up most of the extra K^+ from the extracellular fluid. What becomes of the K^+ taken up by astrocytes? It diffuses rapidly along its concentration gradient within the cytoplasm, and since the astrocytes are coupled by numerous gap junctions, it passes from one astrocyte to another. Taken together, the processes of astrocytes form a three-dimensional maze of conduits that are continuous with one another through the gap junctions (Fig. 6.3). This conduit system leads ultimately to the end-feet at capillaries and at the pial surface of the CNS, where the excess K^+ is discharged into the blood and cerebrospinal fluid by an abundance of K^+-Na^+ exchangers and perhaps by voltage-sensitive K^+ channels in the end-foot membrane. This system allows the part of the astrocyte membrane that is associated with neuronal elements to always favor the uptake of K^+. The structural and functional organization of this astrocyte "drainage" system is thus reminiscent of the lymphatic drainage of the interstitium of body tissues.

The K^+ that escapes from active neurons can, and often does, depolarize adjacent astrocytes. In fact, the resting potential of astrocytes (which is slightly higher than that of neurons) is so sensitive to extracellular K^+ levels that astrocytes can be used as K^+ electrodes to accurately monitor the concentration of K^+ in the extracellular fluid. The depolarization that occurs in one astrocyte readily spreads electrotonically to other astrocytes via the gap junctions that join them. Such a local spread of electrical activity may act as a signal that affects the metabolism of astrocytes in significant, but as yet unknown, ways. One hypothetical, but plausible, effect of this local depolarization spread is the activation of voltage-dependent K^+ channels at the capillary end-feet that would permit the release of excess K^+ from the astrocyte cytoplasm.

Astrocyte processes surround and isolate synapses in the CNS. *This prevents the neurotransmitter that is released from a particular axon ending from activating nearby unrelated receptive surfaces and thus ensures spatially discrete signaling between neurons.* The proximity of the astrocytic processes to the synaptic cleft also enables participation of the astrocytes in neurotransmitter deactivation (uptake). You'll recall from Chapter 3 that, in the case of glutamate, GABA, or 5HT, *astrocytes express high-affinity transporters for the neurotransmitter released by the axon ending with which they are associated.* The astrocytes can then dispose of the neurotransmitter through enzymatic degradation to an inactive form or, in the case of glutamate, at least participate in a recycling mechanism.

It has recently been discovered that *astrocytes synthesize and release* **trophic** *molecules and express cell surface proteoglycans, both of which guide the development of the CNS in many subtle ways.* Even in the mature nervous system, *different astrocytes secrete different trophins that are necessary to sustain specific neuronal populations.* These important functions of glial cells will be more fully discussed in the chapters (30 and 31) that deal with the molecular control of CNS development.

Yet another important function of glia is their response to CNS tissue damage caused by trauma and infection. Through some mechanism yet to be illuminated, *tissue damage within the CNS triggers a characteristic astrocyte reaction.* Reactive astrocytes become enlarged, increase their metabolic activity, invade the injured tissue, and, to a limited extent, phagocytose debris. Presumably, this response is an adaptive one that minimizes detrimental secondary effects due to the uncontrolled release of active substances from the injured neural tissues. The result of this process is often a **glial scar** (for more, see Chapter 32).

Resident Macrophages of the CNS

Another type of cell found in the parenchyma of the CNS, called **microglia,** has been acknowledged by many students of the brain since the early part of the twentieth century. Conventional wisdom holds that the microglia are monocytes that migrate into the CNS during late prenatal and early postnatal development. In the adult, there are about one-tenth as many microglia as true glia, and they are sprinkled mainly throughout the gray matter. Microglia lie dormant until the integrity of the CNS is challenged. They are sensitive to any number of injury-related factors that cause them rapidly (within hours) to differentiate into motile cells that express a number of macrophagelike molecules, including the major histocompatibility class I and II antigens. In the activated state, they scurry about performing a number of damage-control functions. They surround damaged tissue and contribute to the restriction of infectious and toxic materials. They activate astrocytes and perhaps attract peripheral macrophages to the sites of injury. In the presence of neuronal degeneration, they become phagocytotic, extending arms that engulf debris. Unfortunately, when activated to this extreme degree, they also begin to promote inflammation and necrosis through the secretion of reactive oxygen species. On a topical note, microglia are the only cell type in the brain known to be infected by human immunodeficiency virus (HIV) and may be crucial to the development of the acquired immunodeficiency syndrome (AIDS) dementia complex.

Ependymal Cell Functions

The ventricles of the adult CNS are lined by a simple epithelium of cuboidal ciliated cells that are GFAP-positive. This epithelium is called the ependymal layer or simply the **ependyma.** The cilia of the ependymal cells beat actively in life and may assist the flow of cerebrospinal fluid through the ventricular system. Evidence suggests that there is regional specialization of the ependyma and that different types of ependymal cells exist that serve different functional roles in the CNS. For instance, *the ependyma of the choroid plexus is important in the formation of cerebrospinal fluid from blood serum.* The choroid plexuses form in the developing brain ventricles as highly vascularized invaginations of the thin pial and ependymal epithelia (see Chapter 8). The endothelial cells of the capillaries within the choroid plexuses lack tight junctions, a condition referred to as a *fenestrated* capillary bed. All but formed elements and large proteins can escape from the bloodstream into the fluid space between the capillaries and the monolayer of ependymal cells. This serum may provide the raw material that is taken up by the cells of the ependymal epithelium that secrete the cerebrospinal fluid into the ventricles.

Some ependymal cells synthesize and secrete biologically active products into the cerebrospinal fluid, whereas others may be engaged in the active transport of ions, water, and other molecules between the cerebrospinal fluid and the extracellular fluid compartments. Definitive data on the ependymal cells are still too scant to permit unifying concepts about their impact on CNS functions.

Detailed Reviews

Federoff S, Vernadakis A (eds). 1986. *Astrocytes,* vols 1–3. Academic Press, Orlando, FL.
Lemke G. 1988. Unwrapping the genes of myelin. *Neuron* 1: 535.
Perry VH, Gordon S. 1988. Macrophages and microglia in the nervous system. *Trends Neurosci* 11: 273.

MORPHOGENESIS AND TOPOGRAPHIC ANATOMY

II

Armed as you now are with a grasp of neuronal and glial function and faced with an anatomical and functional analysis of brain circuits, it is important that you become familiar with the general anatomical plan of the adult CNS. Whereas the spinal cord is a linear structure that poses little challenge to comprehension, the brain is another matter altogether. It is characterized by differential expansions and foldings of its component parts into a three-dimensionally elaborate organ that rivals the complexity of the joints used in traditional Japanese furniture making. And, like furniture joints, much of the brain structure is hidden from view. One efficient way to grasp the three-dimensional complexity of the adult human brain is by dissection—a method highly recommended to the novice. Another way is by following its embryonic morphogenesis, as we shall do in the three chapters that follow.

The Basic Plan and Ontogeny of the Spinal Cord

7

The CNS consists of the brain and the spinal cord, which are contained entirely within the cranial vault and the vertebral canal. The CNS can be distinguished from the PNS, which is made up of neurons (most often contained in ganglia) and their processes (which bundle together to form nerves) that are distributed throughout the body. Given the delicate consistency of the CNS—something akin to aspic—it is fortunate that Mother Nature has encased this vulnerable organ within a hard bony fortress. Within the fortress, the CNS is further protected by three meningeal coatings. From outside in, the meninges are the thick and leather-tough **dura mater,** the diaphanous **arachnoid mater,** and the **pia mater,** which is everywhere adherent to the surface of the CNS. The CNS is further cushioned by a layer of viscous liquid called the **cerebrospinal fluid.** It fills a space between the arachnoid and the pia mater called the subarachnoid space. The cerebrospinal fluid also fills the hollow core of the brain, the **ventricular system,** increasing further the shock-absorbing capacity of the brain. Although these raiments of the brain serve their wearer well, their indirect relationship to the workings of neuronal circuits places them beyond the scope of this survey. Here, we'll focus on the wearer.

The Basic Plan of the Adult CNS

Let's begin our anatomical overview of the human CNS with a quick look at the end product of morphogenesis. If you isolate the mature CNS from the rest of the body and strip away all its coatings and superficial blood vessels, you are left with a large bulbous brain sitting atop a gracefully slender stalk of spinal cord (Fig. 7.1). Most of what you see of the brain is the outermost covering of one of the paired **cerebral hemispheres:** the highly convoluted **cerebral cortex.** Each of the folds of this cortex is a **gyrus,** and each of the furrows between them is a **sulcus** or, if particularly deep, a **fissure.** Tucked under the occipital end of the cerebral hemispheres you see a smaller sulcate structure. This is the cortical surface of the **cerebellum,** and its folds are called **folia.** You can probably discern a bit of tissue anterior to the cerebellum and inferior to the cerebral hemisphere. It is part of the **brain stem,** most of which is hidden from view behind the cerebral hemisphere. Finally, you notice that the brain stem is continuous below with the spinal cord.

The CNS is bilaterally symmetrical: a perfect slice down the midline would split the CNS into a right and a left half that are mirror images of one another. Whatever structures one identifies on the right side would also be present on the left. If we were to make our sagittal slice just a few millimeters off the midline and examine the flat surface that is exposed, we would see a complicated pattern of gray and white matter—nuclei and fiber tracts—that is representative of either half (Fig. 7.2). Further inspection would reveal several salient features. For one thing, you would appreciate that the cortex is only the outermost portion of the

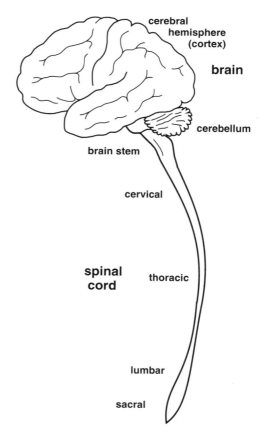

Figure 7.1 The sketch shows a lateral view of the human CNS in isolation.

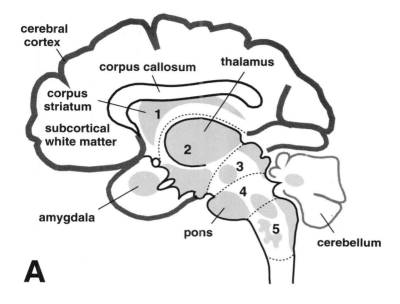

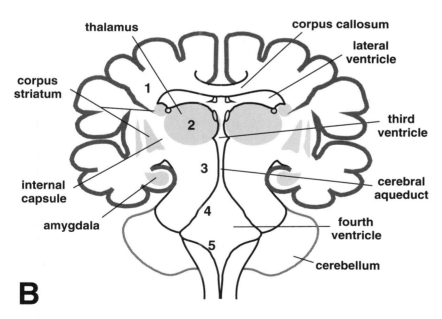

Figure 7.2 Sagittal (**A**) and frontal (**B**) sections through the adult CNS show the five major subdivisions called the (1) telencephalon, (2) diencephalon, (3) mesencephalon, (4) metencephalon, and (5) myelencephalon. Some of the major subcortical nuclear groups are shown in red.

cerebral hemisphere; deep to it lies a massive core of white matter, aptly called subcortical white matter, and a number of **subcortical** nuclei. We can use landmarks visible on this sagittal section to distinguish the *five major subdivisions* that are somewhat arbitrarily identified in the human brain. The cerebral hemispheres—there are two of them, one on either side of the midline—constitute the first subdivision, the **telencephalon** ("end-brain"). It is attached to the second subdivision, the **diencephalon** ("between brain"), so named because it sits astraddle the midline between the left and right hemispheres. The two major parts of the diencephalon are the dorsal **thalamus** and, ventral to it, the **hypothalamus.** Together, the telencephalon and diencephalon are referred to as the **forebrain.** The diencephalon is continuous caudally with the third subdivision, a narrow part of the brain stem called the **mesencephalon** or simply the **midbrain.** Below the midbrain, the brain stem expands due to the presence of two add-ons; on the ventral surface is a mass of neural tissue called the **ventral pons,** which is attached to the dorsally situated **cerebellum** by a massive pair of fiber straps. The ventral pons and cerebellum, together with the core of the brain stem, the **pontine tegmentum,** make up the fourth subdivision, the **metencephalon.** Caudal to the pons is the fifth subdivision, the **myelencephalon** or **medulla,** which finally tapers into the spinal cord.

Three groups of subcortical gray matter are contained within each of the two cerebral hemispheres. Two of them can be seen in Figure 7.2A: the **corpus striatum** and the **amygdala.** The third group, the **ventral forebrain nuclei,** lies alongside the midline in the base of each hemisphere and is not visible in our sagittal section. The cerebral hemisphere of either side is continuous with the diencephalon, which itself caps the uppermost end of the brain stem. In a coronal slice that passes downward through the forebrain and into the brain stem (Fig. 7.2B), you can appreciate that the attachment consists of a prominent fiber bundle called the **internal capsule.** A great many of the axons that make up the internal capsule originate from cells in the cerebral cortex and descend to make synaptic connections in the thalamus and points lower down in the brain stem and spinal cord. Other axons in the internal capsule arise in the thalamus and ascend to synapse in the cortex. Another massive fiber bundle is apparent within the forebrain called the **corpus callosum;** its axons connect the cortices of the two cerebral hemispheres with each other. In the sagittal slice (Fig. 7.2A), the fibers of the corpus callosum are cut in cross section; the plane of the coronal slice runs parallel to the course of the axons as they pass between the hemispheres (Fig. 7.2B). In addition to these more evident nuclei and tracts, a multitude of more or less well-defined nuclear groups and fiber tracts are present within each of the five subdivisions of the brain. We'll learn about many of them in the context of functional systems in the chapters ahead. For now, let's try to better understand the topographic anatomy of the adult CNS by following its elaboration in the embryo.

The Formation of the Neural Tube

The earliest stages in the development of the nervous system happen quickly and are morphologically, if not mechanistically, simple compared to what goes on later (Fig. 7.3). Between days 18 and 24 of gestation, the dorsal midline ectoderm of the embryo folds up on either side of the midline until, beginning at a more or less mid-rostrocaudal site, the folds on either side touch one another and fuse. Zipperlike, the neural folds progressively close over in both the caudal and the rostral directions until a tubular structure, the **neural tube,** is formed (Fig. 7.3). As this folding occurs, a population of cells buds off the ectoderm dorsolaterally near the crest of the neural folds to form a cluster of **neural crest cells** outside the neural tube on either side. The crest cells migrate to form the neurons and supportive cells of the PNS: the nearby **dorsal**

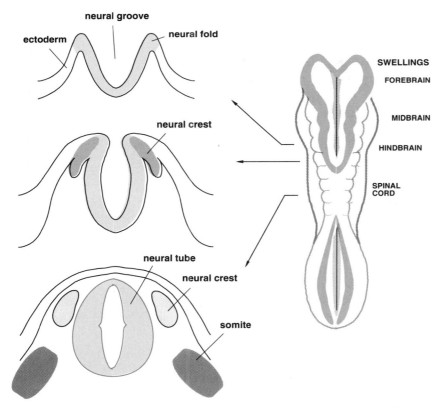

Figure 7.3 A series of sections (left) through different rostrocaudal levels of an embryo at approximately 3 weeks of gestation (shown from a dorsal viewpoint on the right) indicates the early stages in the closure of the neural folds into a tube (red).

root ganglia, the para- and prevertebral sympathetic ganglia, and most distantly, the juxta- and intramural parasympathetic ganglia. The epinephrin-secreting chromaffin cells of the adrenal medulla also derive from the neural crest cells.

Sometimes the neural tube fails to completely close over during development. When this happens, neural tissue remains exposed at the surface of the body. If the failure occurs at the rostral end of the neural tube, a malformation known as *anencephaly* ensues, in which the whole of the forebrain or at least the cerebral hemispheres fail to develop. As you might suspect, this condition rapidly leads to death if the baby comes to term at all. More likely to survive are babies in which the caudal end of the neural tube has not completely closed, leading to a condition known as *spina bifida.* The degree of exposure varies between a relatively mild exposure of the vertebrae to a more serious form in which the spinal cord and spinal nerves are exposed. This extreme condition is called *myelomeningocele* and is often accompanied by *hydrocephalus,* the accumulation of cerebrospinal fluid in the lateral ventricles.

The major subdivisions of the human brain are foreshadowed even before the neural folds begin to meet and fuse to form the neural tube (Fig. 7.3). This foreshadowing takes the form of thickenings at the rostral end of the neural ectoderm. Rostral to caudal, the thickenings are named the forebrain, midbrain, and hindbrain (or rhombic) enlargements. The hindbrain enlargement exhibits three smaller swellings, called rhombomeres, separated by shallow transverse grooves. The longer, caudal end of the neural tube retains a more constant diameter and will give rise to the spinal cord.

Neurogenesis

When the neural tube first closes, it remains as a columnar epithelium in which the bipolar cells have a radial orientation, with one process attached to the outer surface of the tube and the other to the lumenal surface. These cells are the progenitors of all the neurons and glia of the CNS and are therefore exuberantly mitotic (Fig. 7.4). Cell division occurs after a temporary retraction of the distal process, so that during metaphase and anaphase the cell bodies lie adjacent to the lumen. Most of the early cleavages are in the radial plane, so that the two daughter cells retain their attachment to the lumenal surface and both re-extend a process to the surface to re-enter the mitotic cycle. Later on, cell divisions occur with increasing frequency in a plane tangential to the lumen, so that one daughter loses its remaining inner attachment and can freely move away from the lumen of the neural tube and forsake the mitotic cycle. The factors that regulate the number of mitoses that a given progenitor cell will perform are unknown.

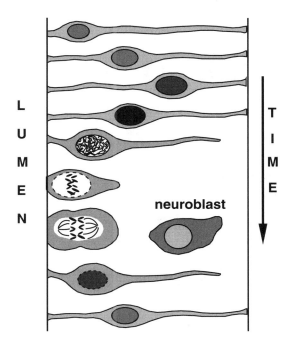

Figure 7.4 The repeated mitosis of progenitor cells in the ventricular zone of the neural tube gives rise to migratory neuroblasts. In this drawing, one progenitor is followed through a single cell division over time.

As increasing numbers of cells leave the mitotic cycle, the neural tube transforms itself from a cellular monolayer into a structure that consists of three concentric layers: a **ventricular zone** where mitosis continues unabated, an **intermediate zone** composed of postmitotic neuroblasts and glioblasts that will eventually form the gray matter core of the spinal cord, and a **marginal zone** where the growing axons of the neuroblasts accumulate to eventually form the white matter surround of the spinal cord (Fig. 7.5). As the intermediate zone thickens by the addition of more and more postmitotic cells, the cells coalesce and differentiate into nuclei according to influences within the local environment that are manifold and incompletely understood at present. The molecular mechanisms that govern ultimate cellular phenotype and nuclear formation will be discussed in Chapter 30.

Spinal Cord Ontogeny

Since the spinal cord remains the simplest part of the CNS, we'll begin our structural analysis by following its development in the remainder of

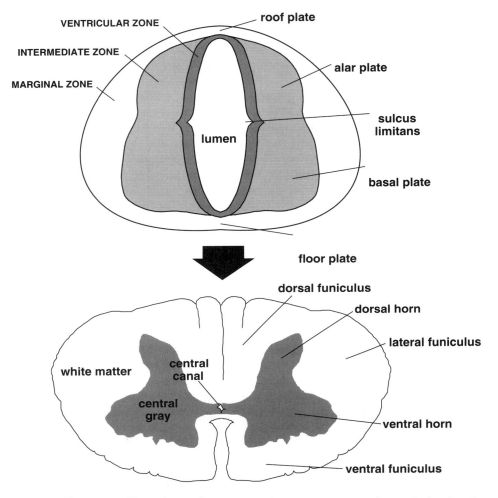

Figure 7.5 Drawings of representative cross sections through the developing (top) and mature (bottom) spinal cord show the relatively simple ontogeny of the cord.

this chapter. The development of the brain stem and the forebrain will follow in, respectively, Chapters 8 and 9. In the lumenal walls of the neural tube, a shallow groove, the **sulcus limitans,** is visible about midway along the dorsoventral axis. An imaginary horizontal plane that passes through the grooves divides the emergent gray matter of the spinal cord into a dorsal **alar plate** and a ventral **basal plate.** The neural crest-derived neurons that congregate as the dorsal root ganglia grow their sensory axons into the alar plate, where they synapse on the differentiating secondary sensory neurons. Neurons destined

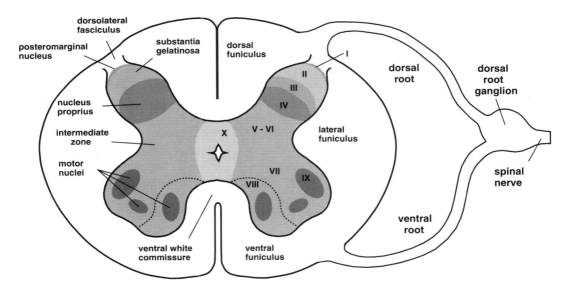

Figure 7.6 A drawing of a cross section through the adult spinal cord shows its basic organization. For reference purposes, the gray matter is divided according to named nuclei (left) or a scheme of layers (right) devised by the Swedish neuroanatomist Bror Rexed.

to send their axons out of the CNS—the motoneurons—coalesce into nuclei of the basal plate, and their axons emerge from the spinal cord ventrally.

The intermediate and marginal layers continue to expand as more and more neurons and glia are added. At some point, mitosis stops, and the ventricular zone remains as a single cell layer of nonmitotic ependymal cells lining the lumen of the neural tube. Eventually, the portion of the lumen that becomes the central canal of the mature spinal cord collapses until it is no longer patent in adult humans.

The segmental organization of the spinal cord is imposed by the incoming sensory fibers. Dorsal root ganglia form in each somite, and the neurons of the ganglia send their axons into the spinal cord as the dorsal roots (Fig. 7.6). Each dorsal root fans out into rootlets over a short rostrocaudal extent of the cord, thus more or less defining a segment. The axons of motoneurons leave the spinal cord ventrally, forming the ventral roots of the spinal nerves. These roots fuse with the dorsal roots at the ganglia, and the sensory and motor fibers together form the peripheral spinal nerves. The spinal nerves emerge from the vertebral column through the intervertebral canals.

Since most of the cellular elements of the spinal cord are in place at birth, there is little subsequent growth in the mass or length of the cord.

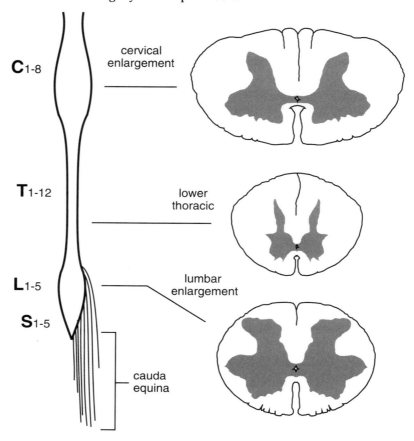

Figure 7.7 Drawings of representative cross sections from different levels of the adult spinal cord (shown from a dorsal point of view at the left) show the regional differences in the size and shape of the gray matter.

The vertebral column, on the other hand, continues to grow for the next 16 to 20 years. As a result of this differential growth, the spinal cord ends at about the level of the L2 vertebra in most humans. Consequently, the spinal nerves from progressively lower levels of the cord are dragged down with the elongating vertebral column, giving rise to a bundle of spinal nerves within the caudal end of the vertebral canal called the **cauda equina** ("pony tail"; Fig. 7.7). Since the dura and the arachnoid mater are carried along with the vertebral growth, the cauda equina is contained in a cerebrospinal fluid-filled subarachnoid space called the lumbar cistern. The neurosurgeon can sample cerebrospinal fluid from this cistern by inserting an aspiration needle below the L3 vertebra.

The Mature Spinal Cord

The spinal cord is the first central site of sensory input from the body and contains all of the motoneurons that innervate the body musculature. A cross section through the spinal cord reveals its consistently simple anatomical appearance—a butterfly-shaped gray matter core surrounded by a perimeter of white matter columns called **funiculi** (Fig. 7.6). The gray matter differentiates into a number of nuclear groups that can be described by name or according to a system of layers devised by the Swedish neuroanatomist Bror Rexed. The funiculi are occupied by long ascending and descending axons that connect the spinal cord with the brain, and shorter axons that interconnect different levels within the spinal cord. The intersegmental, **propriospinal** axons tend to hug the fringes of the gray matter. Since ascending axons accumulate as they are contributed from different levels of the cord, and the number of descending axons diminishes as they distribute along the length of the cord, the ratio of white to gray matter progressively increases as one ascends from sacral to cervical levels (Fig. 7.7).

Most of the synapses of incoming sensory fibers are made on neurons in the nuclei of the dorsal horns of the central gray substance. The motoneurons reside entirely in the ventral horns, where they are assembled into muscle-related clusters called **motor pools.** The dimensions of the dorsal and especially the ventral horns vary considerably along the length of the spinal cord. Since the cervical segments of the spinal cord must innervate the arms and the lumbar segments must innervate the legs, these segments contain a larger number of sensory and motor neurons. As a result, the stretches of the cord from segments C_3 to T_1 and L_1 to S_2 are wider than the other segments and are called, respectively, the **cervical** and **lumbar enlargements** (Fig. 7.7). At thoracic levels, the dorsal and ventral horns are much attenuated. However, a feature of the gray matter occurs here that is not evident at other levels of the cord. It is a small spur on the lateral margin of the ventral horn on either side that extends into the lateral funiculus. The spur is called the **intermediolateral nucleus** and consists of the preganglionic sympathetic neurons that innervate the sympathetic ganglia (see Chapters 15 and 22) of the PNS.

The part of the spinal cord gray matter that extends from the base of the dorsal horn to surround the motoneuron pools of the ventral horn (or Rexed's layers V through VIII) is known as the **intermediate zone** (not to be confused with the intermediate zone of the neural tube). The neurons of the intermediate zone are involved in reflex connections between the "secondary sensory" neurons of the dorsal horns and the motoneurons. They also receive the bulk of the descending axons that carry movement commands from the brain. Thus, the intermediate zone appears as a tangle of axons and neuronal cell bodies and dendrites engaged in an intricate network of synaptic connections. This tangle continues unabated into the brain stem, where it is known as the **reticular formation**—a topic to be taken up in the next chapter.

Detailed Reviews

Jacobson M. 1978. *Developmental Neurobiology.* Plenum Publishing, New York.
Purves D, Lichtman JW. 1985. *Principles of Neural Development.* Sinauer Associates, Sunderland, MA.

The Ontogeny of the Brain Stem 8

In the previous chapter, the general topographic anatomy of the human CNS was described. The brain was somewhat arbitrarily divided into five regions called, proceeding from caudal to rostral, the medulla, pons, midbrain, diencephalon, and telencephalon. As the next installment on learning the anatomical organization of the brain, let's take a closer look at the embryonic emergence of the more prominent structures of the brain stem. In the next chapter, we will do the same for the forebrain. Familiarity with such major landmarks will facilitate orientation in the three-dimensionally complex human brain when you undertake an analysis of functional systems in later chapters.

Remember that the rostral part of the neural tube, which is destined to become the brain, consists of the forebrain, midbrain, and hindbrain enlargements by about 3 weeks' gestation (Fig. 8.1). Two dorsally convex bends soon appear in this early brain. One bend, called the **cephalic flexure,** occurs at the level of the midbrain swelling. The second bend appears at the juncture of the spinal cord with the brain. Call it the **cervical flexure.** A third, ventrally convex bend, the **pontine flexure,** appears a short time later about midway through the rostrocaudal length of the hindbrain (Fig. 8.2). From the time that the pontine flexure appears, we distinguish two major subdivisions of the hindbrain: the metencephalon (or pons) and, caudal to it, the myelencephalon (or medulla). During this time, two pairs of vesicles begin to bud off the forebrain enlargement. The more rostral and larger are the **telencephalic vesicles,** destined to become the cerebral hemispheres of the adult. The smaller **optic vesicles** appear more caudally (Figs. 8.1 and 8.2); they develop into the optic cups, which eventually form the neural retinas and optic tracts. Thus, the retina and optic tract are bona fide extensions of the CNS. A third bud, the **infundibular stalk,** appears on the midline at the base of the developing hypothalamus (Fig. 8.2); it will form the posterior lobe of the pituitary gland. A second midline bud, this one on

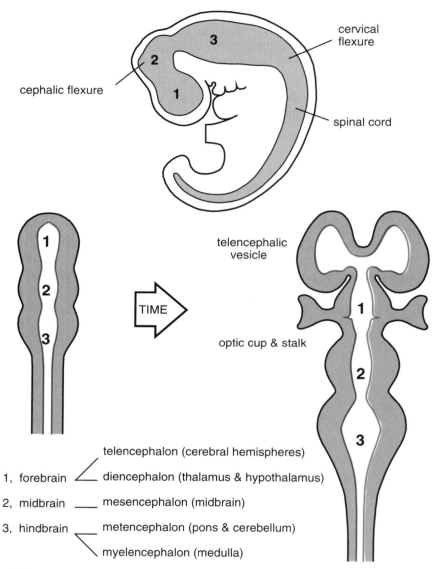

Figure 8.1 The three-enlargement brain of the early embryo is shown in a side view at the top and as a straightened, longitudinal slice to the left. From the fourth through the eighth week of gestation, it undergoes further changes. Paired vesicles (telencephalic and optic) emerge from the forebrain enlargement, as shown in the longitudinal slice to the right. The red lining of the lumen of the neural tube and telencephalic vesicles represents the proliferative, ventricular zone.

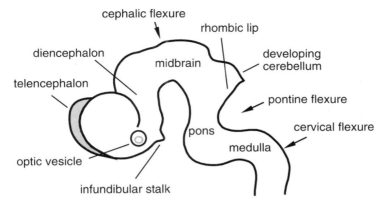

Figure 8.2 The secondary vesicles are evident in a lateral view of an isolated embryonic brain at four weeks of gestation.

the dorsal surface of the developing thalamus, eventually forms the pineal gland.

At the level of the pontine flexure, the lumen of the neural tube widens to form the caudalmost of four such widenings, called **ventricles.** This one is the **fourth** ventricle. It is connected by way of a narrow passage through the midbrain called the **cerebral aqueduct** to the **third** ventricle in the diencephalon. The first two ventricles form the hollow cores of the two cerebral hemispheres; these **lateral ventricles** communicate with the third ventricle through the narrow interventricular foramens. Since the fourth ventricle has a diamond shape when viewed from above, it is also called the **rhomboid fossa** (see Box 8.1). The roof plate above the fourth ventricle is stretched into a thin structure formed by the apposition of an inner layer of ependymal cells and an outer layer of pial cells. This structure, which appears also in the roof of the diencephalon and the medial wall of either telencephalic vesicle (Chapter 9), is referred to as a **choroid membrane** (see Fig. 8.5). As capillaries form in the choroid membrane, it folds inward into the ventricular lumen along a line on either side of the midline. It then refolds repeatedly as it becomes vascularized and invaded by mesenchymal cells, ultimately forming the **choroid plexus** of each ventricle. The choroid plexuses secrete cerebrospinal fluid into the ventricular system (see Chapter 6).

Yet another change takes place in the roof of the fourth ventricle. A large hole opens on the midline at its caudal end, and smaller holes open on either side at the extreme lateral recesses of the rhomboid fossa. It is through these **median** and **lateral apertures** that the cerebrospinal fluid produced by the choroid plexus escapes into the subarachnoid space to bathe the exterior surface of the CNS.

Box 8.1 A Useful Brain Schematic

To diagram functional systems in later chapters, we can use the schematic representation shown below. It is an outline of the brain stem and thalamus from a dorsal point of view capped by a highly stylized rendition of the cerebral hemispheres. The cerebellum has been removed for convenience, and some prominent landmarks have been indicated.

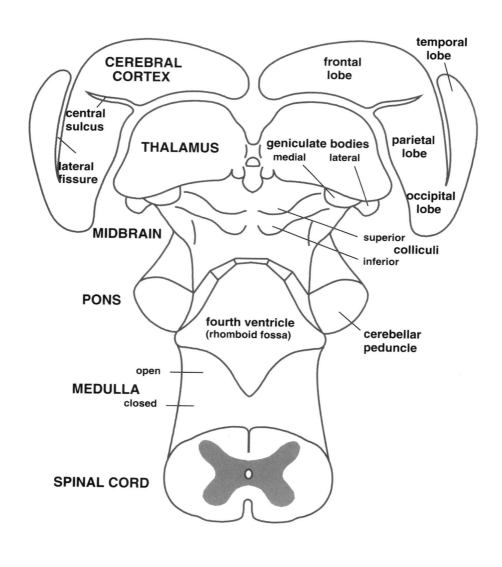

Ontogeny of the Medulla

The pontine flexure allows us to distinguish the medulla from the pons; the approximate dividing line lies at the fundus of the flexure (Fig. 8.2). Since the rostral part of the medulla lies beneath the caudal half of the floor of the fourth ventricle and the caudal part of the medulla continues to surround the central canal, we can distinguish, respectively, an *open* and a *closed* medulla (or "bulb," as the medulla is often called). The mitotic ventricular zone of the medulla gives rise to the blast cells that will form the circumscribed motor and sensory nuclei of the medulla, as well as other nuclei that are neither directly sensory nor motor in function. Other neuroblasts remain as a loose collection at the core of the medulla, forming the caudal part of the brain stem reticular formation.

The adult brain stem shows little resemblance to the spinal cord. Even so, careful analysis reveals many common anatomical principles. In the developing spinal cord ventral horn, we identified an alar plate and a basal plate of gray matter (Fig 8.3). In the brain stem, the sulcus limitans persists in the floor of the fourth ventricle and continues to demarcate an alar (sensory) region from a basal (motor) region, although in the open bulb, the former is shifted to a more dorsolateral position with respect to the latter.

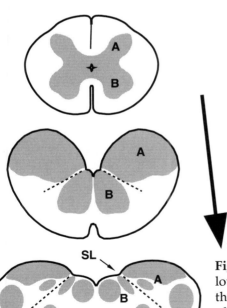

Figure 8.3 The transition from the spinal cord to the lower brain stem (closed and open medulla) preserves the relative relationship of the alar (**A**) and basal (**B**) plates. In the medulla, each plate differentiates into specific secondary sensory and motor nuclei, but other nuclear groups form within the brain stem that are not present in the spinal cord and are neither sensory nor motor. Abbreviation: SL, sulcus limitans.

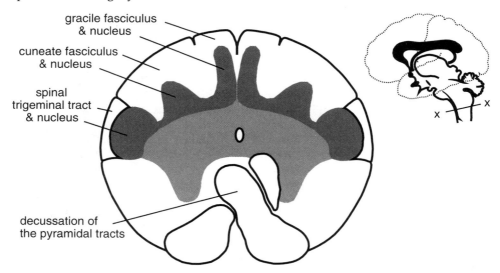

gracile fasciculus
& nucleus

cuneate fasciculus
& nucleus

spinal
trigeminal tract
& nucleus

decussation of
the pyramidal tracts

Figure 8.4 A sketch of a cross section through the closed (lower) medulla of the adult near its transition into the spinal cord shows the changed relationships between the gray and white matter (compare to Fig. B1).

In a cross section of the mature closed bulb, the surround of white matter that was present in the cord is broken up into more discrete ascending and descending bundles (Figs. 8.4 and B1). We define the border between the spinal cord and the medulla by the **decussation** (crossing over) of one of the descending fiber bundles, the **pyramidal tracts,** even though it takes several millimeters to fully accomplish the crossing. The spinal canal and the intermediate zone and ventral horns of the central gray appear largely as they did in the spinal cord. But the dorsal horns have merged with the caudalmost end of a particular brain stem sensory nucleus, the **spinal trigeminal nucleus,** and some new sensory nuclei, the **dorsal column nuclei,** have appeared in place of the dorsal funiculi. Note, however, that as in the cord, the sensory nuclei reside dorsal to the horizontal plane defined by the sulcus limitans.

Ontogeny of the Pons

The rostrolateral lips of the rhomboid fossa thicken rapidly due to a massive proliferation of neuroblasts in the local ventricular zone. As ever more cells are added to these rhombic lips, they accumulate in a mass that expands caudalward above the fourth ventricle. The blasts of the rhombic lips remain mitotic, and the mass continues to grow for the duration of gestation, to eventually form the cerebellum. At the same time, the ventricular zone in this rostral part of the hindbrain generates an enormous number of neurons that migrate ventrally to form the large mass of the ventral pons (Figs. 8.5 and B3–5). The corresponding growth

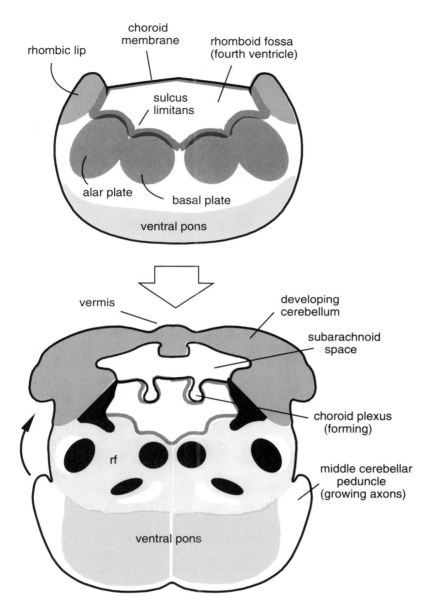

Figure 8.5 The developing pontine nuclei and cerebellum are, respectively, ventral and dorsal add-ons to the metencephalon. The part of the brain stem sandwiched between the two is called the pontine tegmentum to distinguish it from its more glamorous associates. The mitotic ventricular zone (red) lines the fourth ventricle. Abbreviation: rf, reticular formation.

of the cerebellum and ventral pons is not a simple coincidence. The neurons of the ventral pons send their axons to synapse on neurons in the cerebellum, forming a massive pair of fiber straps that adhere to the lateral aspects of the pons. These pontocerebellar bundles are called the **middle cerebellar peduncles.** The middle peduncles and two other pairs—the **superior,** above, and the **inferior,** below the middle one— connect the cerebellum to, respectively, the midbrain and the medulla.

The cerebellum and ventral pons can be considered add-ons to the brain stem that form, respectively, dorsal and ventral to the central mass of tissue of the pontine tegmentum (Fig. 8.6). Unlike the cerebellum and ventral pons, the tegmentum is continuous below with the medulla and above with the midbrain tegmentum. As the pontine tegmentum is formed by migration of postsynaptic neurons from the ventricular zone, some of the neuroblasts eventually coalesce to form circumscribed nuclei. Many, however, remain as a loosely arranged collection of neurons and growing axons that form the pontine portion of the brain stem reticular formation.

Ontogeny of the Midbrain

The midbrain remains relatively small throughout development (Fig. 8.7). The lumen of the neural tube persists as the narrow cerebral aqueduct. The cells that migrate dorsal to the aqueduct form a region called the **tectum** ("roof of the tent"), while those that go south form the **tegmentum** ("carpet"). Eventually, the blasts that remain in the region surrounding the cerebral aqueduct form the tightly packed, small neurons of the even-textured **periaqueductal gray substance.** From an outside view, the midbrain develops two pairs of little hillocks on its tectal surface. The rostral hillocks are the **superior colliculi** (a visuomotor structure) and the caudal ones are the **inferior colliculi** (a structure involved in the processing of auditory signals). Like the pontine tegmentum, much of the core of the mature midbrain tegmentum is occupied by the reticular formation (Figs. 8.8, B6, and B7). Two formed nuclei of the midbrain tegmentum that are conspicuous landmarks are the spherical **red nuclei** and, more ventrally, the **substantia nigras,** which take on a dark appearance with age due to an accumulation of neuromelanin. The only add-on to the midbrain comes in the form of two white matter straps that overgrow the ventral surface of the tegmentum on either side. These **cerebral peduncles** contain axons of mainly cortical origin that are en route to various brain stem and spinal cord targets. The substantia nigras are couched in the cerebral peduncles along their length through the midbrain.

The Brain Stem Reticular Core

The term **reticular formation** refers to the impossibly tangled neurons and fibers that form the core of the brain stem. It is phylogenetically

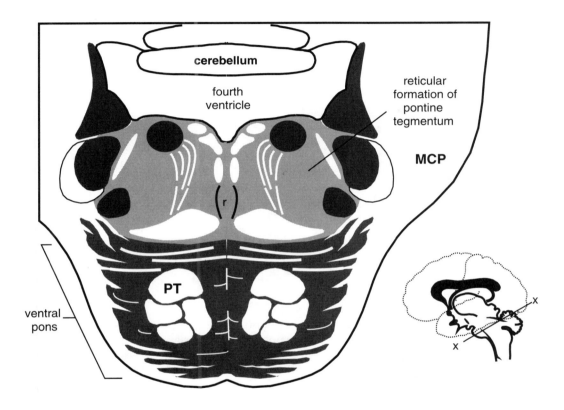

Figure 8.6 A drawing of a cross section through the adult pons shows some of the major nuclei (dark shading) and fiber bundles (white; compare to Figs. B3 and B4). All of the cerebellum, except a small piece of a midline part called the vermis, has been removed for clarity. Abbreviations: PT, pyramidal tract; MCP, middle cerebellar peduncle; r, raphé nucleus.

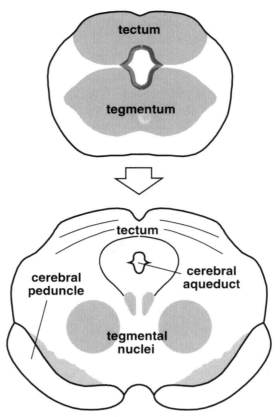

Figure 8.7 The emergence of tectal and tegmental gray matter and the later appearance of the cerebral peduncles are shown on these drawings of cross sections through the midbrain at early and later stages of development. The mitotic ventricular zone (red) lines the part of the neural tube lumen that will become the cerebral aqueduct of the mature midbrain. The axons that form the cerebral peduncles that overgrow the ventral surface of the midbrain are of telencephalic (cortical) origin.

among the oldest parts of the brain. It appears as a parenchyma of interlaced fibers and neurons extending without interruption through the medulla, pontine tegmentum, and midbrain. The brain regions that are caudally and rostrally contiguous with the brain stem reticular core—the intermediate zone of the spinal gray matter and the lateral hypothalamus, respectively—share this reticular structure. In fact, one cannot distinguish any boundaries between these regions at all. Such anatomical similarity suggests that the signal transformations performed by these reticular tissues are similar regardless of their location. Circumscribed cell groups that will concern us greatly in future chapters—the inferior olivary nuclei of the medulla, the red nuclei of the

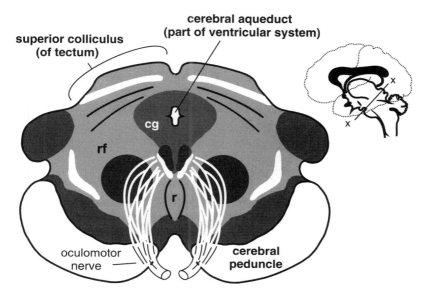

Figure 8.8 A drawing of a cross section through the adult midbrain at the level of the superior colliculi shows the major nuclei (dark shading) and fiber bundles (white; compare to Fig. B7). Abbreviations: cg, central gray substance; r, raphé nucleus; rf, reticular formation.

midbrain, and the cranial nerve nuclei that are scattered throughout the brain stem, to name but a few of the more prominent—are *not* included as part of the reticular formation.

Many of the neurons of the reticular formation have long, straight, poorly ramified dendrites that usually are spread out radially in a plane perpendicular to the long axis of the brain stem. *This peculiar dendritic array allows the reticular neurons to monitor a variety of signal traffic in axons coursing up and down through the brain stem.* The axons of many reticular neurons are long and collateralize extensively both nearby and at more distant sites. In fact, it has been shown that some (perhaps many) individual reticular cells have axons that dichotomize into long ascending and descending branches. One branch can travel down to the spinal cord and the other up to the thalamus. Thus, *the same neuron can signal in both rostral and caudal directions simultaneously.* Because of its structure, connections, and physiological properties, *the brain stem reticular formation as an entity seems to be particularly well suited to carrying out rather global sensorimotor activities that require the integration of many sensory modalities and the orchestration of a variety of skeletomotor and visceromotor responses.* Cardiorespiratory functions, for example, are particularly dependent on the integrity of the reticular formation, making it essential to the survival of the organism. Another function in which the reticular formation plays an important role is generalized brain arousal, the subject of Chapter 13.

Despite this seemingly random appearance and lack of functional specificity, it is becoming increasingly clear that the brain stem reticular core consists of a complex, highly organized interdigitation of numerous individually signatured cell systems. Each has its own histochemistry, input–output relations, and functional roles. These regions within the reticular formation will be singled out in the context of particular functional systems in the chapters ahead. For now, a few general comments on anatomical divisions and neurochemical cell groups are in order.

The Three Reticular Zones

In Nissl-stained sections of the medulla, the reticular core is seen to consist of three zones: a **paramedian zone** made up of the **raphé nuclei** (raphé means "seam"), the ventromedial half of the medulla made up of large cells called, accordingly, the **magnocellular zone,** and the remaining dorsolateral half made up of smaller cells and called the **parvicellular zone** (Fig. 8.9). The magnocellular and parvicellular zones cannot be distinguished in the pontine and midbrain tegmentum, but the raphé nuclei are distinct throughout the rostrocaudal extent of the brain stem.

The raphé nuclei are made up of serotonergic neurons. Although these nuclei are all confined to the narrow midline and paramedian zone, they

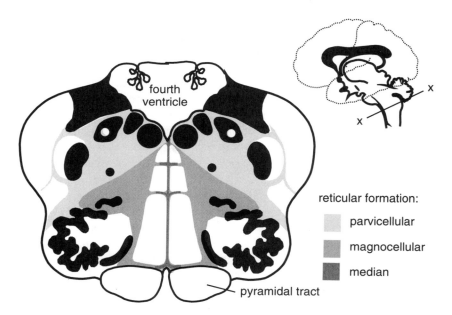

Figure 8.9 The three areas of the brain stem reticular formation (red shading) are identifiable in this cross section through the open medulla of an adult. Some of the formed nuclei are shown in dark shading, and fiber bundles are shown in white.

can be distinguished from one another on architectonic grounds and by virtue of their connections. The rostral raphé nuclei send their axons rostrally to the forebrain. The pontine raphé nuclei project heavily to the cerebellum, and the medullary raphé nuclei are the source of serotonergic axons that end in the spinal cord. Among other functions, the descending serotonergic systems have been implicated in pain sensation in the CNS, and the ascending ones have been implicated in such normal functions as sleep induction and arousal and in disorders of sleep and emotion.

Neurons that synthesize norepinephrin are restricted to the pontine and medullary tegmentum where they are especially concentrated in the **nucleus locus coeruleus.** These noradrenergic neurons give rise to highly branched ascending and descending fiber systems that course the length of the neuraxis and form terminal networks in the spinal cord gray matter, the cerebellar cortex, throughout the brain stem, and in many forebrain structures including the cerebral cortex. The function of such widespread noradrenergic projections is not understood. There are many other neurotransmitter and cotransmitter phenotypes represented at various sites among the neurons of the reticular formation.

Now that you have a general view of the topographic anatomy of the brain stem, we will proceed in the following chapter to pursue the ontogenesis of the forebrain in a similar way. In the meantime, take a moment to peruse the schematic of the brain stem and thalamus shown in Box 8.1 and relate it to what you have learned about the topography of the brain stem. You are also invited to examine the cross sections of the human brain stem provided in Appendix B, now and whenever you need an idea of the relationships of major landmarks to one another.

Detailed Reviews

Nauta WJH. 1986. *Fundamental Neuroanatomy.* WH Freeman & Co, New York.
O'Rahilly R, Müller F. 1994. *The Embryonic Human Brain.* Wiley-Liss, New York.
Purves D, Lichtman JW. 1985. *Principles of Neural Development.* Sinauer Associates, Sunderland, MA.

9

The Ontogeny of the Forebrain

The emergence of the telencephalic vesicles allows us to distinguish the final two major forebrain regions, the telencephalon and the diencephalon, from one another (have a look back at Figs. 8.1 and 8.2). The telencephalic vesicles give rise to the two cerebral hemispheres while the diencephalon remains as the rostral end of the neural tube. In the diencephalon, the lumen of the neural tube forms the tall narrow slit of the third ventricle (see Fig. 9.3). The roof of the third ventricle remains as a thin choroid membrane, which eventually invaginates to form a modest choroid plexus. The sulcus limitans persists in both walls of the third ventricle, where it is called the **hypothalamic sulcus** and delimits the thalamus above from the hypothalamus below. The hypothalamus grows more modestly than the thalamus during development, and each differentiates into several nuclear groups. Neither, however, expands as exuberantly as the cerebral hemispheres. In fact, the hemispheres are by far the most productive parts of the developing CNS in terms of ultimate volume and morphological differentiation.

Ontogeny of the Telencephalon

The hollow telencephalic vesicles expand like balloons. The hollows of the balloons persist in the adult cerebral hemispheres as the lateral ventricles. These ventricles remain confluent with the diencephalic (third) ventricle, each through a narrow passage at the neck of the balloon called the interventricular foramen (see Fig. 9.2). At first, the balloons are the ovate kind common to birthday parties—they expand in all directions equally. Later, they assume a particular shape as they expand. The caudalward expansion outstrips expansion in the other directions so that the balloons grow back over the entire diencephalon and most of the brain stem (Fig. 9.1). At the same time, the lateral expansion turns

108

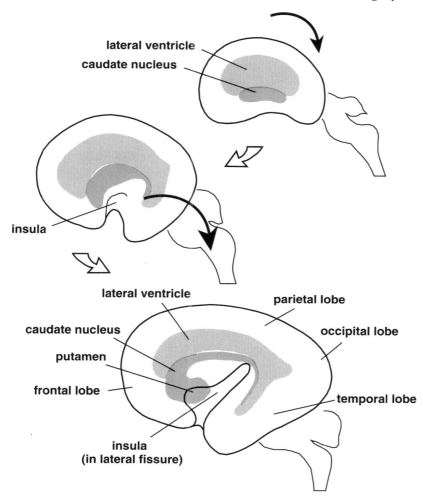

Figure 9.1 The telencephalic vesicles expand into the C-shaped cerebral hemisphere. The cerebral ventricle (gray), some of the subcortical gray matter (shaded red), and some major axon bundles (not shown) follow the C shape.

downward to cover the sides of the diencephalon and the upper part of the brain stem. Next, a peculiar event ensues in which a central disc on the lateral face of the hemisphere ceases to expand while the surrounding regions continue to grow. The surround bulges outward above, behind, and below the disc, revealing for the first time the C-shape of the adult hemisphere. Soon, the bulges overgrow the disc entirely and their rims meet to form the **lateral fissure.** The central disc, now completely buried behind the bulges, persists as the **insular cortex** of the adult. Further expansion of the cortex results in the prominent sulcate pattern of the mature brain.

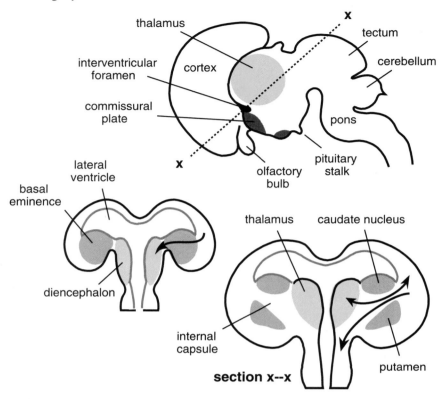

Figure 9.2 The development of deep-lying telencephalic structures includes most prominently the formation of the internal capsule, corpus striatum (shaded dark red), and thalamus (lighter red). The lumens of the telencephalic vesicles and the rostral end of the neural tube are lined by the mitotic ventricular zone (red).

The lateral ventricles and one of the deep-lying nuclei of the telencephalon, the **caudate nucleus,** faithfully follow the expansion and recurvature of the hemispheres (Fig. 9.1). Likewise, many of the fiber tracts that grow out of the lower limb of the C—the temporal lobe of the cerebral hemisphere—to reach the cortex of the upper limb or other parts of the forebrain are obliged to follow this arching route.

Concurrently with this hemispheric expansion, the **olfactory bulbs** and tracts form on the ventral surface of the frontal lobe (Fig. 9.2). Also, a mass of gray matter, the **basal eminence,** grows in the base of each hemisphere and will eventually give rise to the three assemblages of subcortical nuclei on each side (Fig. 9.2; see also Figs. B8–10): the corpus striatum (made up of the **caudate nucleus, putamen,** and **globus pallidus**), the amygdala (in the temporal lobe), and the ventral forebrain nuclei that lie at the ventromedial base of the hemispheres.

During and after the expansion of the telencephalic vesicles, their walls begin to thicken, first by the addition of neurons and glia as a superficial sheet, and later by the accumulation of axons beneath the sheet, until in the adult each hemisphere is dominated by the vast and continuous sheet of cortex that lies at its surface and a cable basement of white matter. The axons that make up the cable basement are of three varieties: those that interconnect cortical areas, both nearby and distant, *within the same hemisphere,* called *association* fibers; those that *cross-link cortical regions of the two hemispheres,* called *commissural* fibers; and those that *join the cortex with subcortical nuclei,* mainly by way of the internal capsule, called *projection* fibers.

As each internal capsule grows downward through the basal eminence, it splits this gray matter into separate nuclear groups (Figs. 9.2, B8, and B9). The caudate nucleus ends up medial to the internal capsule, and the putamen and globus pallidus lateral to it. The internal capsule from each hemisphere continues to grow downward along the lateral surfaces of the diencephalon. Many fibers of the internal capsule enter and synapse in the thalamus. Also, axons emerge from thalamic neurons and enter the internal capsule to travel up to the cerebral cortex. The axons of the internal capsules that continue to descend beyond the diencephalon form the cerebral peduncles covering the ventral aspect of the midbrain tegmentum (see Figs. 8.8 and B7). Many of these axons will terminate in the ventral pons as they penetrate that structure (see Figs. 8.6 and B6). Those that do not, emerge from the caudal end of the pons as the **pyramids** running along the ventral aspect of the medulla (see Figs. 8.9 and B2) until they cross the midline and enter the lateral funiculi of the spinal cord at the spinomedullary junction (see Fig. 8.4).

The first of the commissural fibers to bridge the gap between the two hemispheres do so by piercing the closely apposed medial walls at a point called the **commissural plate** (Fig. 9.2). The smaller **anterior commissure** forms at this rostroventral site. Commissural axons that cross later create a horizontally running, thick slab of white matter called the corpus callosum (Figs. 9.3 and B8–10). The thin remnants of the closely apposed walls of the telencephalic vesicles ventral to the corpus callosum remain as the **septum** of the adult (see Fig. B10). The walls dorsal to the corpus callosum constitute the free medial edges of the cerebral cortex of each hemisphere. This free edge is carried around the C of each hemisphere and onto the medial aspect of the temporal lobe, where it is refolded upon itself to form a special cortex called the **hippocampus.**

The caudomedial wall of each telencephalic vesicle does not contribute to the elaboration of neurons and glia. Instead, it remains a thin apposition of pia and ependyma, another choroid membrane, that is soon infolded to elaborate the choroid plexus of the lateral ventricle (Fig. 9.3). Its first infolding creates the choroid fissure, which is carried down and around with the C-shaped expansion of the hemisphere.

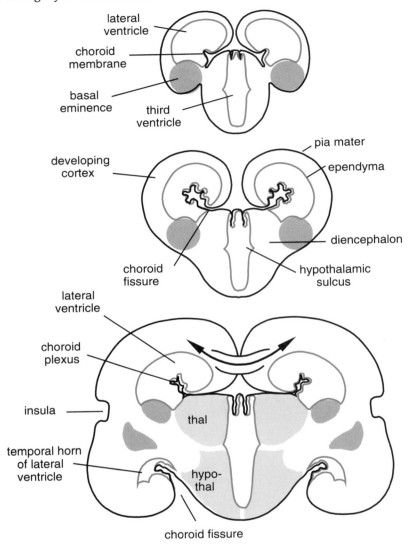

Figure 9.3 The invagination of the thin choroid membrane forms, sequentially, the choroid fissure and the choroid plexus of the lateral and third ventricles. The arrow with two heads indicates the site at which the axons of the corpus callosum will cross between the hemispheres.

The Lobes of the Cerebral Hemispheres

One of the earliest sulcuses to appear in the cortex is the deep **central sulcus,** which runs dorsoventrally about midway along the rostrocaudal extent of the hemisphere. The central sulcus has an important functional

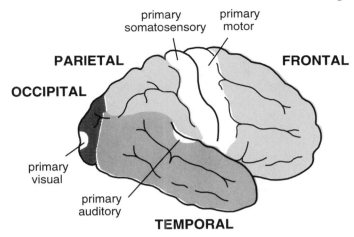

Figure 9.4 The lobes and some major functional areas of the mature cerebral cortex are delimited by prominent and relatively invariant sulcuses.

significance: the gyrus in front of it is destined to become the primary motor cortex, while the one behind it will become the primary somatosensory cortex. Another landmark sulcus that forms early is the **parieto-occipital sulcus** (or fissure). These two sulcuses, together with the lateral fissure, define the major lobes of the cerebral cortex (Fig. 9.4). The **frontal lobe** extends from the central sulcus to the frontal pole of the hemisphere. The **parietal lobe** extends back from the central sulcus to the parieto-occipital sulcus. The **occipital lobe** (the visual cortex resides here) extends the rest of the way back to the occipital pole. The frontal and parietal lobes are bordered ventrally by the lateral fissure. The part of the hemisphere below the lateral fissure is the **temporal lobe** (the primary auditory cortex resides here).

The Nuclear Organization of the Thalamus

Although the individual thalamic nuclei and the cortical areas to which they project will be discussed more fully in the many chapters that follow on functional brain systems, it is helpful to have at least a general idea of the structure and functional organization of the thalamus. The term *thalamus* refers to the large, egg-shaped mass of the dorsal diencephalon on either side of the third ventricle (Figs. 9.5, B8, and B9). The thalamus has always been accorded weighty significance because, like *Cephalonia* to *Ithaca*, it is an obligatory stopover for signals that seek ingress to the cultural capital of the brain—the cerebral cortex. Indeed, *the thalamus is the major source of synaptic input to the cortex.* The thalamocortical projec-

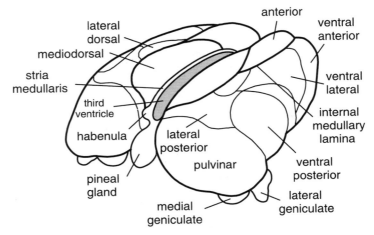

Figure 9.5 The thalamic nuclei are shown on a drawing of an isolated thalamus from a right caudolateral perspective.

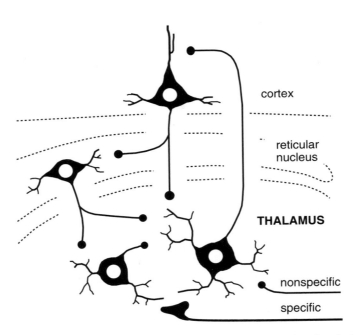

Figure 9.6 The sketch shows the basic processing circuit of a thalamic nucleus and the reciprocity of synaptic connections between the thalamus and the cerebral cortex. Specific axons carry relatively unprocessed sensory data to different thalamic nuclei. Nonspecific axons arrive from the brain stem reticular formation.

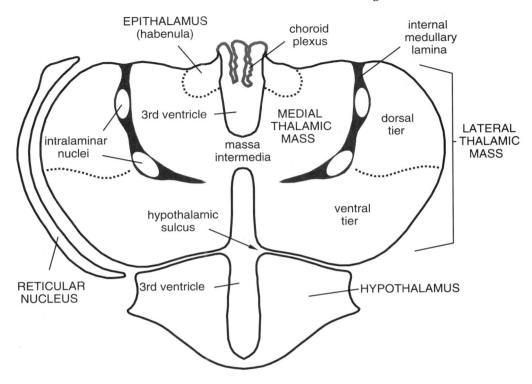

Figure 9.7 This schematic frontal section through the diencephalon shows the basic organizational plan of the thalamus (compare to Fig. B8). The reticular nucleus is drawn on one side only for clarity.

tions convey sensory data arriving from the spinal cord, the brain stem, and other brain regions, motor-related data from the corpus striatum and cerebellum, and other highly processed, multimodal data from several brain regions. The thalamus is not simply a passive relayer of impulses to the cortex. At any synaptic station along a conduction pathway in the CNS, the pathway is opened to the influence of convergent synapses. And, like Odysseus at *Cephalonia,* the signals can even be denied passage. To a degree not yet fully appreciated, the thalamus transforms the data it receives into a format that can be used by the cortical processing circuits. It accomplishes this transformation by way of (1) intrathalamic (local circuit) connections, (2) corticothalamic (feedback) connections (*the cortex reciprocates the thalamocortical projections*), and, in some instances, (3) convergent inputs from more than one source (Fig. 9.6). The thalamus has few, if any, descending projections.

The internal structure of the thalamus is complex, and several nuclei can be identified. For starters, the thalamus can be divided into **medial** and **lateral** nuclear masses with the **internal medullary lamina,** a thin sheet of white matter, as the dividing line between the two (Figs. 9.7 and

B8). Rostrally, the lamina splits, its outstretched arms embracing the **anterior** mass of nuclei (Figs. 9.5 and B9). The anterior nuclei project to the cortex on the medial aspect of the cerebral hemisphere—the cingulate gyrus. The medial thalamic mass is dominated by the **mediodorsal nucleus,** which projects to much of the cortex of the frontal lobe. The internal medullary lamina itself contains several small intralaminar nuclei and one large one called the **centromedian nucleus.** The intralaminar nuclei project to the corpus striatum as well as to the cerebral cortex.

The lateral nuclear mass is the largest of the three masses and is subdivided into more or less ventral and dorsal tiers. All the major nuclei that relay sensory data to the cortex and the nuclei most obviously related to motor control are situated in the ventral tier. Hung on the underside of the ventral tier at its caudal end are the **medial** and **lateral geniculate nuclei** (MGN and LGN), which process, respectively, auditory and visual sensations. The MGN provides input to the primary auditory cortex of the temporal lobe, and the LGN conveys information to the primary visual cortex in the occipital lobe. Near the geniculate nuclei at this caudal end of the ventral tier, we find the **ventral posterior nucleus,** which receives the ascending axons for somatic sensations arising from the face and body. The ventral posterior nucleus sends these data to the primary somatosensory cortex, which forms the **postcentral gyrus.** At the rostral end of the ventral tier are the **ventral anterior** and **ventral lateral** nuclei (often lumped together as the VA-VL complex), which influence the motor cortical areas of the frontal lobe. The dorsal tier of the lateral thalamic mass is made up of nuclei—the **lateroposterior** and **pulvinar** nuclei—that are attractively large, but nonetheless poorly understood from a functional perspective. They project to the cortex of the parietal, occipital, and temporal lobes.

The **reticular nucleus** of the thalamus lies as a thin sheet of cells that envelops the thalamus and is separated from it by the thin **external medullary lamina.** This nucleus does not project to the cortex, but only into the thalamus (Fig. 9.6). Cells of the reticular nucleus receive collaterals of the corticothalamic and thalamocortical axons that are obliged to pass through it. *The reticular nucleus is therefore strategically located to gate the two-way conversation that takes place continuously between the thalamus and the cortex.*

The Warp and Woof of the Cerebral Cortex

The mature cortex is a thin (1.5 to 4.5 millimeters, with an average of about 2.5 millimeters), folded sheet of gray matter. We attribute our highest mental functions, such as language, abstract thought, and creativity, to the cerebral cortex, but it plays a role in virtually all CNS functions. It is especially involved in sensory interpretation (perception), learning and memory, and the control of complex behavior. Two mor-

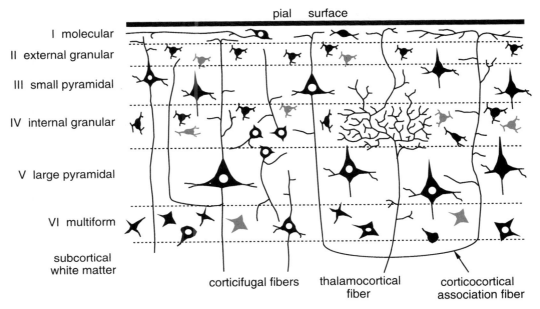

pial   surface

I   molecular

II   external granular

III   small pyramidal

IV   internal granular

V   large pyramidal

VI   multiform

subcortical
white matter

corticifugal fibers   thalamocortical   corticocortical
                                  fiber   association fiber

Figure 9.8 A slab through the depth of the cortex shows the six layers, the cell types that predominate in each, and a simplified version of basic cortical circuitry. Many cortical cells are excitatory, glutamatergic neurons (jet black), whereas others are inhibitory, GABAergic, local circuit neurons (red).

phological types of neuron dominate the cortex: **stellate** (or **granule**) and **pyramidal** cells (Fig. 9.8). The axons of stellate (star-shaped) cells mediate local connections within small regions of the cortex. The pyramidal cells, so called because of the pyramidal or conical shape of their somas, are responsible for long corticocortical axonal projections and projections to subcortical nuclei.

The cortex has both *laminar* (layered) and *columnar* organization. Most of the cortical mantle (all of what is visible from an external view) has *six* layers (Fig. 9.8). We call this part of the cortex *neo*cortex because we suspect it is a relatively *new* product of evolution. It is also called *iso*cortex because it has everywhere the *same* general appearance. We distinguish this six-layered cortex from the *three*-layered *archi*cortex or *allo*cortex of the piriform (olfactory) area and the hippocampus (see Chapters 21 and 24). The layers of the neocortex are labeled with roman numerals I to VI from the outside in. Layer I is called the **molecular layer.** It is rich in fibers and poor in cells. Layer II, the **external granular layer,** has many small granule (stellate) cells. Layer III is the **small pyramidal layer,** IV is the **internal granular layer,** V is the **large pyramidal layer,** and VI is the **multiform layer** which has a variety of cell types. *Layer IV is the main target of thalamocortical axons in the primary sensory cortices. Layers II and III*

Box 9.1 A Schematic Sagittal Section of the Brain

The schematic representation shown below should come in handy for diagramming functional systems in the chapters that lie ahead. Although some liberties have been taken, the outline drawing is generally faithful to the contours one would see in a sagittal section of the brain a few millimeters off the midline. The major districts and some of the more prominent landmarks have been labeled.

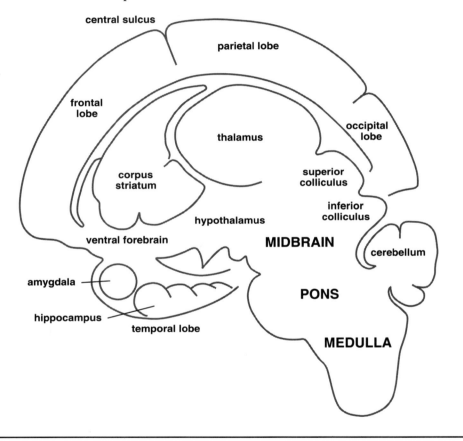

(together called the **supragranular** layers) *are considered to be the source and terminus of most corticocortical* (association and commissural) *projections,* whereas *layers V and VI (the* **infragranular**) *are considered to give rise to axons that leave the cortex* to synapse in the corpus striatum, thalamus, brain stem, and spinal cord.

The columnar organization of the cortex is an evolving story that begins with the observations of Rafael Lorente de Nó in his Golgi-stained material. He noted that every one of the pyramidal neurons sent its towering apical dendrite radially (perpendicular to the plane of the cortical surface) through the thickness of the cortex. None coursed horizontally or even obliquely. He noted further that the orientation of these dark stelae was mimicked by many of the axons in the cortex. This histological observation was given functional teeth by Vernon Mountcastle, who studied the physiology of the somatosensory cortex, and was made positively glamorous by David Hubel and Torsten Weisel in their Nobel Prize-winning work on signal processing in the visual cortex. The cortical column, which is considered by some brain scientists to be a modular component or functional unit, has so far been identified only in the primary sensory and motor cortices and is most easily defined in electrophysiological terms as a vertically oriented column in which all the cells share a particular response property. Further discussion of cortical columns must be deferred until we possess a fuller understanding of functional systems.

As we have seen in the last three chapters, the elaborate complexity of the mature brain is formed from a simple neural tube. Even though each of you now has a reasonable three-dimensional image of the human brain contained within your own mind's eye, you can appreciate that it remains difficult to depict the brain for purposes of instruction. To this end, we can use the schematic representation of the brain shown in Box 9.1 in which the major subdivisions are indicated. It is a more or less sagittal section on which we can diagram functional systems in later chapters.

Detailed Reviews

Cowan WM. 1978. Aspects of neural development. In Porter R (ed), *Neurophysiology III. International Review of Physiology, Vol 17.* University Park Press, Baltimore.

O'Rahilly R, Müller F. 1994. *The Embryonic Human Brain.* Wiley-Liss, New York.

Purves D, Lichtman JW. 1985. *Principles of Neural Development.* Sinauer Associates, Sunderland, MA.

SKELETOMOTOR CONTROL SYSTEMS III

Animals move about and influence their environment in order to obtain nutrients, avoid injury, reproduce, and generally enjoy themselves. The range of behaviors exhibited by an animal is proportional to the complexity of its nervous system. Simple animals depend largely on stereotyped, reflexive movements that are pretty much hard-wired in their nervous systems and relatively immutable. At the other end of the behavioral spectrum are the complicated, volitional behaviors performed by humans. But even many of these become relatively stereotyped—intentionally so—if they are successful in achieving desired goals. That is to say, successful behaviors are duplicated as long as they are successful and modified when they are not.

Movement is guided first and foremost by sensory input, much of it in the form of feedback from the executed behavior itself. Even so, neuroscientists often tend to conceptualize the CNS as having a set of *motor systems* that are distinct from the *sensory systems*. After all, when neurons of the CNS receive relatively direct synaptic input from peripheral sensory fibers (e.g., those in the dorsal horn of the spinal cord), is one not logically compelled to regard them as sensory? Applying the same logic, aren't neurons that are synaptically close to the motoneurons to be regarded as motor in function? The fallacy of this logic is easily exposed. For instance, there is a patch of the cerebral cortex near the calcarine sulcus that reprocesses visual data conveyed to it from the primary visual cortex. When the patch is electrically stimulated, the eyes move to the contralateral side. Is this area sensory or motor? As we shall soon see, many motoneurons themselves receive direct synaptic inputs from primary sensory neurons! Should they be considered secondary sensory neurons? The point of these rhetorical questions is that *no clear distinction can be made between the sensory and motor functions within the confines of the CNS.* For this reason, it is useful to consider the many cell groups involved with movement control as *sensorimotor integrators* rather

121

than acquisitors of sensory data or autonomous generators of movement. The circuits that control movement depend at all levels on sensory data to perform their computations. We will be guided by this concept as we analyze the movement control apparatuses in the following chapters.

Sensory Transduction and Coding   10

Most parts of our bodies are equipped with sensory nerve fibers that monitor the external environment, our internal conditions and the movements we make. Sensations of the eight major **modalities**—*cutaneous sensation, vision, hearing, balance, musculoskeletal sensation, smell, taste, and visceral sensation*—are detected by the peripheral endings of these sensory nerves and carried to the CNS for processing over various circuits. Some sensations are conveyed all the way to the cerebral cortex, where, presumably, they intrude on awareness; when so privileged, we call them **perceptions.** Somewhat counterintuitively, however, the greater share of our sensory experience falls short of the cortex and goes unnoticed. But, as we shall see in the forthcoming chapters, this sensory information is nonetheless essential to many CNS functions. We will deal with the central processing of sensory events in several later chapters. Our only goal in the present chapter is to examine how the energies to which we are exposed are transformed into useful neural impulses.

Transducer Specializations

The endings of sensory nerves vary in structure and often are associated with anatomically specialized cellular or connective tissue apparatuses that are essential to their proper function (Fig. 10.1). The ending of a sensory fiber, together with its accessory structures (if any), constitutes the tissue assembly responsible for *the transformation of naturally occurring stimulus energy into nerve impulses,* and is referred to as a **transducer.** Incidentally, although transducers are often called *receptors,* it seems advisable nowadays to reserve the term receptor for proteins that bind signal molecules.

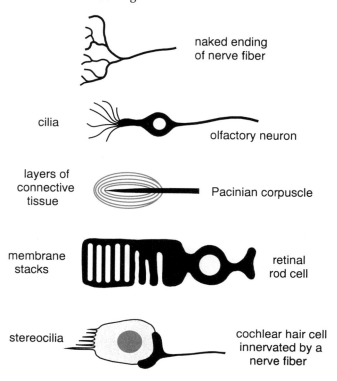

Figure 10.1 A few of the many varieties of sensory transducers present in the human body are shown in these sketches. Complexity increases from a simple naked fiber ending in the skin to the more elaborate transducers of the eye (rods and cones) and ear (hair cells). The neural element is shown in jet black and the nonneural elements are shown in red.

Within each of the eight major sensory modalities, we can identify a few to several **submodalities** (Table 10.1). For example, the submodalities of cutaneous sensation can be described as *fine touch, pressure, pain,* and so on. In taste, we speak of salty and sour, sweet and bitter. We are able to recognize these submodalities because we possess sensory transducers that are dedicated to their detection. Indeed, the very existence of such a wide variety of anatomical specializations exhibited by transducers implies a dedication to the transduction of particular stimulus energies (Fig. 10.1). Take the Pacinian corpuscle found throughout the skin. Its onionlike layers of connective tissue readily transmit an abrupt pressure to the nerve ending that sits at its core. But the compliance inherent in a layered arrangement of soft elements quickly dissipates a steadily applied pressure over the surface of the corpuscle rather than transmitting it to the nerve fiber. Relieve the pressure, and the nerve ending is again momentarily distorted. Thus, the Pacinian corpuscle is

Table 10.1 Classification of sensory modalities and submodalities and the general form of the energies to which the transducers respond

Modality	Submodality	Energy	Transducer
Cutaneous	Fine touch	Mechanical	Meissner's corpuscle
	Pressure	Mechanical	Merkel's discs
	Vibration	Mechanical	Pacinian corpuscle
	Tickle	Mechanical	Unknown
	Fast pain	Mechanochemical	Naked ending
	Slow pain	Chemical	Naked ending
	Itch	Chemical	Unknown
	Heat	Thermal	Naked ending
	Cold	Thermal	Naked ending
Musculoskeletal	Joint angle	Mechanical	Naked ending
	Muscle length	Mechanical	Muscle spindle
	Muscle tension	Mechanical	Golgi tendon organ
Balance	Angular acceleration	Mechanical	Hair cell
	Linear acceleration	Mechanical	Hair cell
Hearing	Tone	Mechanical	Hair cell
Vision	White light	White light	Rod cell
	Color	Red/blue/green light	Cone cell
Smell	Many odors	Chemical	Ciliated neuron
Taste	Sweet	Chemical	Hair cell
	Sour	Chemical	Hair cell
	Bitter	Chemical	Hair cell
	Salty	Chemical	Hair cell
Visceral	Many substances	Chemical	Unknown

able to detect rapid changes in pressure, as might occur with a vibrating stimulus applied to the skin surface. It is one of several different kinds of *mechanoresponsive* endings in the skin. In contrast to the Pacinian corpuscle, a nerve ending that lies naked in the skin may be more suited to detect heat or cold; if so, it is *thermoresponsive*. The stacked membrane structure of the rod and cone cells in the eye enables them to efficiently capture photons of light; such retinal transducers are *photoresponsive*. The transducers for smell, taste, and much of visceral sensation are designed to interact with chemical substances. Call them *chemoresponsive*. And so it goes that transducers are dedicated to a particular type of stimulus energy—a notion referred to as **transducer specificity**. It means that *the stimulus quality that is required to produce discharges in a sensory fiber is determined by the structural specializations of its peripheral transducer.* Thus, for any given transducer (and hence, its corresponding sensory axon), there exists a particular type of stimulus energy that is the one most likely to produce a response. That stimulus is called the **adequate stimulus** for the transducer (or fiber).

Experimental data acquired by recording the impulse activity of peripheral sensory fibers in animals support this concept. A given sensory axon that is found to conduct impulses in response to light stroking of the skin will not respond to, say, pressure or heat applied to the same area of skin. A retinal cone cell responsive to blue light will not respond as well to red light of the same intensity. Often, however, an extreme form of an unnatural stimulus (like electric shock) can set off impulse activity in a sensory fiber. But even then, the brain will interpret the activity as if were elicited by the adequate stimulus. You can demonstrate this on yourself by pressing on the side of your eyeball with your index finger. You *see* patterns of light and dark because the pressure is sufficient to cause discharges in the phototransducers, but their activity is still perceived as light patterns by the brain. Thus, under ordinary circumstances, a *given sensory fiber conducts information in one, and only one, submodality. This principle is the primary means by which sensory submodalities are coded and is referred to as the* **labeled line** *principle.* This principle is reflected in the symptoms that are described by patients who suffer compression damage or disease affecting one or more peripheral nerves. Such *peripheral neuropathy* often leads to a variety of aberrant sensations, depending on the particular fibers within the nerve that are affected. Patients often describe the sensations as numbing, tingling, burning, or pain in the receptive area served by the nerve. Often, normal sensation is restored when the nerve recovers from the inflammation or compression.

Transducer Location Is Important

Some sensory endings respond to stimuli (a **stimulus** *is an energy that can be transduced by a neuron or nervous system*) that originate outside the body (*exterosensitive*), whereas others respond to stimuli from within the tissues of the body (*interosensitive*). The tactile, visual, acoustic, taste, and smell transducers monitor sensory events of external provenance and are necessarily located at or near the surface of the body. Some nerve endings in deep tissues of the body monitor stimuli resulting from internal mechanical or chemical events (*viscerosensory*). Other nerve endings, both superficial and deep-lying, detect stimuli associated with tissue damage (*nocisensory*). Transducers in the muscles, tendons, and joints, along with the balance transducers of the vestibular system, convey information about body positions and movements (*propriosensory*).

The specific location of a transducer within these broad bodily distributions is also important for coding sensory information. Transducers *are responsive to stimulation only within a particular part of the sensory surface or field.* As gratuitous as this statement may seem—after all, a prick to the finger does not elicit a sensation of pain from the toe!—it is nonetheless important in considering the *representational maps* encountered within the CNS and the capability of sensory processing systems to resolve discrete stimuli. Thus, cutaneous transducers can be activated by

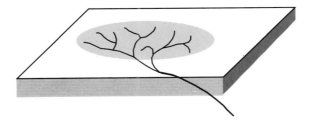

Figure 10.2 The receptive field of a transducer (in this case a naked ending) in the skin is the zone (red-shaded area) in which an applied stimulus will elicit a response (spikes) in the sensory fiber.

stimulation within a certain patch of skin, the size, shape, and location of which will vary from one transducer to the next. This zone is referred to as the **receptive field** of the sensory neuron (Fig. 10.2). Receptive fields can be tiny and closely spaced, as for cutaneous transducers on the tips of the fingers or phototransducers in the fovea, or they can be large and farther apart, as for thermal transducers on the back of the trunk. Later we will see that *neurons within the CNS that receive sensory information also exhibit receptive fields. The size and shape of the receptive fields of central neurons are, however, composites built up from the convergent inputs they receive from several primary sensory axons.*

Mechanisms of Transduction

Transducer endings of nerve fibers respond to impinging energies by opening cation channels in their plasma membrane. The inward movement of Na^+ ions results in a transmembrane potential change called a **generator potential,** which (unlike an action potential) diminishes with distance along the membrane and over time. In this way, generator potentials are analogous to the EPSPs that occur postsynaptically at excitatory central synapses. *The generator potential is proportional to both stimulus intensity and frequency.* A mild stimulus will elicit a smaller generator potential than a more intense stimulus (Fig. 10.3A). Repeated stimulations of equal magnitude that follow rapidly after one another will sum to produce a progressively larger generator potential. The membrane in a segment of the sensory fiber adjacent to the transducer ending, called the **trigger zone,** is functionally equivalent to the initial segment of an axon; it is specialized (by an abundance of voltage-sensitive cation channels) to respond to voltage changes that exceed a certain **threshold** value. The response of the fiber is the firing of one or more action potentials that are conducted nondecrementally along the sensory axon all the way through the sensory ganglion and into the CNS, where the first synaptic contacts are formed.

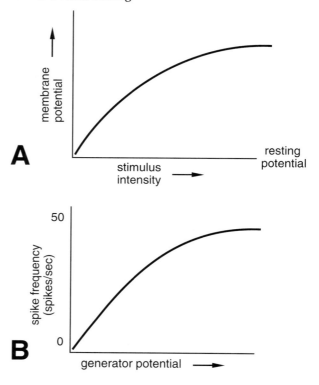

Figure 10.3 The generator potential (**A**) is proportional to the stimulus intensity, and the spike frequency (**B**) is, in turn, proportional to the generator potential.

Sensory Information Is Coded by Spike Frequency

In addition to *submodality* and *location* there are two other stimulus features that the sensory system must encode in order to work effectively: *intensity* and *duration*. Stimulus intensity is encoded in two ways. First, the frequency of impulses in a sensory fiber increases with increasing intensity of the stimulus at its transducer ending. We call this a **frequency code** (Fig. 10.3B). Second, stimuli of progressively increasing intensity will tend to influence a wider area of the sensory surface and thereby cause a proportionally larger number of sensory axons to discharge. This recruitment of additional fibers is called a **population code.** Although the mechanism for such coding is obscure, the *pattern* of discharges across a population of sensory fibers may also provide a useful means of coding sensory information. Such **pattern coding** has been suggested to explain the complexities of taste and odor recognition (see Chapter 21).

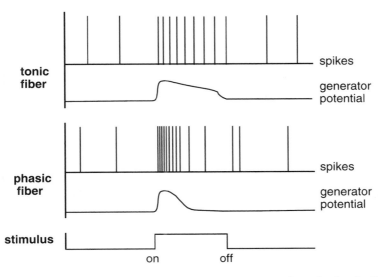

Figure 10.4 The responses of tonic (slow-adapting) and phasic (fast-adapting) fibers to the same stimulus are very different.

All sensory fibers exhibit a more or less gradual reduction of generator potential and hence of spike frequency upon continued application of a stimulus. We call this phenomenon **adaptation.** Importantly, some transducers adapt very slowly while others are quick to adapt (Fig. 10.4). Slow-adapting or **tonic** transducers—the ones that continue to trigger action potentials as long as the stimulus is applied—signal the *duration of stimulation.* Transducers that respond only *transiently* are called fast-adapting or **phasic.** The burst of impulses they discharge at the onset or offset of a stimulus signals a *change* in stimulation rather than the continuing presence of a stimulus. Many transducers are characterized by intermediate adaptation: they respond with a burst of spikes, like a phasic transducer, but follow the burst with a steady and less frequent train of impulses, like a tonic transducer.

Sensory axons come in a wide spectrum of diameters and with varying degrees of myelination (Table 10.2). In general, *larger, more heavily myelinated axons are capable of conducting larger-amplitude spikes at higher frequencies than smaller axons.* Hopefully, you also remember that *larger axons conduct impulses faster than smaller axons.* As it happens, there is a general correlation between sensory submodality (or transducer) and the diameter and degree of myelination of the sensory fiber that carries the data. One of the clearest examples of this correlation involves the axons that convey pain information to the spinal cord. The pain that we describe as *sharp* or *fast* is carried by fast-adapting, fast-conducting, lightly myelinated axons called type Aδ axons. Dull, throbbing pain is conveyed by small-diameter, nonmyelinated axons called type C axons.

Table 10.2 Classification of sensory axons by myelination, diameter, and conduction velocity

	Skin	Muscle	Diameter	Velocity
Myelinated				
Large	—	Ia & Ib	13–20 μm	80–120 m/sec
Medium	Aβ	II	6–12 μm	35–75 m/sec
Small	Aδ	III	1–5 μm	5–30 m/sec
Nonmyelinated	C	IV	0.2–1.5 μm	0.5–2 m/sec

The CNS Uses Sensory Information for Three Main Purposes

The same sensory data that enter the CNS are processed over multiple pathways. In upcoming chapters, we will follow many of these path-ways—they have been mapped using neuroanatomical tracing methods (see Box 1.1)—and discover that they are designed to extract particular *features* from the raw sensory data to serve particular computational purposes of different CNS circuits. One primal purpose that is served by sensory input is the *arousal* of the CNS to an alerted state (see Chapter 13). Imagine, for instance, your response to a sudden rustle of leaves in an otherwise quiet forest glade. Another important use of sensory informa-tion is to guide *movement*. Relatively unprocessed data may do for *reflexes,* whereas more complex behavioral patterns—returning a kickoff for a 98-yard touchdown in the Superbowl—require the construction of composite representations of the environment called *cognitive maps.* Finally, we use sensory input to acquire knowledge about the world called *perceptions.* We can mentally manipulate these perceptions to form *ideas*—sometimes even novel ones—and solve the many problems that confront us on a daily basis.

Detailed Reviews

Burgess PR, Wei JY, Clark FJ, Simon J. 1982. Signaling of kinesthetic information by peripheral sensory receptors. *Annu Rev Neurosci* 5: 171.

Iggo A, Andres KH. 1982. Morphology of cutaneous receptors. *Annu Rev Neurosci* 5: 1.

Mountcastle VB. 1980. Sensory receptors and neural encoding. In Mountcastle VB (ed), *Medical Physiology.* Mosby, New York.

Sathian K. 1989. Tactile sensing of surface features. *Trends Neurosci* 12: 513.

Vallbo ÅB, Hagbarth K-E, Torebjörk HE, Wallin BG. 1979. Somatosensory, pro-prioceptive, and sympathetic activity in human peripheral nerves. *Physiol Rev* 59: 919.

Primary Somatosensory Processing 11

Upon entering the CNS, sensory signals are distributed over several parallel pathways in each of which information is serially processed through multiple synaptic stations. Several chapters in this book are devoted to the processing of information in specific sensory modalities. In the present chapter, we will identify the locations of the nuclei in the spinal cord and brain stem that receive the first synapses of peripheral sensory axons, and analyze in greater detail the intrinsic structural and functional organization of those nuclei that receive sensory information from the skin and musculoskeletal (together, the *somatosensory*) systems. This information should be correlated with details given in the next chapter on the output of these secondary sensory nuclei and further somatosensory processing.

Somatosensory Axons Synapse in the CNS

Impulses from somatosensory transducers are carried to the CNS by sensory (or primary afferent or first-order) nerve fibers whose cell bodies lie in the **dorsal root** (spinal) **ganglia** or in the ganglia of cranial nerves, 5, 7, 9, and 10. Embryologically, each spinal nerve associates with one somite, so that the muscles and skin innervated by individual spinal nerves in the adult reflect the segmental organization of the embryo. Consequently, the somatosensory fibers entering the spinal cord with each spinal nerve innervate a particular strip of skin (or underlying muscle) called a **dermatome** (Box 11.1). Each segment of the spinal cord thus receives its somatosensory data primarily from one dermatome. But because the sensory nerve fibers branch to terminate in segments both above and below the one at which they enter (see below), *adjacent dermatomal representations overlap extensively in the spinal cord.*

Box 11.1 The Segmental Innervation of the Body

The peripheral innervation of the body by spinal nerves reflects the original segmental organization of the embryo. Each spinal nerve is composed of motor and sensory axons that innervate the muscles and skin that develop from one somite. The area of skin that is innervated by sensory axons of one spinal nerve is called a **dermatome.** A crude map of the dermatomes is presented below with a few of the dermatomes labeled. Remember that there is considerable overlap between the neighboring dermatomes; the borders are not as sharp as depicted. The curious posture of the figure represents the way a neuroscientist envisions the human body for the purpose of anatomical reference to the CNS. This is a consequence of our tendency to view the human CNS for the most part through the metaphor of experimental animals. The four-legged attitude is far more popular than the bipedal throughout the mammalian world! Thus, *rostral* means toward the nose and *caudal* toward the buttocks (where a self-respecting mammal would proudly display a tail). *Dorsal* is toward the back of the trunk and *ventral* toward the tummy. The brain stem flexures and overgrowth of the cerebrum make navigational terms problematic within the brain. However, if one simply pretends that the brain has a straight axis, then the terms dorsal/ventral and rostral/caudal do nicely.

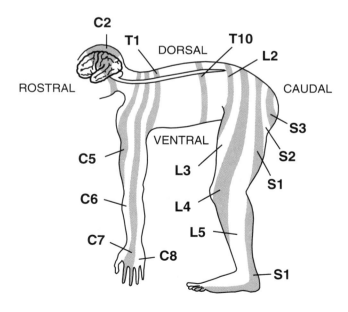

The primary sensory axons that enter the CNS excite the neurons upon which they synapse. They release excitatory amino acid neurotransmitters, primarily glutamate, at their central synapses. Peptide cotransmitters are also present in the presynaptic endings of many primary somatosensory fibers. In general, the region of the CNS in which the primary sensory axon terminals distribute is defined as a *sensory nucleus.* The sensory nuclei themselves contain the cell bodies that give rise to *secondary* (second-order) sensory axons that, in turn, make contacts with other cells within the CNS.

Spinal Cord Somatosensory Nuclei

The primary afferent axons in the dorsal roots enter the spinal cord at the *dorsolateral sulcus.* Incidentally, it is now clearly established that a few primary sensory fibers enter the spinal cord with the ventral roots. These are considered to be some of the propriosensory afferent axons involved in muscle reflexes. We can disregard them for the time being. There is a wide size spectrum of afferent axons entering the spinal cord, but two major groups can be distinguished (Fig. 11.1): a *medial* group of larger,

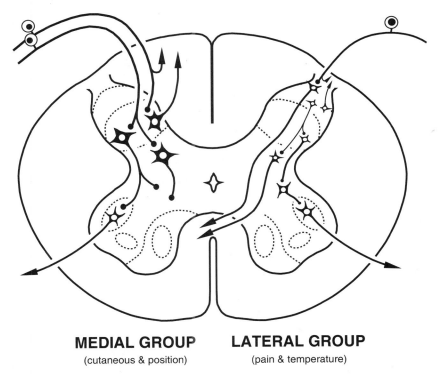

MEDIAL GROUP
(cutaneous & position)

LATERAL GROUP
(pain & temperature)

Figure 11.1 Two general categories of sensory axons enter the spinal cord and engage in different patterns of synaptic connection.

myelinated fibers (types I, II, and Aβ), and a *lateral* group of smaller, myelinated (Aδ) and nonmyelinated (C) fibers. A knife cut (sometimes made by the neurosurgeon in an attempt to relieve otherwise intractable pain in terminal cancer patients) that severs the lateral group produces analgesia and loss of thermal sensation in the ipsilateral dermatome innervated by the segment of the cut. In contrast, *tabes dorsalis,* a now uncommon disease that affects mainly the large-diameter fibers in the dorsal roots, causes a loss of tactile sensation and propriosensation without a loss of pain and temperature sensation. Thus, it appears that *the lateral group of smaller fibers mediates pain and temperature sensibility, and the medial group of larger fibers carries cutaneous (fine touch, pressure, vibration) and propriosensory information.* Notably, many of the type C (pain) fibers contain the neuropeptide substance P, a cotransmitter of demonstrated importance to pain transmission in the spinal cord.

   The two groups of incoming primary fibers terminate in different regions of the dorsal horn (Fig. 11.1). Although almost all primary afferent axons divide into short ascending and descending branches upon entering the cord (one to three segments up and down), those of the lateral group conveying pain and temperature data travel in the **dorsolateral fasciculus** (or tract of Lissauer) before ending mainly in the **posteromarginal nucleus** (Rexed's lamina I) and in the outer part of the **substantia gelatinosa** (laminae II and III). Fibers of the medial group, which convey cutaneous and propriosensory data, on the other hand, branch in the **dorsal funiculus** (or column) and distribute in the **nucleus proprius** (lamina IV) and more deeply in the **intermediate zone** of the spinal gray matter.

   The secondary pain- and temperature-responsive cells of the substantia gelatinosa are small and give off *short* axons that travel in the dorsolateral fasciculus and make contacts with other small cells and larger cells in Rexed's laminae I, IV, V, and VI. Notably, a shallow cut in the dorsolateral fasciculus will produce analgesia in the corresponding ipsilateral dermatome. The *larger* cells of the dorsal horn give rise to the ascending axons that convey the already somewhat processed pain and temperature data up to the brain over a pathway known as the **anterolateral tract** (see Fig. 12.2). The functional organization of the anterolateral tract will be considered in Chapter 12.

   Some incoming axons are involved in the local, spinal *reflex* connections with cells in the ventral horn of the spinal cord. Many such neurons in the ventral horn, especially those in Rexed's laminae VII and VIII, have short axons that contact motoneurons in Rexed's lamina IX. We'll have a look at these reflex connections in Chapter 15.

   Some of the larger fibers in the medial incoming group, believed to be mainly propriosensory ones, synapse on cells in a special nucleus called the **nucleus dorsalis** (of Clarke or Clarke's column), which is located medially at the base of the dorsal horn, but only in segments T_1 down to L_2. It is most prominent near its lumbar end. The nucleus dorsalis receives primary sensory fibers conveying propriosensory signals from muscle spindles, tendon organs, and joint endings and some touch and pressure information from the leg and lower trunk. Many of the larger cells of the nucleus dorsalis send their axons into the lateral funiculus of the spinal cord white matter, where they form the **dorsal spinocerebellar tract** and ascend to terminate in the cerebellar cortex (see Chapter 19). The functional counterpart of the nucleus dorsalis serving the arm and upper trunk is the **external cuneate nucleus** of the medulla (see below).

Local Processing in the Dorsal Horn

There is a rich local circuitry in the dorsal horn of the spinal cord that uses both excitatory (mainly glutamate) *and inhibitory* (mainly glycine) *neurotransmitters to modulate the activity patterns produced by the incoming sensory axons.* Although much of the local circuit processing remains unknown, it is clear that *the dorsal horn of the spinal cord is not simply a relay for incoming sensory data.* The complex local circuitry provides mechanisms whereby the incoming sensory information is processed, transformed, and rerouted for different purposes. A salient example of the local transformations that can occur involves our ability to override pain by intense tactile stimulation—vigorously rubbing a freshly barked shin seems to mask a good bit of the agony. This phenomenon is known to depend on a local mechanism in the dorsal horn where there is a convergence and interaction of pain and tactile transmissions.

The dorsal horn of the spinal cord also receives descending inputs from brain stem structures, such as the reticular formation, and even from regions of the cerebral cortex. Such *descending influences can modulate the responses of dorsal horn cells to sensory input* from the periphery. For example, serotonin axons originating in a caudal part of the brain stem raphé nuclei have been especially implicated in the modulation of pain information. Such neurons send axons to the dorsal horn of the spinal cord, where the serotonin they release suppresses pain transmission. Interestingly, the serotonergic cells of this raphé nucleus are themselves responsive to opiate compounds, which are known to be potent analgesics.

Brain Stem Sensory Nuclei

The sensory nuclei of the brain stem can be classified in a number of ways. An easy way is to identify *three* categories: somatosensory, viscerosensory, and special (Fig. 11.2). The last category refers to the **vestibular nuclei** and the **cochlear nuclei** which receive their input from the vestibulocochlear (8th cranial) nerve (see Figs. B2 and B3). The **nucleus of the solitary tract** represents the taste and viscerosensory complex that receives axons of the 7th, 9th, and 10th cranial nerves (see Fig. B2). Box 11.2 summarizes the association of cranial nerves with the brain

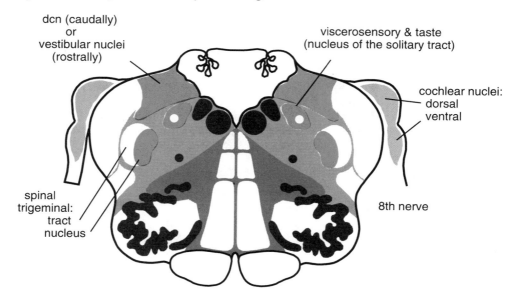

Figure 11.2 A schematic frontal section through the rostral medulla shows the locations of sensory nuclei (red) in the lower brain stem (compare to Fig. B2). Abbreviation: dcn, dorsal column nuclei.

stem afferent nuclei. The viscerosensory and special nuclei will be dealt with in chapters to come. For now, let's concentrate on the somatosensory nuclei. Included in the somatosensory group are the **dorsal column nuclei,** located in the medulla and the **trigeminal complex,** which extends throughout the medulla, pons, and midbrain (Fig. 11.3).

The Dorsal Column Nuclei Mediate Somatosensory Discrimination

The term dorsal column nuclei refers to the **gracile, cuneate,** and **external** (or lateral) **cuneate nuclei** and is derived from the position of these nuclei as the terminal nuclei for fibers ascending in the dorsal columns of the spinal cord (Fig. B1). Many of the larger primary sensory fibers that enter the spinal cord turn rostrally in the **dorsal funiculi.** Such fibers from the leg are located most medially near the dorsal median septum of the spinal cord, and fibers coming in at progressively higher segments of the spinal cord are layered on laterally so that the fibers in the dorsal columns are arranged according to the dermatomes. Thus, sensory axons from the leg occupy the **gracile fasciculus,** and those from the arm occupy the **cuneate fasciculus,** with the smaller contingent of trunk fibers more or less evenly divided between these two fasciculi (see Figs. 12.2 and B1).

Box 11.2 The Sensory Cranial Nerves and Nuclei

Since a look-up table sometimes comes in handy, one for the sensory cranial nerves is offered below.

Sensory Nucleus	Transducers	Nerves
Trigeminal	Cutaneous, propriosensory: face, oral cavity, dura, anterior two thirds of tongue	5
	External auditory meatus	7, 9, 10
	Caudal one third of tongue	9
Spinal nucleus	Pain and temperature	5
Main sensory nucleus	Largely fine touch and pressure	5
Mesencephalic	Pressure and propriosensory from teeth, periodontium, muscles of mastication	5
Solitary		
Rostral (taste)	Anterior two thirds of tongue	7
	Posterior one third of tongue	9
	Epiglottis	10
Caudal (viscerosensory)	Heart, lungs, gastrointestinal tract, carotid sinus, etc.	9,10
Special		
Cochlear nuclei	Hearing (organ of Corti)	8
Vestibular nuclei	Head position with respect to gravity, angular acceleration: maculae of utricle and saccule and cristae of semicircular canals	8

The fibers in the dorsal columns reorganize themselves as they ascend so that when they terminate in the dorsal column nuclei they are no longer organized according to the segmental (dermatomal) innervation of the skin, but instead are organized so that the afferent axons carrying information from adjacent parts of the body surface terminate in adjacent parts of the dorsal column nuclei. Thus, *there is a continuous, point-to-point representation of the body surface contained in the dorsal column nuclei* (Fig. 11.4). We call this representation a *somatotopic map*.

Early descriptions maintained that the dorsal columns contained only branches of primary sensory axons. More recent work has shown that *many* (perhaps a majority) *of the ascending dorsal column axons are* secondary *sensory fibers that originate from neurons of the dorsal horn of the spinal cord*. The classical view, based mainly on clinical observations, holds that the dorsal column nuclei mediate sensations of fine touch, pressure, vibration, and position, and that pain and temperature are mediated by another ascending system. More recent clinical and experimental evidence has challenged this view. Today, rather than attempting to correlate a particular somatic submodality with a limited central pathway, it is

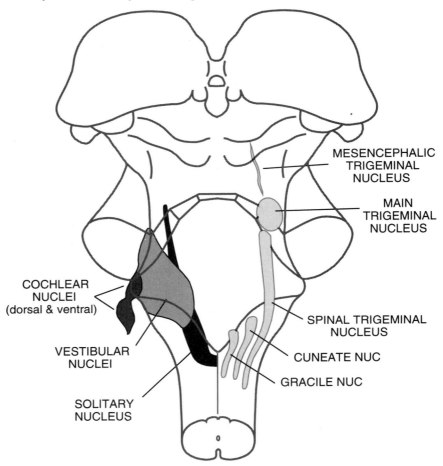

Figure 11.3 The distribution of sensory nuclei is shown on this outline of the brain stem. Although all of the nuclei are bilaterally paired, the somatosensory are shown on the right (in red) and the special and viscerosensory on the left for clarity.

more useful to consider that *somatosensory data of all or many submodalities are channeled along a number of parallel central pathways that extract information and use it for different computational purposes.* Thus, *the dorsal column system is better viewed as mediating somatosensory signals necessary for rather complex discrimination tasks, such as the identification of objects by feel.* Accordingly, the receptive field sizes of neurons in the dorsal column nuclei, especially those of the hand representation in the cuneate nucleus, are small (Fig. 11.5). There is a much higher density of transducers in the fingertips than most other parts of the integument. The representation of the hand in the cuneate nucleus is large in proportion to this high density of transducers. The axons of several transducers converge on

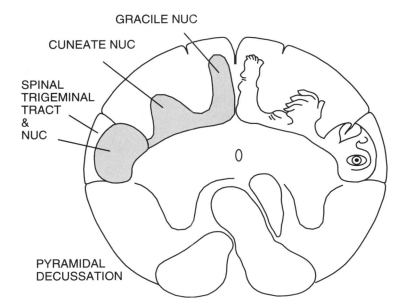

Figure 11.4 A schematic cross section through the closed medulla shows the somatotopic organization of the dorsal column nuclei (shaded red; compare to Fig. B1).

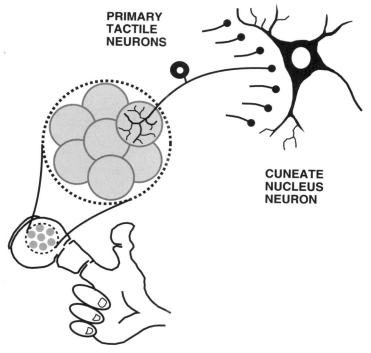

Figure 11.5 The receptive field properties of a sensory neuron in the cuneate nucleus (shown in the upper right) are established by convergence of primary sensory axons on it.

single cuneate neurons and thereby impose the parameters of their receptive fields. Because of the high density of transducers with small receptive fields and the generous numbers of correlated central neurons, fine discrimination is enhanced on the finger tips. We shall see in Chapter 12 that this *disproportionate representation,* which is first evident in the dorsal column nuclei, is sustained all the way to the somatosensory cortex.

The secondary sensory neurons of the gracile and cuneate nuclei give rise to the crossed **medial lemniscus,** which ends in the thalamus (see Chapter 12). The external cuneate nucleus on each side receives propriosensory information from the ipsilateral arm and relays that input to the cerebellar cortex by way of the **cuneocerebellar tract** (see Chapter 19).

Like the dorsal horns of the spinal cord, the dorsal column nuclei have complicated intrinsic connections made by local circuit neurons, and they receive descending inputs from higher levels of the brain. One important descending input comes from the somatosensory cortex—an ultimate target of signals from the dorsal column nuclei. Thus, the ascending impulses carried to the cells of the dorsal column nuclei are subject to modification at this synaptic way-station.

The Trigeminal Nuclei Mediate Somatic Sensations from the Face

The trigeminal complex receives primary somatosensory axons of the 5th cranial nerve, whose cell bodies lie in the trigeminal (semilunar, Gasserian) ganglion. There are three sensory nuclei of the trigeminal complex (Fig. 11.3). The most caudal is the **spinal trigeminal** nucleus (see Figs. B1–3). In the pons at the rostral end of the spinal nucleus is the **principal** (or main or chief) nucleus (Fig. B4). The most rostral, coursing throughout the midbrain, is the narrow **mesencephalic** nucleus. The 5th cranial nerve enters the brain stem by penetrating the middle cerebellar peduncle (Fig. B4). Many of these primary sensory axons synapse directly on neurons of the main nucleus, others descend along the lateral margin of the brain stem as the **spinal trigeminal tract** (Fig. 11.7A). Terminal axon branches are given off all along this tract; they enter the medially adjacent spinal trigeminal nucleus to synapse.

The axons in the spinal trigeminal tract are arranged so that those of the ophthalmic branch are located most ventrally, those of the mandibular branch most dorsally, and those of the maxillary branch in between. The fibers terminate in the spinal nucleus in this same ordered pattern (Fig. 11.6). The spinal trigeminal nucleus also receives the small contingent of somatic sensory fibers that travel in cranial nerves 7, 9, and 10.

Functionally, it would appear that the principal nucleus and the rostral end of the spinal nucleus (called the *oral* part) are involved in discrimi-

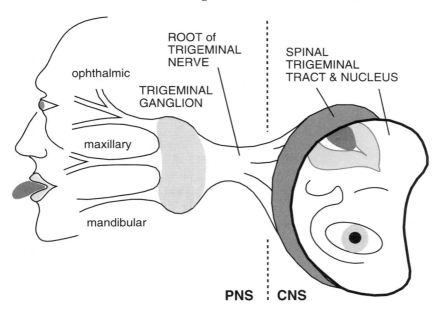

Figure 11.6 The representation of facial somatotopy is inverted in the spinal trigeminal nucleus. Abbreviations: CNS, central nervous system; PNS, peripheral nervous system.

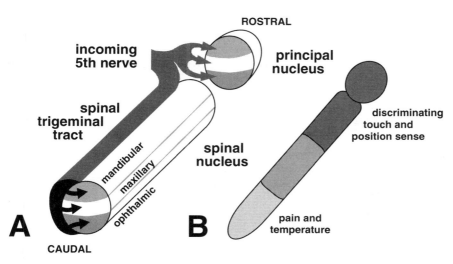

Figure 11.7 The principal and spinal trigeminal nuclei contain an upside-down, ipsilateral representation of the face (**A**). The somatosensory submodalities are represented in different parts of the trigeminal complex (**B**).

natory touch sensation for the face (Fig. 11.7B). The middle portion, going rostrally to caudally, is called the *interpolar* part, and although its functions are less well understood than those of the oral and caudal parts, some evidence suggests it may be involved with the processing of propriosensory data; it sends a prominent projection to the cerebellum. It is especially important to note that the most caudal end of the spinal trigeminal nucleus, called the *caudal* part, looks histologically very much like the substantia gelatinosa in the dorsal horn of the spinal cord. In fact, it is not possible to tell with certainty where the spinal trigeminal nucleus ends and the dorsal horn of the spinal cord begins; they simply fuse at some point. It is now quite clear that the caudal part of the spinal nucleus is biased toward the mediation of pain and temperature transmission. For these reasons, it is plausible to *regard the main nucleus and the oral part of the spinal nucleus as serving the same function of discriminating touch for the face that the dorsal column nuclei do for the body. The caudal part of the spinal nucleus serves the same function of pain and temperature processing as the dorsal horn of the spinal cord.* Keep in mind, however, that this is surely an oversimplification; much remains to be learned about the functional organization of the trigeminal complex.

The mesencephalic trigeminal nucleus is made up of *primary* sensory cell bodies—that's right!—the primary cell bodies rest within the bounds of the CNS rather than in a peripheral ganglion. Their peripheral axons carry propriosensory signals from the muscles of mastication and pressure sensations from the base of the teeth. Their central axonal segments terminate in the **trigeminal motor nucleus.** The mesencephalic trigeminal axons are considered to provide the afferent limb of a very fast reflex arc that is designed to inhibit the contraction of the muscles of mastication (innervated by the trigeminal motor nucleus) when the teeth encounter a hard object during biting or chewing. The reflex may have appeared as an evolutionary adaptation to prevent damage to the teeth—a fortunate circumstance for early hominids who were forced to dine frequently on pebble-laced tubers.

Detailed Reviews

Kass JH. 1990. Somatosensory system. In Paxinos G (ed), *The Human Nervous System*. Academic Press, New York.

McCloskey DI. 1978. Kinesthetic sensibility. *Physiol Rev* 58: 763.

Willis WD, Coggeshall RE. 1978. *Sensory Mechanisms of the Spinal Cord*. Plenum Press, New York.

Ascending Somatosensory Systems

12

An ascending somatosensory pathway can be considered literally as any axonal conduction route from the spinal cord or brain stem to higher levels of the brain. *There are a number of such pathways, which convey somatosensory information to the brain for three purposes: arousal and attention, movement control, and the perception of somatic sensation.* In this chapter, we will concentrate mainly, though not exclusively, on the ascending pathways that are generally believed to enable the organism to experience a *conscious* appreciation of energies impinging on transducers of the body surface and deep tissues. These sensory modalities are classified as *tactile*, *nocisensory*, and *propriosensory* and are typically described by terms such as touch, pressure, vibration, pain, heat, cold, position, itch, and tickle. The pathways extend from the peripheral transducers as far as the cerebral cortex and involve a *minimum* of three synaptic connections to get there (Fig. 12.1).

We learned in the preceding chapter that the incoming sensory axons of the spinal cord can be divided into two groups: a medial group of larger, myelinated fibers and a lateral group of smaller, myelinated and nonmyelinated fibers. Pain and temperature sensibility are conveyed preferentially by the lateral group, tactile and propriosensory data by the medial group. From this point onward, we can distinguish two major ascending conduction routes for somatic sensation, one concerned mainly (though not exclusively) with pain and temperature sensation, and another concerned with discriminating touch and propriosensation. Remember further that it is the larger neurons of the dorsal horn that give rise to the long ascending axons generally referred to as *secondary sensory fibers*.

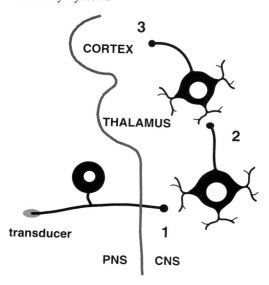

Figure 12.1 The minimal synaptic path from the peripheral transducer to the cerebral cortex involves three neurons (or synapses). Abbreviations: CNS, central nervous system; PNS, peripheral nervous system.

The Anterolateral System Conveys Pain and Temperature Data

Secondary sensory axons originate from the larger cells of Rexed's laminae I through VI (mainly I, IV, V, and VI) that receive incoming nocisensory and temperature data. Many such axons cross the midline in the **anterior white commissure** of the spinal cord within one or two segments of their origin (see Fig. 11.1). These fibers collect in the contralateral white matter as the **anterolateral tract** and ascend without interruption to the brain (Fig. 12.2). A unilateral transection of the anterolateral fasciculus results in the loss of pain and temperature sensation on the contralateral body surface, beginning a few segments below the level of the cut. Since no other cuts of the spinal cord produce this effect, the anterolateral tract is considered to mediate pain and temperature impulses. This notion is further supported by the clinical observation that electrical stimulation of the anterolateral tract leads patients to report experiences of pain and temperature changes. Upon entering the brain stem, the anterolateral tract is commonly referred to as the **spinothalamic tract** (Fig. 12.3 and see Figs. B1–6). This is a bit of a misnomer, since most of its axons do not reach the thalamus. Instead, they terminate on cells of the brain stem reticular formation and in the central gray substance. For this reason, a **spinoreticular** and **spinothalamic** component are distinguished within the anterolateral system.

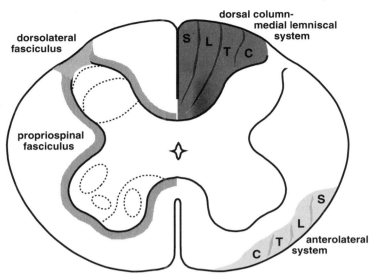

Figure 12.2 The ascending and local somatosensory tracts of the spinal cord are shown on a representative cross section. Fiber groups related to pain and temperature sensation are shown in red. Abbreviations: C, cervical; L, lumbar; S, sacral; T, thoracic.

The reticular formation makes up much of the core of the brain stem, and its neurons receive a wide variety of sensory and other types of input in addition to input from the spinoreticular axons. The projections of the reticular formation are as diverse as its inputs; it sends axons to virtually all other parts of the CNS, from the spinal cord to the cerebral cortex. This extensive connectivity together with the electrophysiological observations that individual reticular neurons are often multimodal and have large receptive fields, indicates that *the reticular formation is well-suited to the tasks of arousal and the maintenance of different levels of vigilance* (see Chapter 13). It should surprise no one that a neuronal system involved with arousal should receive ample data concerning painful stimulation.

The lesser spinothalamic contingent of axons of the anterolateral system synapses on cells in the **ventral posterolateral nucleus** (VPL), the amorphous **posterior group** of nuclei, and some of the rostrally located **intralaminar nuclei** (Fig. 12.3). Around mid-century, Vernon Mountcastle and his collaborators explored these thalamic areas with recording electrodes. They found that many neurons in the posterior group respond only to pain-producing stimulation, whereas only a few neurons in the VPL seemed responsive to painful stimuli. However, this notion has been challenged by more recent experiments, and it is important to be aware that the thalamic representation of pain remains controversial to this day. The third-order sensory axons from the VPL and the posterior group project to the cerebral cortex. But, before considering

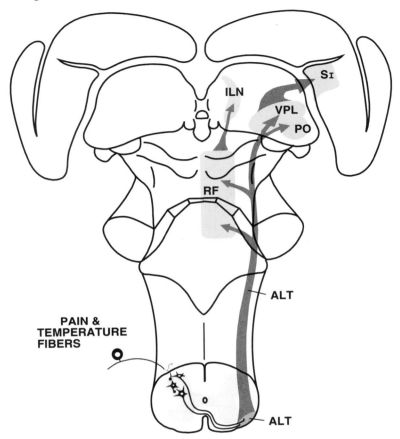

Figure 12.3 The anterolateral tract (ALT) mediates sensations of pain and temperature. Abbreviations: ILN, intralaminar nuclei; PO, posterior group; RF, reticular formation; S$_I$, primary somatosensory cortex; VPL, ventral posterolateral nucleus.

these thalamocortical projections, let's examine the ascending pathway for discriminating touch.

The Lemniscal System Conveys Data Used for Discrimination

The *lemniscal system* is considered to carry sensations that are important for discrimination of the shape, size, texture, and hardness of a tactile stimulus object. It is this system that enables us to identify a letter traced on our back by the finger of a friend or to choose a quarter rather than a dime from our pocket. This information is carried to the thalamus, which in turn sends it on to the cerebral cortex. Axons arising from neurons of

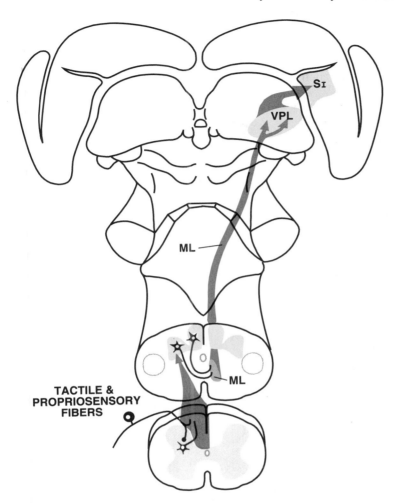

Figure 12.4 The dorsal column-medial lemniscal (ML) system mediates discriminative touch and conscious awareness of limb position. Other abbreviations as in Figure 12.3.

the gracile and cuneate nuclei on either side *cross the midline* as the **internal arcuate axons** to form the **medial lemniscus** of either side (Figs. 12.4 and B1). The medial lemniscuses ascend to the VPL without further crossing (Figs. B2–7). Unlike the anterolateral tracts, the majority of axons that form the medial lemniscuses reach the thalamus, and only a few synapse elsewhere along the way through the brain stem. Thus, each medial lemniscus conveys a systematic representation of the *contralateral* body surface. Electrophysiological studies have shown that the cells of the VPL respond in a place-specific manner and have rather small receptive fields—conditions requisite to the detection of discrete stimuli (Fig. 12.6).

The Trigeminothalamic System

Most of the axons that leave the main nucleus and rostral part of the spinal trigeminal nucleus cross the midline to join the contralateral medial lemniscus and ascend as its medialmost contingent to the thalamus, where they terminate on cells of the **ventral posteromedial** nucleus (VPM) (Fig. 12.5). The VPM is medially contiguous with the VPL, and thus, these two thalamic nuclei together contain a complete somatotopic representation of the contralateral body surface and face (Fig. 12.6). The cells in the VPM respond to stimulation of the face in a manner analogous to the response of VPL cells to stimulation of the body surface. There may also be a diminutive ipsilateral trigeminothalamic pathway arising from the principal trigeminal nucleus (Fig. 12.5). Recall that the caudal one-third of the spinal trigeminal nucleus is specialized to transmit pain and

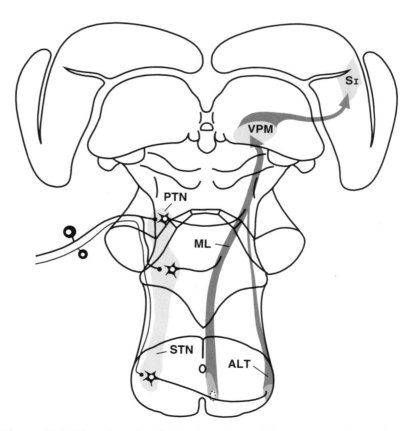

Figure 12.5 The trigeminothalamic pathway(s) carry somatic sensations from the face. Abbreviations: PTN, principal trigeminal nucleus; STN, spinal trigeminal nucleus; VPM, ventral posteromedial nucleus; other abbreviations as in Figures 12.3 and 12.4.

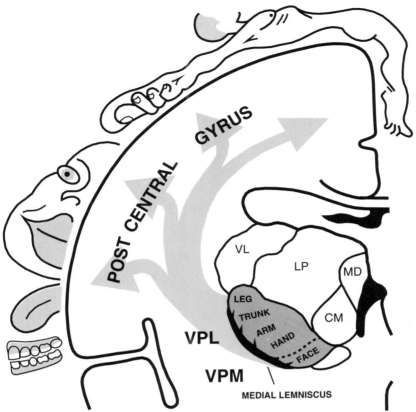

Figure 12.6 The neurons of the ventral posterior thalamus and the primary somatosensory cortex are somatotopically organized. Abbreviations: CM, centromedian nucleus; LP, lateral posterior nucleus; MD, mediodorsal nucleus; VL, ventrolateral nucleus; VPL, ventral posterolateral nucleus; VPM, ventral posteromedial nucleus.

temperature sensation. Axons from this caudal part, like those of the anterolateral system, ascend on the contralateral side and distribute mainly in the reticular formation.

The Somatosensory Thalamocortical System

The third-order somatosensory fibers originate in the thalamic nuclei VPL, VPM, and the posterior group and ascend to the cerebral cortex by way of *thalamic radiations*. The major cortical target of VPL-VPM is the caudal bank of the central sulcus and much of the crown of the

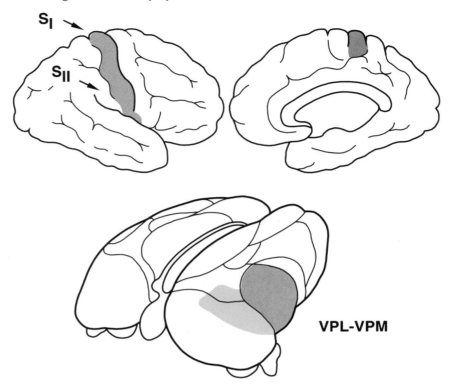

Figure 12.7 The diagram shows the location of the ventral posterior nuclear complex of the thalamus and the S$_I$ cortex to which they project. Abbreviations: VPL, ventral posterolateral nucleus; VPM, ventral posteromedial nucleus.

postcentral gyrus (Fig. 12.7). This cortical area corresponds to Brodmann's areas 3, 1, and 2; call it the primary somatosensory area (S$_I$). The thalamic projection to S$_I$ is topographically organized so that it imposes a point-to-point representational map of the contralateral half of the body. On this map, the zone that responds to the leg is located dorsomedially, the zone that responds to the arm and hand is located ventrolaterally, and the smaller zone that responds to the trunk is located in between (Fig. 12.6). The face is represented below the arm-hand zone, followed by the intraoral cavity and pharynx representations. The somatosensory thalamocortical axons have a strong excitatory effect on neurons in S$_I$ mediated by excitatory amino acid neurotransmitters. Very few, if any, cells can be found in S$_I$ that respond to noxious stimulation. Instead, *most of the S$_I$ neurons exhibit response properties and receptive fields that are appropriate for discriminating touch.* Not surprisingly, then, a stroke that produces damage in a circumscribed region of S$_I$ usually causes a diminution in two-point resolution and a reduced ability to recognize the texture and shape of objects applied to the affected area.

Three major points should be understood regarding the S$_I$ cortex. First, *its neurons are modality- and place-specific.* That is, a given cell is maximally responsive to only one kind of stimulus (e.g., light touch, vibration, pressure) on only one part of the body surface. Second, *the different parts of the body are represented by cortical areas of different size.* This reflects a general principle of all sensory cortices, to wit, *the size of a cortical zone of representation for a given unit of body (receptive) surface is proportional to the density of peripheral receptors in that unit, not to the absolute size of the unit.* Thus, the hand, which has a total surface area much smaller than the entire trunk, has a much larger cortical representation in S$_I$ than the trunk, because its receptor density is much higher. The third point to remember is that *there are at least three body maps in S$_I$* lying parallel to the central sulcus and corresponding roughly to Brodmann's areas 3, 1, and 2. Area 3, which lies in the fundus of the central sulcus, contains a complete contralateral body representation, in which the cells are responsive to stimulation of superficial tactile receptors. The receptive fields of these neurons are relatively small. Area 1, on the shoulder of the central sulcus, contains a second complete map, in which the cells are responsive to stimulation of receptors at an intermediate depth in the skin. Area 2, on the crown of the postcentral gyrus, contains a third complete map in which the cells respond best to stimulation of deep-lying receptors, especially by stimuli that are in motion across the skin. The receptive fields of cells in area 2 are much larger than those in area 3, and the response properties of the neurons indicate preferences for the particular direction of stimulus movement. Thus, distinct features of the same stimulus are extracted along the ascending and intracortical processing circuits and distributed to the three different regions of the S$_I$ cortex.

Using electrophysiological methods, it can be shown that neurons with common response properties within each of the three regions are grouped together in units that extend through the depth of the S$_I$ cortex. These cortical *columns,* which are considered by some brain scientists to be modular components or functional units, have so far been identified only in the primary sensory and motor cortices. Columns are most easily defined in electrophysiological terms as follows: *a column is a small cortical zone (1 or 2 millimeters in diameter) extending through the depth of the cortex that is responsible for a single class of sensory transducer coming from one part of the sensory field.* This means that a column in the S$_I$ cortex, for example, is specific to both modality (e.g., pressure) and place (e.g., knee). A corollary to this concept is that the computational processing of signals within each cortical module is largely the same regardless of the cortical area in which it resides. *Differences in the function of cortical modules are determined by their inputs rather than by local processing circuits.* We will revisit this notion of columnar organization several times in the chapters that lie ahead.

Other Cortical Areas Receive Somatosensory Data

A second somatosensory cortex, the S$_{II}$ cortex, is located ventrocaudal to S$_I$ and dorsal to the lateral fissure (Fig. 12.7). It also receives axons from VPL-VPM. Although a body surface representation can be demonstrated in S$_{II}$, this area appears to be less precisely organized than S$_I$, and its cells are responsive to stimulation on both sides of the body. This bilaterality is a consequence of commissural axons that cross through the corpus callosum. Several other cortical regions exist in which the resident neurons respond to somatic stimulation, but we know far less about them.

Various nuclei within the posterior group have been shown to project to all of the somatosensory cortical areas as well as to association areas. The function and organization of these thalamocortical projections are less well understood than those of VPL-VPM. Whether they communicate pain and temperature information to the cortex is unknown, and how pain is represented in the cerebral cortex is still far from clear.

The Transformation of Ascending Sensory Information

When you consider the ascending sensory systems and the other sensory systems to be described later, it is important to remember that signals are not always transmitted with high fidelity at each of the synaptic stations along the sensory pathway. At each of the so-called relay nuclei, there is convergence and divergence of ascending axons to the postsynaptic cells, which allows for the reintegration of the signals. There is also an influence of local circuitry and of descending feedback from stations higher up in the pathway. *These neural mechanisms serve to gate and transform the ascending signals in an orderly and systematic fashion that provides for extraction of information in a form more useful to the computational goals* (e.g., movement control) *of the different brain systems that receive the information.* Because of the multiple parallel pathways over which somatic information is conveyed to the cortex, lesions of one or another of the conduction routes often fail to give rise to entirely definitive sensory deficits. Damage to the dorsal columns, such as occurs in *tabes dorsalis,* for instance, does not eliminate fine tactile sensibility, such as might be measured by two-point resolution, but it does impede the recognition of objects by touch. Likewise, transection of the anterolateral tract does not always fully eliminate pain sensibility.

Detailed Reviews

Pons TP, Garraghty PE, Friedman DP, Mishkin M. 1987. Physiological evidence for serial processing in somatosensory cortex. *Science* 237: 417.

Rustioni A, Weinberg RJ. 1989. The somatosensory system. In Björklund A, Hökfelt T, Swanson LW (eds), *Handbook of Chemical Neuroanatomy, Vol 7: Integrated Systems of the CNS*. Elsevier, Amsterdam.

13 Arousal and Sleep

A living nervous system is inherently active. Spontaneous signaling between neurons begins as soon as the first synapses are formed in the womb. In addition to a widespread tendency for spontaneous activity, the fully developed human brain contains populations of neurons that are dedicated to a *pacemakerlike* function; they can sustain a generalized level of brain activity or selectively augment the activity in specific brain circuits. Such neuronal populations reside mainly in the brain stem reticular formation and are relatively immediate recipients of incoming sensory signals. The sensory data serve to modulate the baseline activity in these systems in order to elevate the responsiveness of the brain. This knowledge that brains are intrinsically active leads us as neuroscientists to reject Cartesian rationalism and instead declare to the contrary, *"sum ergo cogito"*!

Levels of Consciousness Are Reflected in the Electroencephalogram

When we are awake, we have a general awareness of ourselves as independent beings, of the passage of time, and of events in the space around us. This awareness is almost completely absent when we are deeply asleep. When awake, awareness can range from a very low level, when, for instance, we are drowsy, to a heightened state, when, for instance, we are scared. Scalp macroelectrodes can be used to monitor the summed electrical activity of the underlying cerebral cortex during these states of consciousness. The recorded activity—the electroencephalogram (EEG)—is a reflection of the *extracellular current flow caused mainly by postsynaptic potentials of cortical pyramidal neurons* (Fig. 13.1). The frequency and amplitude of the cortical EEG range from about 1 to 30 hertz and 20 to 100 microvolts, respectively. Although the EEG record exhibits wide variations within these ranges at any period of time, four main patterns are typically observed (Fig. 13.2). An *alpha wave* pattern of

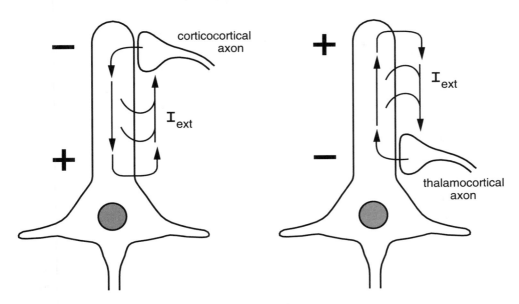

Figure 13.1 The EEG registers the extracellular current flow (volume conduction) created by excitatory synapses on pyramidal cells of the cortex. The macroelectrode detects the summated activity of several thousand such cells whose apical dendrites are all oriented parallel to one another.

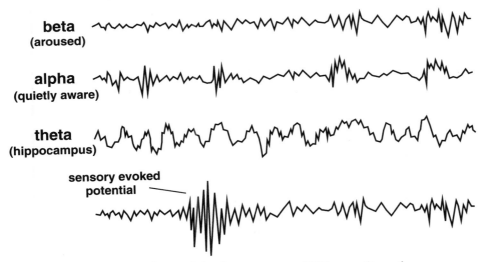

Figure 13.2 Shown are three of the four common EEG recordings from a normal human adult in the waking state.

8 to 13 hertz typifies the state of relaxed wakefulness. The alpha wave gives way to the *beta* pattern when a person engages in intense thought or is more highly aroused. The beta wave has a higher frequency (13–30 hertz) and a lower amplitude. The *delta* (0.5–4 hertz) and *theta* (4–7 hertz) waves are usually associated with sleep in the normal adult. They have the highest amplitudes and the lowest frequencies. Interestingly, the theta wave dominates in the hippocampal cortex when the neocortex is in the desynchronized beta pattern. The EEG patterns produced during specific sensory stimulations are called *sensory evoked potentials*. The EEG is clinically useful for the detection of abnormal patterns that might indicate epileptic activity and other central pathologies.

The Reticular Formation Contributes to General Brain Arousal

The so-called *ascending reticular activating system* has been a widely accepted concept. After entering the CNS, sensory signals can reach the cerebral cortex rather directly by way of modality-specific *lemniscal* pathways, such as the one described for somatic sensation in the last chapter. But the same raw data also can ascend to the cortex by way of a parallel route involving way-stations in the brain stem reticular formation. The spinoreticulotha-lamic pathway of the last chapter is an example of such a parallel avenue over which somatic sensation is conveyed to the cortex.

It is common for neurons of the reticular formation to extend four or five major dendrites that are long, relatively straight, and poorly ramified (Fig. 13.3). This dendritic tree is usually arrayed in a single plane—like the spokes of a wheel—perpendicular to the long axis of the brain stem. This morphol-ogy enables reticular neurons to intercept axon collaterals from several ascending and descending sources. In addition to spinal input, the reticular formation receives information conveyed in sensory cranial nerves. The auditory and vestibular nuclei project directly to the reticular formation. Visual input arrives indirectly by way of complex circuits. Finally, informa-tion of an emotional nature reaches the reticular formation by way of descending connections from the hypothalamus (see Chapters 23 and 24). Thus, not only is the reticular formation as an entity a site of high conver-gence, but individual reticular neurons can receive convergent data from several sources. How the reticular formation transforms these diverse signals is unknown, but it is clear that the information is sent to several destinations in the CNS over long and highly branched axons (Fig. 13.4).

One important destination for neurons mediating cortical arousal is the thalamus, where reticular axons terminate mainly, though not exclu-sively, in the intralaminar nuclei. The intralaminar nuclei, in turn, give rise to thalamocortical projections that are diffusely distributed through-out virtually all parts of the cerebral cortex. In other words, the cortical recipients of signals ascending through this reticular pathway are not

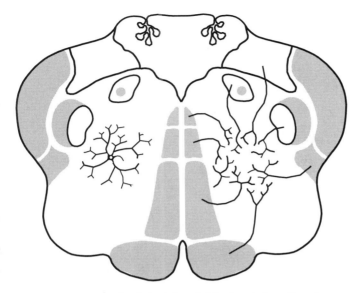

Figure 13.3 Drawn on the left side of this frontal section is a neuron of the reticular formation, which, like many of its fellows, extends its long, poorly branched dendrites in a radial pattern. This radial dendritic array usually defines a plane perpendicular to the long axis of the brain stem. Imagine the neuron superimposed on the right side of the section, where an assortment of axon branches are drawn. The axons, which arise from a variety of sources, can readily access the reticular cell because of its radial dendritic pattern.

as specific as those that receive signals via the lemniscal routes. For this reason, the reticular pathways are often referred to as *nonspecific*. They are also less specific with regard to sensory modality. The same reticular neurons that mediate somatosensory signals, for instance, may also relay auditory and visual signals upward to the thalamus. *Electrical stimulation of the reticular formation or of the intralaminar thalamic nuclei causes a desynchronization of the cortical EEG that is associated with behavioral arousal.* Clinical observations in cases of irreversible *coma*—the essential absence of arousal—support the notion that the brain stem reticular formation is necessary for the maintenance of arousal. Postmortem examination of patients who have suffered coma following head injury usually reveals extensive damage in the cerebral cortex, the brain stem reticular formation, or both.

To Sleep . . .

The EEG recorded from young adult sleepers can be divided into *four slow-wave stages* (Fig. 13.5). Sleep is characterized by regular cycling

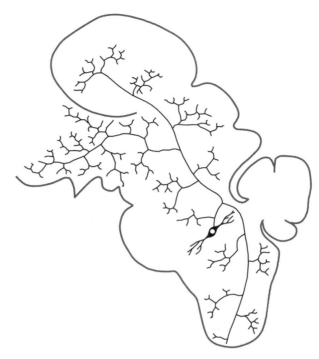

Figure 13.4 Many individual reticular neurons extend axons that branch into long ascending and descending collaterals. Secondary collaterals distribute the neuron's activity to multiple destinations throughout the spinal cord, brain stem, and diencephalon.

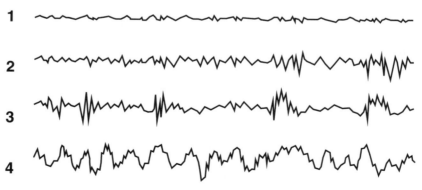

Figure 13.5 The typical EEG patterns recorded during the four identifiable stages of sleep are distinct from one another.

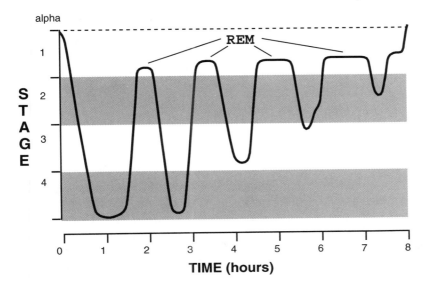

Figure 13.6 Recording the EEG of a normal adult sleeper through the night indicates that the sleep stages are periodic. Abbreviation: REM, rapid eye movement.

through these well-defined stages. After a person initially falls asleep, the EEG pattern passes sequentially through all four stages over about 45 minutes, then retraces the same stages in reverse order over a similar time period (Fig. 13.6). This cycle usually repeats itself a second time, after which the cycles become attenuated, first by the omission of stage 4, and later by the omission of stage 3. During slow-wave sleep, the muscles are relaxed, but motor commands can reach the motoneurons to produce periodic adjustments of body position. Likewise, sensory impulses from the periphery can arrive at the cortex. The threshold for arousal to a waking state varies according to the stages, however, so that stage 4 is the most difficult to interrupt. Curiously, time spent in stage 4 of slow-wave sleep declines steadily with age and may disappear completely after age sixty or so.

Several physiological changes occur during slow-wave sleep. Parasympathetic activity predominates. Alveolar ventilation is slightly reduced, leading to a slight rise in venous and alveolar carbon dioxide. Heart rate and blood pressure decline and gastrointestinal motility increases. Kidney filtration is diminished and body temperature falls by as much as 2°F. Basal metabolic rate decreases by 10% to 30%.

Sleep induction remains a difficult and controversial research subject to this day. One school holds that sleep induction is a passive process that results from an absence of arousing activity conveyed upward through the reticular formation. Other investigators hold that sleep induction is an active process. They argue that although the tonically active ascending reticular activating system, spurred on by sensory input, can arouse a sleeper or elevate vigilance in a waking person, it is clearly not the absence of activity in this system that induces sleep. There is some evidence to support both theories but none that is conclusive. It seems certain, however, that at least one *stage* of sleep is actively induced—the stage in which we dream.

. . . Perchance to Dream

As a sleeper returns to stage 1 after each cycle of slow-wave sleep (about 90 minutes after first falling asleep), the EEG enters a desynchronized pattern reminiscent of the waking *beta* wave: the neocortex exhibits a low-voltage, high-frequency pattern and the hippocampus exhibits a *theta* rhythm. Muscle tone is greatly reduced throughout the body, with the exception of the muscles that control the eyes, the middle ear ossicles, and breathing. The most conspicuous corollary of this paradoxical, desynchronized EEG pattern is an abundance of eye movements. Since the eye movements are often rapid, this phase of sleep is commonly called *rapid eye movement* or REM sleep. REM sleep is accompanied by physiological changes that include variations in blood pressure, heart rate, respiration, and body temperature and an increase in brain oxygen consumption. The duration of each REM episode increases, and the intervals between them decrease throughout the night.

During REM sleep, compound monophasic waves of electrical activity originate in the pontine reticular formation and propagate rostralward to reach especially the LGN and then the visual cortex. Because of this spread of activity from the pons to the geniculate nucleus to the occipital cortex, the waves are termed *PGO spikes*. More recent evidence shows that several thalamic nuclei are influenced by the PGO waves, including prominently the intralaminar group. These phasic PGO waves are thought to generate the rapid eye movements and other phasic events of REM sleep. A part of the pontine reticular formation that contains a dense population of noradrenergic neurons—the locus coeruleus—may be responsible for the atonia that accompanies REM sleep. If the locus coeruleus of a cat is destroyed, the cat will freely engage in coordinated movements during REM sleep. The movements have a goal-directed appearance, as though the cat were acting out its dreams. Could the noradrenergic suppression of spinal motoneurons during REM sleep be an adaptive mechanism that prevents us from harming ourselves or wasting energy by acting out our dreams?

The PGO spikes that herald entry into periods of REM sleep may be triggered by cholinergic neurons whose axons synapse in the pontine reticular formation. Such axons arise from nearby parts of the brain stem reticular core and from cholinergic cell groups in the basal forebrain. Injection of ACh agonists into the pontine reticular formation has been shown to elicit PGO waves and REM sleep in mammals. A distinct population of neurons—in this case, serotonergic neurons of the mid-brain raphé nuclei—appear to suppress PGO waves by tonically active axons that distribute in the pontine reticular formation. These serotonin neurons become quiescent through an unknown mechanism just prior to the onset of PGO spikes. Thus, REM sleep can be turned on by a certain population of neurons and turned off by another.

By the criterion of arousability, REM sleep is the deepest phase of sleep. However, sleepers awakened from REM sleep are far more likely to recall dreaming than when they are roused from other stages of the sleep cycle. Moreover, the dreams that occur during REM sleep are more vivid and emotionally charged than those that occur at other stages. Curiously, the time spent in REM sleep varies with age. REM occupies over 60% of the total sleep time in babies, progressively declines to about 20% by age ten, and remains stable for the remainder of an individual's lifetime.

Sleep Is Important to Most People

Although a few individuals do not sleep at all and seem none the worse for it, most people require regular bouts of sleep to function properly. Sleep disorders, especially insomnias (difficulty falling asleep or remaining asleep) and excessive somnolence (daytime sleepiness), are among the most common discomforts that drive people to visit their doctors. Sleep apnea (the cessation of breathing), the parasomnias (bed-wetting, sleep-walking, and nightmares), and narcolepsy (sudden involuntary lapses into a sleep state) are sleep-associated complaints. Insomnias and excessive somnolence can be caused by any number of factors, both organic and psychological, and take several forms. Sleep apnea is characterized by frequent periods of cessation of breathing, usually resulting from airway obstruction. The associated decrease in blood oxygen and increase in carbon dioxide levels eventually activate the brain stem respiratory centers and initiate inspiration. Narcolepsy is a mysterious syndrome characterized by irresistible sleep attacks that occur without warning and last from 5 to 30 minutes. Amazingly, sufferers of narcolepsy can pass from full consciousness directly into REM sleep.

The biological importance of REM sleep and dreaming seems to be anyone's guess; there is no shortage of provocative theories! Some neuroscientists have proposed that dreaming provides a psychological *release* from the emotional stress of daily events. Others have suggested that dreaming plays some role in memory consolidation. The abundance of REM sleep in neonatal animals has prompted some to surmise that dreaming serves as a mental rehearsal of the species-specific behaviors that will be essential later in life. There is little, if any, evidence for any of these speculations. Nonetheless, when a person is purposely deprived of REM sleep, she experiences a period of *rebound* when allowed to sleep without interruption; REM sleep occurs more frequently and for longer

periods. This rebound phenomenon would seem to indicate a physiological imperative for REM sleep.

How Is the Sleep–Waking Cycle
Entrained to the Day–Night Cycle?

Regardless of the exact biological role of sleep and dreaming, we certainly seem to benefit from the restorative value derived from regular
bouts of rest. We shall see later that the hypothalamus is a part of the
brain that plays a central role in sustaining our bodily well-being; it
regulates vital and near-vital functions like food consumption and body
temperature. Not surprisingly, it is critically involved in the sleep cycle
as well. Specifically, the hypothalamus contains two cell groups that are
essential to sleep induction and the diurnal sleep cycle: the **preoptic area**
and the **suprachiasmatic nuclei.** Insomnia results when the preoptic area
is destroyed in a cat. Conversely, application of serotonin agonists to the
preoptic area induces slow-wave sleep. The suprachiasmatic nuclei are
a pair of small, round nuclei that sit atop the optic chiasm. *Dendrodendritic
synapses are abundant between suprachiasmatic neurons and serve to synchro-
nize their activity.* The activity of neurons in the suprachiasmatic nuclei
can be monitored by their uptake of radiolabeled deoxyglucose at different times of day and night. Such a study in rats shows that the activity
occurs with an intrinsic periodicity; the neurons take up more glucose
during the period of behavioral activity than during periods of sleep.
Thus, the suprachiasmatic nuclei appear to serve as an internal circadian
pacemaker. The period of this pacemaker activity can be entrained to the
light–dark cycle by retinal axons that enter the suprachiasmatic nuclei
from the underlying optic chiasm. Such retinohypothalamic axons are
activated by light and are quiescent in the dark. Destruction of the
suprachiasmatic nuclei leads to disruption of the daily sleep–waking
cycle. The preoptic area receives a dense axonal projection from the
nearby suprachiasmatic nuclei and mediates their influence on the
sleep–waking cycle.

Detailed Reviews

Hobson JA. 1988. *The Dreaming Brain.* Basic Books, New York.
Steriade M, McCarley RW. 1990. *Brainstem Control of Wakefulness and Sleep.*
    Plenum Press, New York.

The Machinery of Movement 14

Skeletal movements occur around joints due to the selective contraction and relaxation of striated muscles. Since muscles can only pull, never push, they must work in opposing groups. Operationally, groups of muscles that contribute to a movement in a particular direction at a particular joint (i.e., pull together) are called **agonists,** and those that *oppose* the movement are called **antagonists.** Much of the time, however, appropriate movements are accomplished by *synergistic* contractions of muscles, even to the extent that antagonistic groups of muscles cooperate to produce the movements. Moreover, the movements that are made during practical behaviors occur around more than one joint at a time and involve several sets of agonist and antagonist muscles in a highly choreographed pattern of contractions and relaxations. To produce effective behaviors, the central systems that control skeletal movements must reflect the anatomical organization of the musculoskeletal apparatuses. Further, the computation of effective movement commands requires that the central control systems receive a continuous stream of sensory data about the *spatial layout of the local environment,* the *position of the body and limbs in space,* and the *load changes* (tension) *on the muscles.* Since the motor systems are arranged in parallel as well as hierarchically (see below), the sensory information delivered to each level must be formated appropriately for the contribution of that particular level to movement control (Fig. 14.1).

Classes of Movement and Levels of Control

Many elementary movements (*reflexes* and *rhythmic movements*) are coded in the spatiotemporal discharge patterns of the sensorimotor circuits contained entirely within the spinal cord and brain stem (Fig. 14.1). In these lower motor circuits, particular sets of sensory signals

163

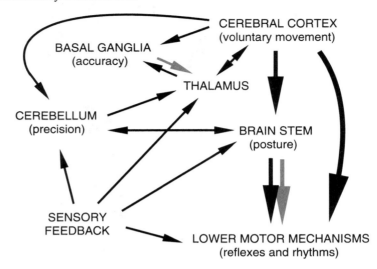

Figure 14.1 Movement is controlled by interactive, parallel systems that work hierarchically. Inhibitory connections (red arrows) play a significant role in movement control.

are conveyed with little synaptic delay to the motoneurons, where they trigger a stereotyped set of contractions and relaxations. When we elect to make more complex, volitional movements, the higher regions of our brains control the execution of the actual movements by selective activation of these same lower reflex and rhythm-generating circuits.

Essentially all of the movements we make are superimposed on a background of *postural stability*. Posture refers to the stability of the trunk over the legs and the head over the trunk. Most of us at some point in our early years were exposed to a stern taskmistress who frequently reminded us to maintain a proper "posture"—the "chest out, shoulders back" sort of comment should evoke fond memories. But postural control is also essential while we are in brisk motion—say, while driving the lane for a lay-up—or we would tumble over every time we began to move. Since humans walk on two legs rather than four and negotiate a terrestrial surface that is, more often than not, of irregular contour and strewn with obstacles, postural control is all the more challenging for the human CNS. Our upright attitude is maintained in part by lower reflex circuits, but it also depends heavily on *longer-loop* reflexes through especially the reticular formation and the vestibular system (Chapter 16). The lower and descending systems for postural control focus their signaling mainly on the axial or trunk muscles and on the leg extensors. The legs work with the trunk to stabilize the body in space.

In addition to the maintenance of an upright posture, the legs are used for locomotion, one of a number of repetitive movements organized at the level of the lower motor mechanisms (Chapter 15). Breathing is another common example of a rhythmic behavior that is patterned by the circuitry of the spinal cord. These rhythmic movements, like the reflexes, are subject to control by the descending systems, which can alter the basic patterns (goose-stepping versus shuffling), the rate (walking to sprinting), or whether they are performed at all (parade rest).

Although the arms can also contribute to posture by adjusting the body's center of gravity, their main use in humans is to reach for and move objects in the environment. Arm positioning (reaching) brings the hands (fine manipulators) to bear on the objects to be manipulated. The output of the arm orientation systems is directed mainly to the motoneurons that innervate shoulder and proximal limb muscles. The output of the systems that control fine manipulative movements is directed to smaller distal muscles in the forearm and hand that control movements of the wrist and fingers. As we shall see in Chapter 18 on voluntary movement control, the red nuclei play an important role in moving the arms, while hand and finger manipulations are controlled much moreso by the cerebral cortex.

Thus, postural, orientational (facing and reaching), and fine manipulative movements of the human body are controlled by partially overlapping and interactive circuits of the brain and spinal cord. Generally speaking, the brain systems that control head orientation and fine orofacial and eye movements are regulated by mechanisms similar to those that control the trunk and limbs. In humans, voluntary movements almost always involve wielding of the arms and head by contractions and relaxations of proximal arm and neck muscles to bring to bear the hands and fingers, the lips and tongue, and, in the case of gaze shifting, the eyes. We shall consider each of these levels of movement control in the chapters that follow.

Two Demands on Motor Systems

Successful behaviors—those that achieve goals—exhibit two important features called *precision* and *accuracy*. Precision refers to the *timing of muscle contraction* to produce spatially appropriate movement, while accuracy refers to the *selection of a movement sequence* appropriate to the task to be accomplished. Precision is often necessary for accuracy, but does not alone ensure it. To illustrate this distinction, consider a budding tennis player in the process of learning to serve. She can learn the proper *form* for serving a tennis ball—a sequence of smoothly blended individual movements—without ever setting foot on a tennis court. However, learning to serve without faulting requires the presence of at least a net and lines, and probably would not suffer from an opponent as well. In other words, one can make the proper movements in a precise manner, but that alone does not ensure that the ball will fall accurately within the service rectangle.

For some simple behaviors such as reflexes or rhythmic patterns of movement, precision is built in to the relatively simple neural circuits that generate the behavior, and the constraints on accuracy are minimal. The precise and accurate execution of a withdrawal reflex (Chapter 15), for example, is organized by the neural circuitry contained in the spinal cord. When the fingertip is pricked, the afferent impulse volley to the spinal cord produces relatively direct excitation of the flexor moto-neurons that serve the arm and hand, evoking a withdrawal of the insulted digit. The pattern of synaptic connections between neurons within the spinal cord imposes the appropriate parameters of force and duration of contraction on the relevant muscles so that the movement is made efficiently and in the appropriate (injury-reducing) direction.

In the case of more complicated behaviors, like serving a tennis ball, a more elaborate set of neuronal circuits is required. Precision becomes dependent on *computing* a set of motor commands and *comparing* the sequence of signals sent to the muscles (motor commands) with ongoing *sensory feedback* about the *performance* (timing and force of muscle con-tractions) of the movement sequence. The brain is equipped with a special device, the cerebellum, which is used by all of the supraspinal motor control systems to increase the precision of movement and smoothly blend one phase of a movement into the next. Accuracy be-comes more difficult to attain and depends on comparing the *intended* outcome of the behavioral sequence with sensory feedback about its execution, and recomputing of the motor commands accordingly. Much of this comparison occurs in the cerebral cortex. But the cortex has access to a large and mysterious mechanism, the basal ganglia system, which enhances the execution of complex movement sequences and may be important in refining motor skills.

The Motoneurons

There are three major types of neuron that reside in the spinal cord and brain stem and send their axons into the periphery: α motoneurons, γ motoneurons, and **visceromotor** (autonomic) neurons. Incidentally, the singularity of β motoneurons was apparently more perceived than real; nobody talks about them anymore. *All three types of efferent neuron utilize* ACh *as an excitatory neurotransmitter,* but they differ drastically in their peripheral targets and functions. Visceromotor neurons are smaller than α motoneurons and are found apart from the groups of α and γ motoneurons. They *innervate other neurons* in the peripheral sympathetic and parasympathetic ganglia or *secretory* cells in the adrenal medulla. The neurons in the peripheral ganglia, in turn, innervate effectors such as smooth and cardiac muscle and secretory glands throughout the body. Further discussion of the visceromotor neurons will be deferred to Chapter 22 on visceromotor reflexes.

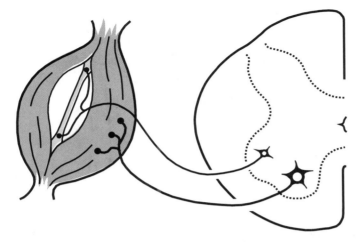

Figure 14.2   Two types of motoneuron innervate striated skeletal muscle. The size of the spindle organ has been exaggerated, and only one of the usual half-dozen or so intrafusal fibers is drawn for clarity.

The α motoneurons are the prime movers; they form neuromuscular junctions on the fibers of striated skeletal muscles and trigger the contractions that produce overt skeletal movements (Fig. 14.2). If the α motoneurons that innervate a muscle are destroyed, as happens in the viral disease *poliomyelitis*, the muscle is incapable of voluntary or reflex contraction; flaccid paralysis and atrophy will occur in the muscle. The cell bodies of α motoneurons are typically large and multipolar and contain a substantial amount of Nissl substance. Their dendrites are highly branched and receive many thousands of synapses. *Each α motoneuron, together with all the skeletal muscle fibers it innervates, is defined as a* **motor unit.** In the largest motor units, a single large α motoneuron innervates upwards of a thousand muscle fibers. This is typical of large limb (antigravity) muscles. Smaller motor units innervate fewer fibers, allowing for graded contractions and finer movements. The very smallest motor units are associated with the hand and tongue muscles, where fine control is needed for precise manipulations. A motoneuron **pool** comprises the motoneurons that innervate a single muscle. *The force production of a muscle is regulated by the number of motor units recruited within the motoneuron pool and by modulation of the firing frequency of the recruited motor units. Therefore, the more motor units dedicated to a muscle, the broader the range of recruitment available to the CNS movement control systems and, hence, the finer the control that the CNS has over the speed and power of muscle contraction.*

Gamma motoneurons are smaller than α motoneurons but are otherwise similar in appearance. They are found intermingled with α motoneurons and, like α motoneurons, send their axons to striated skeletal muscles (Fig. 14.2). But the γ motoneurons synapse on special muscle

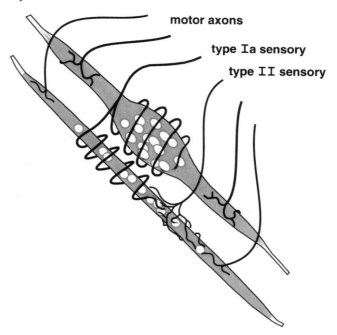

Figure 14.3 There are two types of intrafusal muscle fiber, the nuclear bag type with its bulging midriff and the slender nuclear chain type, which have distinct mechanical properties.

fibers that themselves do not produce overt bodily movements. The neuromuscular contacts of the γ motoneurons occur at the polar ends of *intrafusal* muscle fibers contained in **muscle spindle organs,** where they play an important role in the sensory feedback from the muscle spindles (Fig. 14.3).

The Muscle Spindle Organ

The muscle spindle organ is a specialized sensory apparatus that is important for the maintenance of muscle tone, for postural reflexes, and for movement control in general (Chapter 16). Within each striated muscle, there are a number of spindle-shaped connective tissue capsules that have their long axes in parallel to the fibers in the belly of the muscle. Each capsule is filled with a gelatinous fluid and contains about six to ten individual muscle fibers of two types called *nuclear bag* and *nuclear chain* (Fig. 14.3). These intrafusal fibers insert in the apical ends of the connective tissue capsule. *Only the ends of these intrafusal muscle fibers near their insertions are contractile.* As mentioned above, these contractile ends are innervated by γ motoneurons. The central parts of the fibers are innervated by sensory axons of two types, Ia and II, which respond when the central region of the fiber is stretched (Fig. 14.3). Such stretch occurs when the polar ends contract due to activity of the γ motoneurons, or it can occur when the whole muscle is stretched by, for instance, a change in load.

The two types of fiber in the muscle spindle organ differ in their mechanical properties so that they respond differently to stretch and hence trigger different response patterns in the afferent fibers that innervate them. The type Ia fibers respond with a burst of spikes when the intrafusal muscle fiber is initially stretched, but cease firing when a new constant length is attained. This enables them to inform the CNS about the *dynamic* phase of stretch, which reflects the *velocity* of muscle contraction. The type II axons modulate their frequency of discharge according to the constant degree of stretch of the nuclear chain fibers. In this way, they provide information on the *length* of the muscle at any given instant. It should be obvious that the motor systems of the CNS require information about the velocity of muscle contractions and the instantaneous length of a muscle in order to effectively control movement.

What is the role of the γ motor axons? It was once suggested that they were used by the CNS to trigger muscular contractions. The idea was that the firing of γ axons would stretch the muscle spindle fibers and set up an afferent volley in the type Ia and II sensory axons that would in turn excite the α motoneurons that innervate the same muscle. This hypothesis has been abandoned. Instead, it seems the γ axons' main purpose is to adjust the length of intrafusal fibers *during* muscle contraction so that the integrity of their sensory function is preserved during the execution of movements. That way, *the muscle can detect load changes even during an ongoing contraction* and so signal the CNS.

Local Circuit Interneurons

Although α and γ motoneurons are always intermixed, there are many more α than γ motoneurons. Living cheek-by-jowl with these motoneurons are many interneurons whose axons do not leave the CNS, but instead are involved in local synaptic circuits. Such local-circuit interneurons are found in the intermediate zone of the spinal cord (mainly in Rexed's laminae VII and VIII). Some lamina VII interneurons, called **Renshaw cells** (Fig. 14.4), receive *recurrent collaterals* of the α motoneuron axons and mediate a feedback inhibition (called *recurrent inhibition*) to the same population (pool) of α motoneurons. This circuit is believed to introduce a phasic component to the firing patterns of motoneurons, limit their top-end firing frequency, and perhaps assist in the termination of contraction when activated by descending motor control axons.

The motoneurons of the CNS are said to constitute the *final common pathway* for all CNS output. The term was coined by the founder of contemporary neurophysiology, Charles Sherrington, to emphasize the

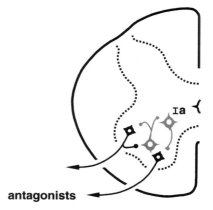

Figure 14.4 The Renshaw neuron mediates a feedback inhibition of α motoneurons and inhibits the Ia inhibitory interneuron (see Chapter 15). Excitatory neurons are jet black; inhibitory are red.

fact that all of the varied activities of the CNS (with its 10^{15} synapses) must ultimately be channeled through the axons of these few million α motoneurons to the effectors (muscles) of the body to produce behavioral expression. Thus, motoneurons receive impulses from many sources: primary sensory fibers synapse on them and several descending axons end directly on them. Much more frequently, however, the sensory and descending signals gain access to the motoneurons only by way of the many interneurons that are found among and make synaptic contacts with the motoneurons.

The Neuromuscular Junction

The business end of the α motoneuron—the neuromuscular junction at the motor endplate—is shown in Figure 14.5. The motor axon ends in a typical widening that is chock-full of ACh-containing vesicles. The vesicles tend to gather near a series of densities arrayed along the presynaptic membrane presumed to represent the active zones for neurotransmitter release. Curiously, the active zones are in good alignment with the apices of the postjunctional folds that characterize the postsynaptic (muscle) plasma membrane at the point of synaptic contact. Not surprisingly, then, one can clearly demonstrate with radiolabeled ligands for the nicotinic ACh receptor (the snake venom α-bungarotoxin does nicely for this purpose) that the receptors are clustered in the apical membrane and nowhere else on the muscle membrane. The cleft at the neuromuscular junction differs in two ways from that of the central synapse: the gap, at its narrowest, is slightly wider than in the synapse, and there is an intervening basal lamina that faithfully follows the contours of the muscle plasma membrane.

After being released into the cleft, ACh binds to the nicotinic receptors, which are themselves cation channels. The ACh is rapidly degraded by the acetylcholinesterase that abounds in the cleft, but not before it causes

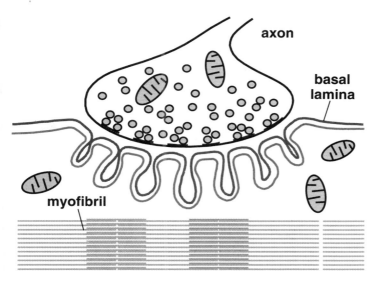

axon

basal lamina

myofibril

Figure 14.5 A sketch of the neuromuscular junction (motor endplate) shows that the ACh-containing vesicles tend to cluster near the active zones, which are spaced along the presynaptic membrane. The nicotinic ACh receptors cluster at the apices of the postjunctional folds.

the channels to open, allowing mainly a Na^+ influx that depolarizes the muscle membrane. This initial depolarization is called the **endplate potential.** It triggers an action potential that propagates along the membrane of the muscle fiber and into the system of T tubules, producing a contraction of the fiber called a *twitch*. This *excitation–contraction coupling* is mediated by voltage-gated Ca^{2+} channels that abound in the membrane of the sarcoplasmic reticulum. When depolarized by the spreading action potential, they open and allow Ca^{2+} to escape from the interior of the sarcoplasmic reticulum, where it is sequestered at high concentration. Some of the cytosolic Ca^{2+} binds to **troponin** which produces a conformational change in **actin** that enables it to interact with **myosin** and produce the transient shortening of the myofibrils.

Detailed Reviews

Evarts EV. 1979. Brain mechanisms of movement. *Sci Am* 241: 164.

Kuypers HGJM. 1985. The anatomical and functional organization of the motor system. In Swash M, Kennard C (eds), *Scientific Basis of Clinical Neurology.* Churchill Livingstone, New York.

15

The Motor Nuclei and the Descending Systems

A distinction can be made between the *lower* sensorimotor apparatuses—the sensory axons, the local circuit interneurons, the motoneurons, and the muscles they innervate—and the *descending* motor systems that exploit them. Clinical neurologists are especially interested in this distinction because of the differential signs that occur when one or the other is damaged by trauma or an interruption of the arterial blood supply (stroke; see Appendix A). Before we begin a functional analysis of movement control, let's briefly survey the anatomical organization of the lower motor nuclei and the descending systems.

The Motor Nuclei Are Arranged in Columns

Motoneurons cluster together into nuclei. Such nuclei are arranged in more or less continuous longitudinal columns in the ventral horns of the spinal cord gray matter and through the brain stem. To help conceptualize the functional organization of the motor nuclei, we can designate *three* columns according to the embryonic derivation of their peripheral targets (Fig. 15.1). The first column contains the α and γ motoneurons that innervate striated skeletal muscle derived from *somite* mesoderm of the head and body. Since most of this segmentally derived muscle ends up controlling the skeletal elements near the midline of the body, the motor nuclei of this column can be called *axial*. The second column is composed of two homologous groups: one in the spinal cord that innervates striated muscles of the body and limbs that are derived from *non*segmented *lateral plate* mesoderm, and one in the brain stem that innervates striated muscles of the head that are derived from *non*segmented *lateral* mesoderm of the *pharyngeal* (branchial) *arches*. Since these motor nuclei control muscles removed from the axis, we can reasonably refer to them collec-

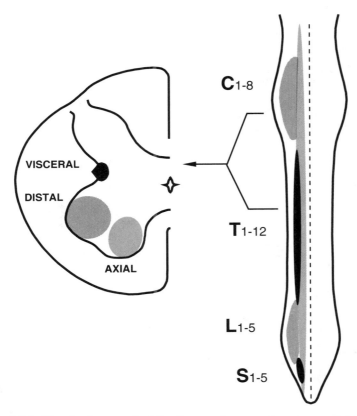

Figure 15.1 The deployment of the three columns of motor nuclei in the spinal cord is represented on a schematic longitudinal section. The hemisection to the left is a fictitious composite of the cervical enlargement and a midthoracic level.

tively as *distal* motor nuclei. The third column consists of efferent neurons that actuate (indirectly, via peripheral ganglia) *smooth* and *cardiac* muscle and *glands*. Call this the *visceromotor* column of nuclei. In the brain and in the sacral segments of the spinal cord, the visceromotor nuclei give rise to the *parasympathetic* outflow; in the thoracic and lumbar segments of the spinal cord, they give rise to the *sympathetic* outflow.

The Spinal Cord Motor Nuclei

The motor nuclei of the axial column are present medially throughout the entire length of the ventral horn of the spinal cord. The distal column, in contrast, is present only at enlargement (C_5 to T_1 and L_2 to S_3) levels where limb-innervating motoneurons are found. The localization of motor nuclei within the these two columns has been facilitated by cell

Figure 15.2 The motoneuron pools at enlargement levels of the spinal cord are logically arranged with respect to the location of the muscles they innervate.

labeling after the injection of retrograde tracer substances into individual muscles (see Box 1.1). The results of such experiments indicate that neurons aggregate into clusters called pools that are associated with single muscles. Thankfully, it is less important to know the individual pools than it is to know their general topographic pattern relevant to the muscles they innervate. As already noted above, within the enlargement levels (Fig. 15.2), *motoneurons that supply the axial muscles are located medially and those that innervate distal muscles are located laterally.* In addition, *the motor pools that innervate flexor muscles are situated dorsal to those that innervate extensor muscles.* At upper cervical, thoracic, and lower sacral levels, the lateral group is, of course, absent.

In the thoracic and upper lumbar segments of the spinal cord, the third column, the visceromotor neurons, forms a nucleus that appears as a distinct spur of the central gray matter extending into the white matter of the lateral funiculus. This intermediolateral nucleus, which extends only from T_1 to about L_2 (Fig. 15.1), is the location of the *preganglionic sympathetic* neurons. The axons of these cells terminate in the chain of sympathetic ganglia, the prevertebral ganglia, or the adrenal medulla. In sacral segments of the cord, in a position analogous to the intermediolateral column, but not readily visible as a discrete spur, are the *preganglionic parasympathetic* neurons of the spinal cord (see Chapter 22).

The Brain Stem Motor Nuclei

Remember that the sulcus limitans persists in the floor of the fourth ventricle and continues to demarcate a dorsolateral sensory region from a ventromedial motor region. With some allowances for the effects of the widening of the fourth ventricle, the three columns of motor nuclei remain in roughly the same positions relative to one another in the brain stem that they occupied in the ventral horn of the spinal cord (Fig. 15.3). Figure 15.4 shows the interruptions (sometimes quite lengthy) that occur in the motor columns and the locations of the motor nuclei with respect to the three subdivisions of the brain stem. Remember that no motor nuclei are present rostral to the midbrain. The correlations of the brain stem motor nuclei with the cranial nerves are summarized in Box 15.1.

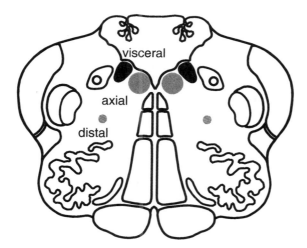

Figure 15.3 A schematic diagram of the motor nuclei in the lower brain stem shows that they retain the same general relationships that occur in the spinal cord (compare to Fig. B2).

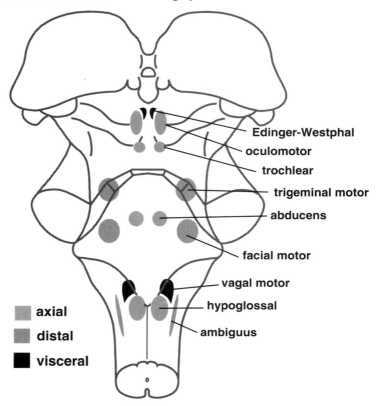

Figure 15.4 The motor nuclei in the brain stem are arranged into discontinuous columns.

The axial column is broken into *four* distinct nuclei. The rostral three nuclei, the **oculomotor, trochlear,** and **abducens,** innervate the extraocular muscles, which are striated and derived from the myotome of the first preotic somite (see Figs. B3, B6, and B7). The axons of these motoneurons form the oculomotor (3rd), trochlear (4th), and abducens (6th) cranial nerves. The oculomotor nucleus is the most rostral; it and the smaller trochlear nucleus are located in the midbrain. The abducens nucleus is located in the pontine tegmentum. The caudalmost member of the medial column is the **hypoglossal nucleus** which is located in the medulla (Fig. B2); its axons form the 12th cranial nerve and innervate the intrinsic tongue muscles, which are derived from the myotome of the last preotic somite. The 3rd, 6th, and 12th cranial nerves exit the brain stem ventrally, not too far off the midline, and ipsilateral to their cells of origin. The intrathecal roots of the paired 4th cranial nerves cross to the contralateral side dorsal to the cerebral aqueduct before exiting the brain stem at the dorsal surface of the pontomesencephalic junction.

Box 15.1 The Motor Cranial Nerves and Nuclei

You may want a handy look-up table for motor cranial nerves as well.

Motor Nucleus	Muscle and Gland	Nerves
Somite somatic		
Hypoglossal	Intrinsic tongue	12
Abducens	Lateral rectus	6
Trochlear	Superior oblique	4
Oculomotor	Remaining extraocular muscles and levator palpebrae	3
Branchial somatic		
Ambiguus	Pharynx and larynx	9, 10
Facial	Muscles of facial expression, stapedius, stylohyoid, posterior belly of digastric	7
Motor V	Muscles of mastication, tensor palatini, tensor tympani, anterior belly of digastric	5
Parasympathetic		
Dorsal motor	Cardiac muscle, smooth muscle and glands of lungs and gastrointestinal tract (always via ganglia)	10
Salivatory	Salivary and lacrimal glands via submandibular and pterygopalatine ganglia (7), and salivary glands via otic ganglion (9)	7, 9
Edinger–Westphal	Ciliary body and sphincter of iris via ciliary ganglion	3

The column of distal (branchiomeric) motoneurons is located ventrolateral to the axial column. There are three nuclei in the distal column. The most rostral is the **trigeminal motor nucleus** located in the rostral pons (Fig. B4). It sits medially adjacent to the main sensory nucleus of the trigeminus. More caudally, close to the pontomedullary junction is the **facial motor nucleus** (Fig. B3) and, in the caudal medulla, is the **nucleus ambiguus** (Fig. B2). Axons from these three nuclei innervate, respectively, the muscles of mastication, the mimetic musculature, and the laryngeal muscles.

The visceromotor column lies close by the axial column for the most part. The rostralmost nucleus of this column is closely associated with the oculomotor nucleus and is called the **Edinger–Westphal nucleus.** It provides the parasympathetic outflow to the ciliary ganglion of the eye by way of the 3rd cranial nerve. After a long interruption in the column, the **vagal motor nucleus** (formerly, the dorsal motor nucleus of the vagus) appears in the medulla, spanning roughly the same rostrocaudal levels as the hypoglossal nucleus (Fig. B2). The axons of the vagal nucleus travel in the 10th cranial nerve and provide parasympathetic outflow to the head, the thorax, and most of the abdomen. Associated with the vagal motor nucleus are the so-called *superior* and *inferior salivary nuclei*. In reality, the preganglionic parasympathetic neurons that innervate the salivary, mucosal, and lacrimal glands are not aggregated as discrete nuclei, as these names imply. Instead, they are rather loosely arranged within the rostral medullary and caudal pontine reticular formation.

The Descending Pathways

One can liken the lower apparatuses to the keyboard of a concert piano upon which the pianist (the brain) plays her score. Instead of ten fingers, the brain uses *five* major descending pathways to strike the right chords in the melody of movement. These descending systems originate from neurons in the *cerebral cortex, red nucleus, superior colliculus, vestibular nuclei,* and *reticular formation* (Fig. 15.5). Since many patients come to the clinic with movement disorders, it is important for diagnostic purposes to know the origin, course, terminal distribution, and contribution to movement of each system. Neurologists often make a clinically useful distinction between the consequences of damage to the motor nuclei and those that result from damage to the motor cortex or other descending

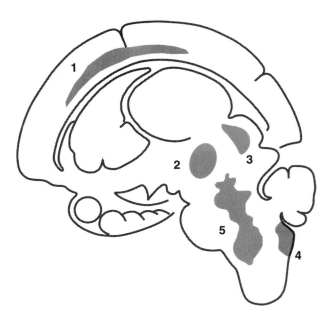

Figure 15.5 The origins of the major descending motor control systems on each side of the brain are the (1) cerebral cortex, (2) red nucleus, (3) superior colliculus, (4) vestibular nuclei, and (5) reticular formation.

motor control pathways. The terms *lower* and *upper motoneuron syndromes* are used in the clinic as shorthand for the two different constellations of motor deficits. The major differences between the two are that lower motoneuron disease affects specific muscles and results in decreased muscle tone or even paralysis, fasciculations (twitchings), and atrophy of the affected muscles. Upper motoneuron disease, in contrast, is more complicated, and its exact manifestations depend on the site of the lesion. Upper motor neuron disease affects groups of muscles and results in spasticity (increased muscle tone), and hyperreflexia. Atrophy is postponed. The vestibular and reticular systems have a major impact on reflexes, rhythms, and posture, functions that will be discussed in Chapter 16. The superior colliculi are concerned entirely with eye and head coordination during gaze shifts, the topic of Chapter 17. The cortical and rubral contribution to *voluntary* movement is particularly important in humans and also merits its own detailed analysis (Chapter 18). In the present chapter, our aim is to overview the anatomical organization of the five descending pathways.

The Pyramidal Tract

The pyramidal tract takes its name from the medullary pyramids. All of the fibers in the medullary pyramids come from the cerebral cortex. In fact, although the common terminology is coincidental, they arise from *pyramidal* cells in layer V. It was originally thought that the corticospinal neurons were all located in the cortex that forms the rostral bank of the central sulcus and extends onto the crown of the precentral gyrus—the primary motor cortex (M_I, Brodmann's area 4). But we now know that cells in cortical areas outside of M_I contribute axons to the pyramidal tract as well (Fig. 15.6). Retrograde tracing studies of the pyramidal tracts indicate that only about 30% of the axons arise from cells in M_I. Another 30% arise from cells rostral to the precentral gyrus in an area of the frontal lobe known as the **premotor cortex** (Brodmann's area 6), and the remaining 40% arise from neurons in the parietal lobe, especially S_I.

The pyramidal tract axons of either hemisphere descend from the cortex first through the subcortical white matter to converge on the internal capsule, where they collect as a component of its *posterior limb* between the thalamus medially and the lentiform nucleus laterally (Figs. 15.7 and B8). The remainder of the internal capsule consists of many other descending fibers, such as corticostriatal, -thalamic, -rubral, -pontine, and -reticular, as well as ascending thalamocortical axons. On each side, the pyramidal tract continues into the cerebral peduncle of the midbrain, where it occupies the middle one-third (Figs. 15.8 and B7). The far medial and lateral thirds of each peduncle are occupied by the corticopontine axons, which synapse on neurons of the ventral pons when the peduncles enter it. The pyramidal fibers, however, split into several bundles, which penetrate the cell masses of the pons (Figs. B3

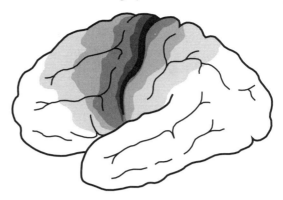

Figure 15.6 Pyramidal neurons that contribute axons to the corticospinal system are expansively distributed in the frontal and parietal lobes. The degree of shading reflects the different density of corticospinal neurons.

and B4), only to regroup as the distinct medullary pyramids as they emerge from the caudal end of the pons (Figs. B1 and B2).

Along the course of this descending pathway through the brain stem, many fibers leave to supply the motor nuclei of the cranial nerves and other motor-related structures such as the red nuclei, superior colliculi, and reticular formation. Since many such axons synapse in or near the motor nuclei of the medulla, they are often said to comprise a *corticobulbar tract*. Some such fibers cross the midline to end on the contralateral side, whereas others end ipsilaterally. Therefore, some of the cranial nerve motor nuclei are ultimately influenced by axons from both cerebral hemispheres.

The *corticospinal* tract represents what is left of each pyramid after it has passed below the brain stem and into the spinal cord. Most such fibers—more than 90%—cross the midline in the pyramidal decussation. In the spinal cord, these axons take up a position in the lateral funiculus of either side (Fig. 15.9), where they are arranged so that the axons that will end at upper levels of the cord are situated most medially and those destined for caudal segments are located most laterally. A much smaller contingent of the corticospinal fibers descends uncrossed as the *ventral* corticospinal tract. Many of these fibers also cross the midline shortly before terminating in the spinal cord gray matter.

The axons of the corticospinal tract vary in diameter and conduction velocity. They can generally be divided into larger, fast-conducting and smaller, slow-conducting axons. The former synapse mostly in a ventral part of the intermediate zone of the spinal gray matter, the latter terminate more dorsally in the intermediate zone and in the dorsal horn. *Some of the corticospinal axons establish direct synaptic contact with motoneurons, whereas most synapse on local circuit interneurons in Rexed's laminae IV through VIII.* Even though there is some overlap between their respective terminal fields, *the fibers from motor and sensory cortices have preferential sites of termination* so that those from motor cortex terminate mainly in

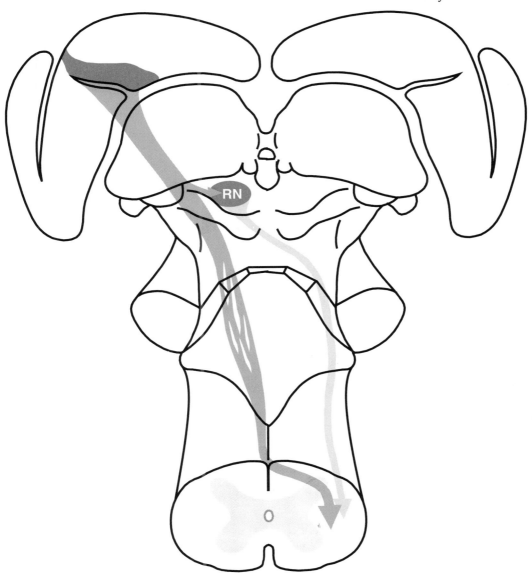

Figure 15.7 The corticospinal axons remain ipsilateral to their side of origin as they descend through the brain. Only at the spinomedullary junction do the vast majority cross as the pyramidal decussation. In contrast, the rubrospinal axons cross the midline shortly after emerging from the caudal end of the red nucleus (RN). They continue laterally to the lateral margin of the brain stem before turning caudalward. Although the pathways exist on both sides, only one is illustrated for the sake of clarity.

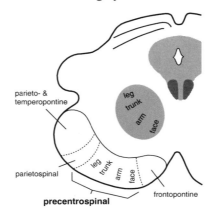

Figure 15.8 A drawing of a cross section through the midbrain shows the organization of corticospinal and corticopontine fibers in the cerebral peduncle and the somatotopy of the red nucleus (compare to Fig. B7).

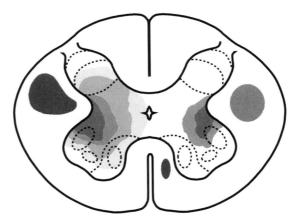

Figure 15.9 The locations of the corticospinal (left) and rubrospinal (right) tracts and their respective terminal distributions are indicated on a spinal cord cross section.

layers VI, VII, and VIII and those from sensory cortex terminate more dorsally in layers IV and V. Because of its termination pattern, the parietospinal system is suspected to be more involved in the modulation of sensory impulses and spinal reflexes than in the conveyance of motor commands to the ventral horn.

The Rubrospinal Tract

The red nuclei are roughly spherical cell groups located in the midbrain tegmentum (Figs. 15.7 and B7). The red nucleus of either side has two major subdivisions: a caudal *magnocellular* part and a rostral *parvicellular* part. Most of the rubrospinal axons arise from magnocellular neurons. Essentially all the rubrospinal axons cross the midline immediately caudal to the red nucleus and course caudolaterally to a position in the ventrolateral margin of the brain stem (see Figs. B1–6). In the spinal cord, the axons continue their descent in the lateral funiculus just ventral and lateral to the corticospinal fibers.

The red nuclei are somatotopically organized, with the neck and upper limb represented rostromedially (these fibers end in the cervical cord), the lower limb represented caudolaterally (these fibers end in the lumbosacral cord), and the trunk in between. The termination sites of the rubrospinal axons in the spinal cord correspond closely to those from the motor cortex (Fig. 15.9).

The red nuclei project to other regions besides the spinal cord. These targets of rubral efferent axons are primarily the lateral reticular nuclei, the inferior olivary nuclei, and the external cuneate nuclei, all of which send their output to the cerebellum. Correspondingly, a major source of input to the red nuclei comes from the cerebellum by way of the superior cerebellar peduncles. The motor area of the cerebral cortex is another major source of afferent connections to the red nuclei, but some corticorubral fibers also originate in the postcentral cortex. This input from the cerebral cortex is excitatory to rubral neurons. Thus, *the red nuclei are a way-station in an indirect corticospinal pathway that can carry cortical commands to the spinal cord in the event of damage to the medullary pyramids.*

The Tectospinal Tract

This modest pathway originates from cells in the deeper layers of either superior colliculus (Fig. B7). Its fibers cross ventral to the central gray substance and descend near the midline in a position ventral to the medial longitudinal fasciculus (Figs. 15.10 and B2). In the cord, the fibers travel in the ventral funiculus and do not reach beyond the upper cervical segments (Fig. 15.11). *These fibers, along with those of the medial vestibulospinal* (see below), *play a critical role in orienting head movements and eye–head coordination* (see Chapter 17). The deeper layers of the superior colliculus are heavily influenced by the cerebral cortex, cerebellum, and basal ganglia, as well as by several other structures.

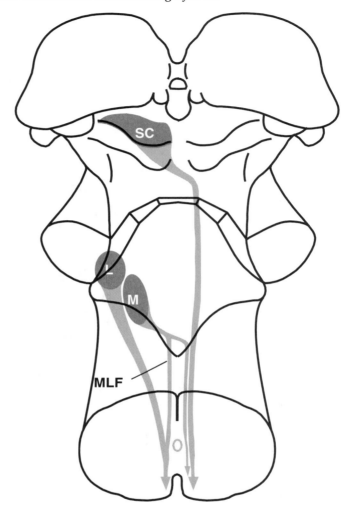

Figure 15.10 The tectospinal tract of either side originates from neurons in the deeper layers of the superior colliculus (SC) and crosses immediately to descend near the midline. The lateral and medial vestibulospinal tracts arise from, respectively, the lateral (L) and medial (M) vestibular nuclei. The lateral tract remains on the ipsilateral side, whereas the medial is a bilateral tract descending through the lower brain stem in company with non-vestibular fibers of the medial longitudinal fasciculi (MLF).

The Vestibulospinal Tracts

The 8th nerve axons transmit impulses to neurons in the **vestibular nuclei** (Fig. 15.12). There are four vestibular nuclei on each side of the brain stem named the **superior, inferior, lateral,** and **medial,** according to their positions relative to one another. Only neurons in the lateral and

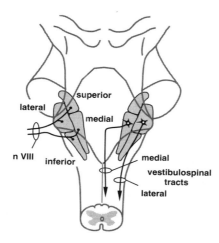

Figure 15.11 This brain stem schematic shows the four vestibular nuclei and the origin of the two vestibulospinal tracts.

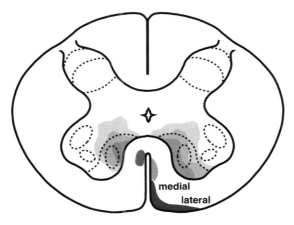

Figure 15.12 The locations of the tectospinal (left) and vestibulospinal (right) tracts and their respective terminal distributions are indicated on a spinal cord cross section.

medial nuclei send axons to the spinal cord (Fig. 15.10). The vestibulospinal system can be divided into two tracts: medial and lateral. The bilateral *medial* tract travels with the medial longitudinal fasciculus in the brain stem and descends in the ventral funiculus in the spinal cord. The *lateral* tract descends ipsilaterally in the ventrolateral white matter of the cord (Fig. 15.12). The lateral vestibulospinal tract originates in the lateral vestibular nucleus, which is characterized by a number of large cells (Fig. B3). Although it is not sharp, there is a somatotopic arrangement in the lateral vestibular nucleus. The termination of these fibers differs from that of the rubro- and corticospinal axons. Lateral vestibulospinal axons end mainly in Rexed's layer VIII and the ventral part of VII, where they strongly influence axial and extensor motoneurons.

The medial vestibulospinal pathway is more modest and takes origin from the medial vestibular nucleus (Figs. 15.10 and B2). The terminal area of this medial pathway is similar to that of the lateral, but does not extend lower than the upper thoracic levels of the cord (Fig. 15.12). Thus, *the medial tract influences musculature of the neck and upper limb only.* It plays a significant role in head posture and eye–head coordination during active movements of the body. Like the lateral, the medial vestibular nucleus receives dense cerebellar input but no direct cerebral cortical input.

The Reticulospinal Tracts

Fibers from large and small cells at all levels of the medullary and pontine reticular formation can be traced into the spinal cord, but there are two maximal areas of origin, one in the pontine tegmentum and one in the medulla (Fig. 15.13). Both areas are medially located and contain many large cells. The pontine fibers descend ipsilaterally, whereas those from the medulla are both crossed and uncrossed. The medullary region from which reticulospinal axons originate is called the **gigantocellular reticular nucleus** because of its ample population of very large neurons. There appears to be *no somatotopic organization in the reticulospinal projections,* and single fibers are known to send branches to upper and lower levels of the cord. The medullary fibers descend in the lateral funiculi, whereas those of pontine origin descend in the ventral funiculi (Fig. 15.14). The two groups also differ in their termination zones: the pontine axons terminate more ventrally in layers VII and VIII, and the medullary end in more dorsal parts of layer VII. Both reticulospinal systems influence muscle tone and posture (Chapter 16). Both also play a role in gating sensory transmission that triggers spinal reflexes.

It is important to keep in mind that the reticular formation is acted upon by diverse ascending and descending systems. There is a considerable input of fibers from the cord, the vestibular nuclei, the cerebellum, and the cerebral cortex, to name but a few.

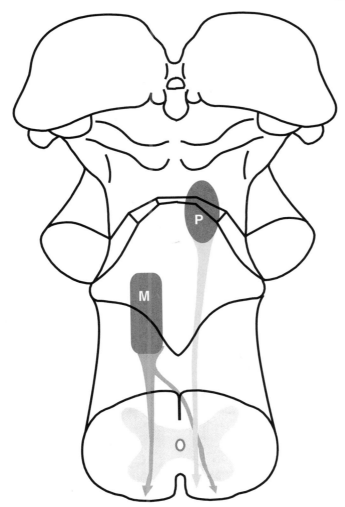

Figure 15.13 The reticulospinal pathways originate mainly from two focal regions, one in the pontine tegmentum (P) and the other in the medulla (M). Each is shown on one side only for clarity. Although the pontine reticulospinal tract remains ipsilateral to its origin, the medullary contains a small number of crossed fibers.

Over the past several years, more or less circumscribed cell groups have been identified within the brain stem reticular formation on the basis of neurotransmitter phenotype. Two of the more prominent are the *serotonergic* neurons of the raphé nuclei and the *noradrenergic* neurons of the **nucleus locus coeruleus.** The neurons in these two monoaminergic cell groups send axons to the spinal cord as well as to several other parts of the brain (Fig. 15.14). Although the raphé–spinal serotonin system is believed to be involved in spinal cord pain mechanisms, little else is known about its functional role or that of the noradrenergic system.

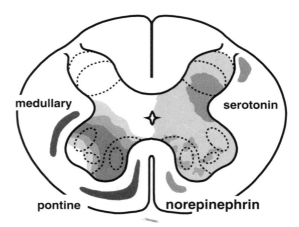

Figure 15.14 The locations of the reticulospinal (left) and monoaminergic (right) tracts and their distributions are indicated on a spinal cord cross section.

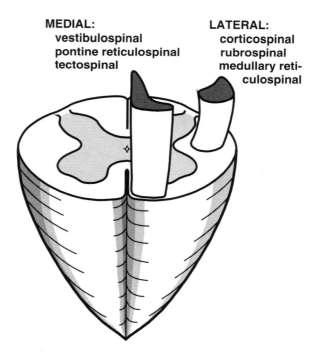

Figure 15.15 The descending movement control projections can be grouped into medial and lateral systems.

Medial Versus Lateral Descending Systems

A dichotomy exists in the descending supraspinal systems (Fig. 15.15) such that the corticospinal, rubrospinal, and certain of the reticulospinal axons descend in the *lateral* funiculus of the spinal cord and terminate on interneurons in *lateral* parts of the intermediate zone (Rexed's laminae V, VI, and VII). In contrast, the vestibulospinal, tectospinal, and others of the reticulospinal axons descend in the *ventral* funiculus and distribute to a more *medial* region of the intermediate zone (laminae VII and VIII). The functional significance of this arrangement emerges when one recalls the organization of the motoneuronal pools in the ventral horn of the spinal cord. That is, the motoneurons that innervate axial muscles are situated medially, and those that innervate limb muscles are located laterally. Thus, *the distribution of the lateral descending system is strategically located so as to influence the motoneurons of limb muscles especially involved in fine, manipulative performances. The medial descending system, on the other hand, is strategically located so as to influence the motoneurons of axial, postural muscles and, to a lesser extent, some of the larger extensor* (antigravity) *muscles of the limbs.* This concept is consistent with both experimental and clinical observations of motor performance. It should *not* be assumed, however, that the cerebral cortex has no control over axial muscle motoneurons. It almost certainly does so by way of its input to the reticular formation, the superior colliculi, and, less directly, the vestibular nuclei.

Detailed Reviews

Kuypers HGJM. 1981. Anatomy of the descending pathways. In Brooks VB (ed), *Handbook of Physiology, Section 1: The Nervous System, Vol 2: The Motor System, Part 2.* American Physiological Society, Washington, DC.

Lundberg A. 1975. Control of spinal mechanisms from the brain. In Tower DB (ed), *The Nervous System, Vol 1: The Basic Neurosciences.* Raven Press, New York.

16 Reflexes, Rhythms, and Posture

One important function of sensory input to the CNS is to rapidly trigger relatively stereotyped movements designed to maintain posture or avoid injury. As we shall see, the proper *spatial* and *temporal* coordination of agonist and antagonist muscle contractions leading to reflex movements is encoded by synaptic circuits contained entirely within the spinal cord and brain stem. Although these lower circuits are activated by sensory input, they are also under the influence of descending axons from the brain. In fact, most reflexes depend for their full execution on contributions from brain stem structures. In such *long-loop* reflexes, the sensory data are conveyed to the brain stem and processed to generate the appropriate motor signals, which are then sent back down to the lower motor apparatuses to augment the excitation and inhibition of motoneurons.

Clinically, examination of the reflexes is fundamental to the analysis of neurological disease. It is essential to know the arrangements of motoneurons in the brain stem and spinal cord and to learn the locations of nuclei that give rise to the efferent axons in cranial and spinal nerves. This knowledge, coupled with observations of the different types of functional disorders, can give valuable information about the nature and location of the pathological process responsible for a particular neurological disease. In this chapter, we'll first explore some examples of simple reflex arcs. Next, we'll examine a special case of rhythmic movements—walking—that can outlast the sensory or descending trigger. Finally, we'll turn our attention to the important postural reflexes mediated by the vestibular system.

The Stretch Reflex

The stretch reflex is the most fundamental of the postural reflexes. The *knee-jerk* reflex, often used by the physician in the neurological examina-

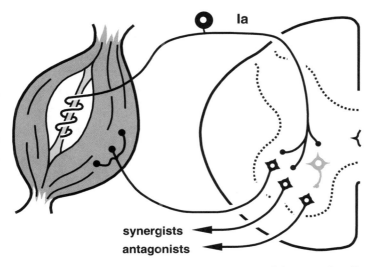

Figure 16.1 The diagram shows the basic circuit of the stretch reflex. The red neuron is inhibitory.

tion, is an example of a stretch reflex. The basic circuit involves a monosynaptic connection between Ia (muscle spindle) sensory axons and α motoneurons to the same muscle and to its agonists, in parallel with a *di*synaptic connection to the α motoneurons that innervate the antagonists of the muscle (Fig. 16.1). When a muscle is passively stretched in an abrupt manner, as with a light hammer blow to the patellar tendon, an afferent volley is established in the type Ia muscle spindle axons that innervate the central regions of the intrafusal muscle fibers. The Ia axons monosynaptically excite the α motoneurons that innervate the same muscle and its agonists, causing contraction of these muscles. The Ia axons also excite a population of so-called **Ia inhibitory interneurons** which in turn inhibit the α motoneurons that innervate the antagonist muscles, thus preventing the antagonists from opposing the reflexive contraction of the stretched muscle. This *reciprocal inhibition* represents an elemental mechanism that is essential to the proper functioning of the lower motor circuits.

   The stretch reflex is in constant operation whether we stand motionless or move about. Whenever we encounter slight variations in the load on a muscle, due to, for example, a tilt of the deck beneath our feet or a strong gust of wind, the resultant stretch of muscles triggers this compensatory postural reflex. The descending systems that activate motoneurons during voluntary movements must also address the mechanisms that govern the stretch reflex. Although some of the descending axons that control voluntary movements synapse directly on α motoneurons, their influence on γ motoneurons of the same muscles is never monosynaptic (see Fig. 18.5). Nevertheless, it is essential that the descending systems *coactivate* the γ motoneurons so that the intrafusal muscle fibers shorten along with the belly of the muscle in which they reside. In this way, the muscle spindles remain sensitive to changes in muscle length even when the muscles are shortened by voluntary contraction; the stretch reflex remains operational, and compensations for load changes proceed accordingly.

The Tendon Reflex

The tendon reflex, or *inverse myotatic reflex,* is a good example of a somewhat more complicated circuit involving additional interneurons (Fig. 16.2). The reflex arc starts with sensory axons (type Ib muscle afferents) arising from the **Golgi tendon organs.** In the tendon organ, the transducer elements of the neural fibers are interlaced with collagen fibers of the tendons. When the tendon is pulled by muscle contraction, the collagen fibers *pinch* the interwoven neural segments to generate a discharge of spikes. Thus, *the Golgi tendon organs monitor the tension produced in their associated muscles.* Upon reaching the spinal cord, the Ib afferent volley inhibits contraction of the muscle associated with the innervated tendon while exciting the contraction of its antagonists. Unlike the stretch reflex, however, there is no monosynaptic connection to the motoneurons; the influence of the Ib afferent axons on the motoneurons is mediated entirely by interneurons. This reflex arc serves to prevent overcontraction of muscles, which would cause sprains or even tears of the muscle and tendon tissues. The spinal cord portion of the circuit may also be engaged by descending motor command axons to signal a halt to a voluntary movement.

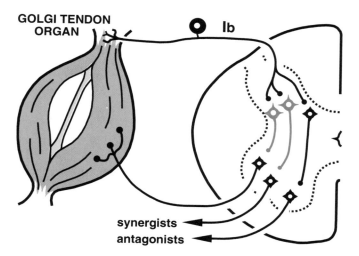

Figure 16.2 The circuit of the inverse myotatic (tendon) reflex is diagrammed in this drawing of the spinal cord. Red neurons are inhibitory.

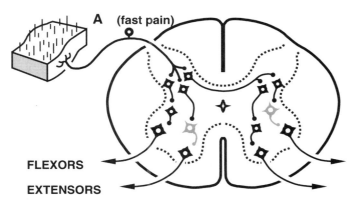

A   (fast pain)

FLEXORS

EXTENSORS

Figure 16.3 The diagram shows the circuitry of the withdrawal and crossed-extensor reflexes. The red neurons are inhibitory.

A Compound Reflex

An example of a reflex circuit at a still higher level of complexity is the *withdrawal* reflex (Fig. 16.3). The afferent limb of this reflex arc involves not muscle afferents, but cutaneous sensory input. Axons that conduct fast pain impulses (the small, myelinated Aδ fibers) make oligosynaptic excitatory contacts with *flexor* motoneurons and inhibitory contacts with *extensor* motoneurons to the same limb, so that the limb is forcefully withdrawn from the pain-inducing and potentially damaging stimulus. Often the limb withdrawal causes a sudden change in the center of gravity of the body, especially if the withdrawn limb is a leg. In this case, another circuit is engaged that mediates a *crossed-extensor* reflex (Fig. 16.3). This intersegmental circuit causes the contraction of extensors and the relaxation of flexors of the leg contralateral to the painful sensory input. Thus, the withdrawal and crossed-extensor reflexes often work in tandem and involve both sides and often several levels of the spinal cord.

The Parvicellular Reticular Formation Serves Bulbar Reflexes

Reflex arcs similar to those described above for the spinal cord are at work in the brain stem as well. The blink of the eyelid when something touches the cornea, the gagging that results from aspiration of fluid into the trachea, and the wincing expression at a loud sound are common examples of bulbar reflexes. In some cases, reflex circuits provide communication between the cranial sensory nerves and the spinal motor neurons by way of the reticulospinal systems. Withdrawal of the head from a punch in the nose is an example of such a bulbospinal reflex.

Studies of the parvicellular reticular formation suggest that its neurons function as intermediaries for bulbar reflexes much as the neurons of the intermediate zone of the spinal cord do for spinal reflexes. The sensory trigeminal complex, the vestibular nuclei, and the nucleus of the solitary tract lie adjacent to the parvicellular reticular formation and send many synaptic inputs to it. A few primary sensory axons also make direct synaptic contacts on neurons of the parvicellular reticular formation. The reticular cells, in turn, project to the brain stem motor nuclei and to the magnocellular reticular formation (which is the origin of an important descending motor control pathway to the spinal cord; see Chapter 15). Also like the intermediate zone of the spinal cord, the parvicellular neurons receive modulatory inputs from higher motor control structures such as the primary motor cortex, the cerebellum, and the superior colliculus.

I Got Rhythm . . .

As in the case of the reflexes described above, the basic neural pattern of activity underlying locomotion is generated by neurons intrinsic to the spinal cord. There is a central program for locomotion that does not require sensory feedback or descending commands to maintain it. If the dorsal roots of a cat are cut and the muscles are paralyzed with curare to prevent any trace of sensory input, the extensor and flexor motoneurons to the hindleg muscles can be made to fire alternately and with the same precise timing they exhibit during actual walking. Similarly a cat can continue to walk normally on a treadmill after its spinal cord has been surgically isolated from its brain if the animal is otherwise intact. Interestingly, late-stage Parkinsonian patients, who initiate movements only with great difficulty and are generally unable to walk on command, will walk normally for long distances if gently shoved by the physician. However, as in the case of reflexes, it is important to bear in mind that even though the *pattern* for locomotion is generated almost entirely by spinal circuits, sensory and descending input are both important for successful goal-directed locomotion.

Breathing is another example of a rhythmic activity that is organized at the level of the lower apparatuses. Although one does not typically think of breathing as a skeletal movement, it is accomplished by the action of striated skeletal muscle (the diaphragm and intercostal muscles) pulling against bone (the sternum, ribs, and second lumbar vertebra). The α motoneurons that control the diaphragm are contained in cervical segments 3 through 5 and those that innervate the intercostal muscles are located in the thoracic segments. Unlike walking, however, the spinal apparatus for breathing is dependent on regions in the brain stem reticular formation. There are regions of the brain stem reticular formation that spontaneously generate the pace of breathing; some regions program the inhalation phase and coordinate their activity through synaptic connections with other regions that generate the exhalation phase. If these regions are destroyed, natural breathing ceases. Although it is under some voluntary control, the *pacemaker* activity of these reticular regions appears to be autonomous to the extent that one cannot hold one's breath indefinitely.

Chewing, swallowing (closure of the glottis and peristaltic contractions of the esophagus), and vomiting (reverse peristalsis) are other commonplace examples of rhythmic behaviors associated with the lower motor circuits of the brain stem and spinal cord. All are dependent on the integrity of the reticular formation to coordinate the activity in the bulbar and spinal circuits.

Descending Systems Control the Lower Circuits

The preceding list of reflexes and rhythms is intended to offer a few examples from among many that are organized mainly by the lower motor circuits of the spinal cord and brain stem, and to convey a sense of the high level of complexity that is attained already at that early level of movement control. It is also important to remember that these simple reflex arcs act in concert with one another much of the time. As you have seen, all of them rely on local-circuit interneurons. Finally, be aware that *the reflex and rhythm circuits, especially the local interneuronal components, are accessible to axons that descend from the motor control nuclei of the brain.* That is to say, *the same circuits of the spinal cord and brain stem that mediate reflexes and rhythms are used by descending motor control systems to bring about voluntary, goal-directed movements by enhancing or suppressing these circuits.*

Clinical observations support this notion. Damage to the corticospinal system, as might occur with stroke, results in a constellation of negative and positive motor signs. Negative signs, such as spastic paralysis and fractionation of movement, result from the loss of the capacity normally controlled by the corticospinal system. Positive signs—movements that do not occur in undamaged people—emerge as well. A common example is the *Babinski sign* in which stroking the sole of the foot from heel to toe with a blunt object leads to a fanning and dorsiflexion of the toes rather than the typical plantar flexion shown by normal adults. Other spinal reflexes become exaggerated (hyperreflexia). These abnormal reflexes are frequently referred to as "release phenomena" because they emerge when descending control is withdrawn from the spinal mechanisms.

Reticulospinal Regulation of Posture and Muscle Tone

The descending reticulospinal systems are multifunctional. They influence visceromotor as well as skeletomotor behavior, regulate lower sensory transmission, and modulate the overall level of responsiveness of spinal neurons. In the present context of movement control, the reticulospinal axons are important for the maintenance of muscle tone and posture through long-loop reflexes and by integrating inputs from the vestibular system with those from the cerebral cortex.

As mentioned in the last chapter, reticulospinal axons originate mainly from two sites: the pontine tegmentum and the magnocellular zone of the medulla. *The pontine system descends ipsilaterally with the medial systems and has a strong excitatory effect on motoneurons that innervate axial muscles and limb extensors.* Electrical stimulation of the pontine system clearly augments spinal extension reflexes. The medullary system, though largely ipsilateral, has a small crossed projection. The effects of the medullary reticulospinal system are more complex than those of the pontine system. It is clear, however, that both reticulospinal systems influence the α and γ motoneurons that innervate extensor muscles. *The general effect of the medullary reticulospinal projection is to inhibit extensor muscle tone and that of the pontine reticulospinal projection is to facilitate such tone* (Fig. 16.4). Both the pontine and the medullary systems are influenced by inputs from the vestibular nuclei (below).

The reticulospinal systems are exploited by the voluntary movement control systems during goal-directed movements. When a voluntary movement is to be executed, it inevitably involves correlative postural adjustments. For instance, if the arm is extended to pluck an apple from a low-hanging limb, the contralateral axial muscles and ipsilateral leg extensors need to be contracted to keep from falling over. The voluntary movement control systems that are organized by the cerebral cortex and that initiate the goal-directed arm movement also send correlative commands to the brain stem reticulospinal regions to produce the requisite postural adjustments. Thus, the reticulospinal systems are engaged by cortico- and rubrospinal axons to control posture during active voluntary movement.

Observations from the clinic support this view of reticulospinal functions. When a person suffers damage to the brain stem between the pons and midbrain, as occasionally happens due to hemorrhage or a tumor, the spinal connections from the cerebral cortex and red nucleus are interrupted, as are their connections to the reticular formation (Fig. 16.4). Since the cortico- and rubrospinal systems facilitate flexor motoneurons and, by way of their projections to the medullary reticular formation inhibit extensors, their interruption leaves a domination by the pontine reticulospinal and vestibulospinal systems. A condition known as *decerebrate rigidity* ensues, in which the limbs become tonically extended and extensor reflexes are exaggerated. Since the ponto- and vestibulospinal axons activate γ motoneurons (as well as α motoneurons), extensor tone is amplified by way of feedback from the muscle spindle afferents, and the muscle spindles become acutely sensitive to stretch.

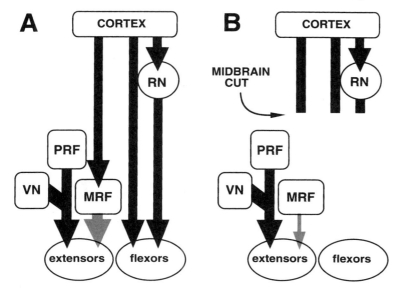

Figure 16.4 The reticular formation contributes to muscle tone and posture. Part **A** shows the reticulospinal systems in the context of the descending motor control systems from the cerebral cortex and the red nucleus (RN). The system that arises from the pontine reticular formation (PRF) and the vestibular nuclei (VN) has an excitatory effect on extensor motoneurons (jet black arrows). In contrast, the system from the medullary reticular formation (MRF) has an inhibitory effect on extensor motoneurons (red arrow) and is itself augmented by an excitatory input from the cerebral cortex. If the midbrain is transected behind the red nucleus, as in part **B,** flexor tone is reduced and extensor tone is exaggerated, leading to a condition of extensor rigidity of the body and limbs.

Vestibulospinal Reflexes and Balance

Displacements of the head in the horizontal and vertical planes are detected by the two otolith organs of the membranous labyrinth. The *utricle* and *saccule* are designed to detect *linear acceleration* of the head and changes of head position with respect to gravity. The other transducer element of the membranous labyrinth, composed of the three mutually perpendicular *semicircular canals,* detects *angular accelerations* of the head and is involved more in eye–head coordination than in posture (see Chapter 17). The two otolith organs appear as widenings of the labyrinth, each containing an epithelial patch, called a *macula,* that consists of hair cells. The hair cells are peculiar in two ways: their basal compartments are full of synaptic vesicles, and their stereocilia ("hairs") are highly specialized and arrayed in a characteristic geometric pattern. The stereocilia on the apical surfaces of each hair cell occur in two forms (Fig. 16.5).

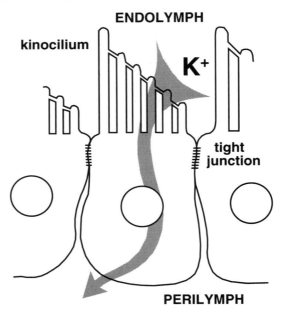

Figure 16.5 The hair cells of the membranous labyrinth form tight junctions with one another near their apices. The tight junctions form an impermeable barrier that separates the high-K^+ endolymph from the low-K^+ perilymph. When channels are opened in the stereocilia, K^+ flows through the hair cell cytoplasm and out of the cell through channels in the basal membrane. This K^+ current depolarizes the hair cell.

There is a single, tall *kinocilium* near one edge of the apical surface of each hair cell and a larger number of flexible *stereocilia* that progressively diminish in height with distance from the kinocilium. The tips of the stereocilia are linked together by membranous filaments called *tiplinks*. The stereocilia are embedded in a gelatinous matrix that contains a cluster of tiny calcium carbonate crystals (the otoliths), and they are bathed in a special kind of extracellular fluid called **endolymph** that fills the lumen of the membranous labyrinth. Unlike the run-of-the-mill perilymph that bathes the lower parts of each hair cell, the endolymph is high in K^+ (about 150 millimolar) and low in Na^+ (about 2 millimolar). This ionic difference establishes a potential difference of some 130 millivolts across the barrier of impermeable tight junctions formed between the hair cells near their apical ends. If the head is displaced, the inertia of the otolith–gelatin matrix causes the embedded stereocilia to bend. If they bend toward the kinocilium, the tiplinks are pulled in such a way that the ciliary membrane is mechanically distorted (Fig. 16.6). The distortion opens mechanically gated K^+ channels in the membrane, allowing an *influx* of K^+ from the endolymph that depolarizes the hair cell membranes. The K^+ that flows into the hair cells at the stereocilia diffuses through the cytoplasm of the hair cells and

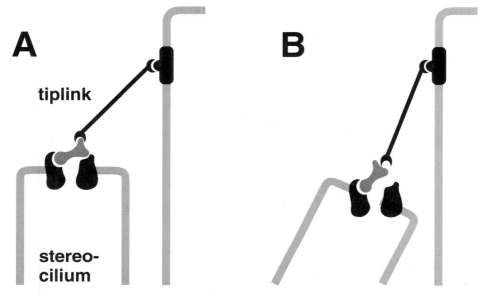

A

tiplink

stereo-
cilium

B

Figure 16.6 The K$^+$ channels (jet black) in the membrane at the apex of each stereocilium are mechanically gated. When the stereocilium is in the upright attitude, as in part **A,** the gate (red) seals off the ionopore. When the stereocilium is bent toward the larger stereocilium, as shown in part **B,** the tiplink pulls the gate open, allowing K$^+$ to flow through the ionopore.

escapes into the perilymph through voltage-gated K$^+$ channels in the basal portion of the hair cell membranes. Since there is a steady leak of K$^+$ across the cell, even when the stereocilia are stable, bending of the stereocilia away from the kinocilium tends to close the leaky K$^+$ channels and hyperpolarizes the hair cells.

The hair cells are innervated at their basal surfaces by large endings of single sensory fibers that emanate from neuronal cell bodies in the vestibular ganglion (Fig. 16.7). The central processes of these bipolar neurons form the vestibular part of the 8th cranial nerve. The hair cells continuously drip neurotransmitter from their basal vesicles onto the nerve fiber endings to drive a relatively high tonic spike frequency in the primary vestibular axons. Depolarization of the hair cells increases, and hyperpolarization decreases, the exocytosis of neurotransmitter from the hair cells. Thus, the activity of the vestibular axons is modulated up or down depending on the deflection of the stereocilia of the hair cells.

The saccule is oriented in the vertical plane, and the kinocilium of each hair cell is located at the superior edge of the apical surface. Thus, the stereocilia of all the saccular hair cells are bent toward the kinocilia when the body (and the head) is accelerated toward the surface of the earth. The utricle is oriented in the horizontal plane and its macula is more complex (Fig. 16.8). The axes of polarity of the stereocilia on the hair cells

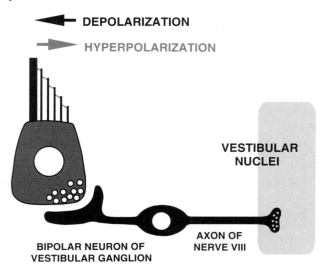

Figure 16.7 Sensory transduction in the membranous labyrinth is accomplished by hair cells whose stereocilia are bent when the head is displaced. The hair cells contain synaptic vesicles in their bases that release neurotransmitter onto the nerve fiber ending.

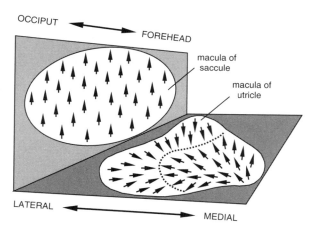

Figure 16.8 The axes of polarity of the cilia on hair cells in the otolith organs are arranged in an orderly fashion. In this view from above and behind the head, the macula of the saccule is oriented in the vertical plane (lightly shaded) and that of the utricle in the horizontal plane (darker shading). The striola of the utricular macula is represented by the dashed line.

are oriented toward a curved line, the *striola,* so that horizontal displacements of the head in any direction will activate some hair cells and deactivate others. The composite *pattern* of activity signals generated in the total population of 8th nerve fibers informs the vestibular nuclei about the exact direction of head displacement.

Axons of the vestibular nerve terminate in all four vestibular nuclei and excite the secondary neurons. Some primary vestibular axons also end in the cerebellum (see Chapter 19). You'll recall that two of the vestibular nuclei, the medial and the lateral, give rise to descending projections to the spinal cord. The medial vestibulospinal tract terminates in the cervical and upper thoracic segments of the cord to influence neck muscles and upper trunk axial muscles. The medial vestibular nucleus, therefore, seems to be more important for head postural reflexes. The lateral vestibulospinal tract extends through the length of the cord and activates intersegmental propriospinal circuits to produce coordinated postural responses of the trunk and limbs. The net effect of vestibulospinal axons is to *excite* ipsilateral α and γ motoneurons that innervate neck and trunk axial muscles and limb extensors and to *inhibit* flexor motoneurons. Both the medial and the lateral vestibular nuclei also contribute to postural control through connections with the reticulospinal pathways.

Detailed Reviews

Friesen WO, Stent GS. 1978. Neural circuits for generating rhythmic movements. *Annu Rev Biophys Bioeng* 7: 37.

Grillner S, Wallen P, Viana di Prisco G. 1990. Cellular network underlying locomotion as revealed in a lower vertebrate model: transmitters, membrane properties, circuitry and stimulation. *Cold Spring Harbor Symp Quant Biol* 55: 779.

Wilson VJ, Peterson BW. 1981. Vestibulospinal and reticulospinal systems. In Brooks VB (ed), *Handbook of Physiology.* American Physiological Society, Washington, DC.

17

Gaze

Humans are highly visual creatures. Like many other animals, we depend heavily on the detection of light energy to guide our essential behaviors in the environment. But as aesthetic beings, we rely on vision for many activities that are at least as important as the elemental functions of survival; we use our eyes to read printed words, to appreciate art, and to gaze wistfully at the displays of nature. So that we may wield our eyes most effectively, our brains are equipped with special sensorimotor circuitries that control eye and head movements to visual targets in space. We can refer to these circuits as *gaze control systems*. The gaze control systems bring the images of interesting visual targets onto the part of the retina that has the highest visual acuity: the **fovea.** Since many such objects are in motion and our bodies may be in motion, the gaze control system also serves to keep particularly engaging targets foveated despite such movements. Further, we often shift our gaze to objects that we hear or feel rather than to objects that intrude on our visual field. Therefore, *the gaze control systems must be able to respond rapidly and accurately to auditory and somatic as well as visual stimulation.*

Basic Features of Eye Movement

The eyeballs differ from other mobile parts of our body in two important ways: they represent a constant load, and the movements they can make are constrained and predictable. Because the mass of the eyeball is a constant and rather light load, eye movements can be much faster than skeletal movements. Moreover, the α motoneurons that innervate the extraocular muscles are not subject to recurrent inhibition, as are the spinal motoneurons, and therefore can fire spikes at much higher frequencies (100 to 600 hertz compared to the 50 to 100 hertz for spinal motoneurons). All the basic movements that the eyes can make are

CONJUGATE

DISJUNCTIVE (VERGENCE)

Figure 17.1 Two types of eye movements can occur in which the eyes move together or in opposite directions.

rotational, with the preferred directions around the vertical and horizontal axes; oblique and torsional rotations occur less frequently. These rotational movements are controlled by three antagonistic pairs of extraocular muscles for each eye: the lateral and medial rectuses, the superior and inferior rectuses, and the superior and inferior obliques. Although all of the extraocular muscles contribute to all basic movements by either contracting or relaxing, each movement in one plane is determined mainly by two sets of muscles.

In most movements the eyes move together. Such movements are referred to as **conjugate** eye movements (Fig. 17.1). However, when we fixate on a visual target that moves toward or away from us, our eyes must converge or diverge. Such movements of the eyeballs in opposite directions are called **disjunctive.** The muscles in each eye that contract together during a particular eye movement are called *yoked pairs* (e.g., the lateral rectus of one eye and the medial rectus of the other during a horizontal movement). Muscles that oppose one another are called *antagonist pairs* (e.g., the lateral rectuses of both eyes during a horizontal movement). Obviously, the pairs of muscles that are yoked and antagonistic in a conjugate eye movement are not the same pairs when a disjunctive movement is made. From these basic properties of eye movements, you might already surmise that *strong inhibitory interactions are required between the motor nuclei that innervate the extraocular muscles.*

The Ocular Motor Nuclei

You'll recollect from gross anatomy that the extraocular muscles are innervated by three cranial nerves: the oculomotor, trochlear, and abducens. Each of these nerves arises from a nucleus or nuclear group in the brain stem that has the same name as the nerve (respectively visible in Figs. B7, B6, and B3). Together we refer to these three pairs of motoneuron pools as the **ocular motor nuclei.** The motoneurons of the ocular motor nuclei constitute the final common pathway over which the gaze control circuits bring the images of interesting visual targets onto the foveas and keep them there. *Direct inputs to the ocular motor nuclei come only from the vestibular nuclei and the reticular formation. All other inputs arrive indirectly by way of cells in the reticular formation.*

There are several interneurons that lie close by and among the oculomotor, trochlear, and abducens nuclei and connect these nuclei with one another. These so-called internuclear neurons come in both inhibitory and excitatory varieties and enable the ocular motor nuclei to coordinate their output for precise gaze control. The internuclear axons travel as part of the **medial longitudinal fasciculi.** Since the vestibular axons that reach the ocular motor nuclei also travel in the medial longitudinal fasciculi, these fiber bundles figure prominently in gaze control. The medial longitudinal fasciculi are critical in orchestrating the relative discharges of the motoneurons in the ocular motor nuclei. Disruption of this pathway in the brain stem due to vascular accident or trauma has devastating consequences for eye movements.

We can identify *four* types of eye movements that are mediated by separate gaze control systems. Each of the four types appears to depend in part upon separate, and in part upon overlapping, neural circuitry. They must all, however, share access to the ocular motor nuclei. The four types of eye movements are termed **saccadic, smooth pursuit, vestibulo-oculomotor,** and **vergence.** Let's consider them each in turn.

What Is a Saccade?

We use saccadic movements to direct our foveas rapidly to a target of interest in visual space. A saccade is a fast, ballistic, conjugate movement of the eyes. Such movements typically take place at speeds approaching 700°/second. They are said to be ballistic because once initiated (about 200 milliseconds after the triggering target appears) they cannot be interrupted. After a saccade is completed, another cannot be initiated for about 200 milliseconds. This latent period represents the time it takes for the saccade-generating system to compute the direction and distance of the saccade. *Two stimulus parameters are used by the saccade-generating brain stem circuits to compute saccade commands.* One is the retinal position of the target with respect to the fovea, or *retinal error signal.* The other is the *position* of the eyeball in the orbit. That the extraocular muscles are richly endowed with muscle spindle organs is no coincidence; they probably signal much to the brain about eye position. After these signals are computed by the central circuits, a command is issued to the ocular motor nuclei that specifies the direction and amplitude of the saccade. *The direction is coded by excitation of the appropriate groups of motoneurons—a*

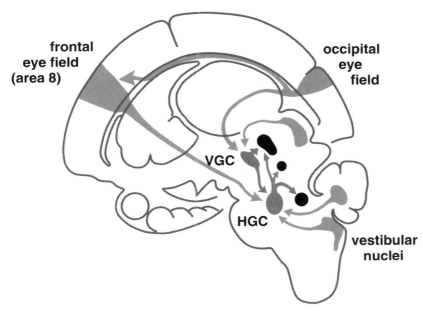

Figure 17.2 The nuclei and circuits that control saccadic eye movements are widespread. The horizontal gaze control center (HGC) and vertical gaze control center (VGC) are shown in red. The ocular motor nuclei are jet black.

population code—and the amplitude is coded by the frequency and duration of the motoneuronal discharge. Since saccades can be made in the absence of a visual target (total darkness), the control circuitry appears to involve a projection from the cerebral cortex to the brain stem that places the saccade system under voluntary control.

Saccades are controlled by three areas in the brain stem that influence either directly or indirectly the motoneurons in the ocular motor nuclei (Fig. 17.2). They are the superior colliculi and two areas in the brain stem reticular formation: one in the pons near the abducens nucleus identified as a *horizontal* (or *lateral*) *gaze control center,* and another, functionally analogous center for vertical gaze control in the midbrain. Neurons in the horizontal gaze control center receive visual information by way of axons from the superior colliculus, and in turn contact motoneurons and interneurons in the abducens nucleus to help coordinate conjugate horizontal eye movements (Fig. 17.2). Let's see how they work.

The Superior Colliculi

The superior colliculi are layered structures forming the rostral part of the midbrain tectum (Fig. B7). Each colliculus receives input from the contralateral half of the visual field (Fig. 17.3). Incoming axons from the

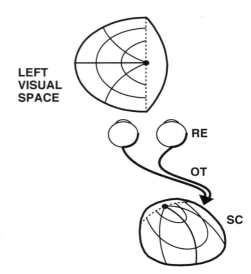

**LEFT
VISUAL
SPACE**

RE

OT

SC

Figure 17.3 The coordinate system of visual space in the right superior colliculus is imposed by the orderly retinotopic terminations of optic tract axons. Abbreviations: RE, right eye; OT, optic tract; SC, superior cell colliculus.

retinas form a horizontal fiber layer (the **stratum opticum**) within each colliculus that separates the colliculus into **superficial** (dorsal to the stratum opticum) and **deep** layers. The cells in the superficial layer receive the direct synaptic input of retinal ganglion cell axons that travel in the optic tract, as well as a dense input from the visual processing areas of the cerebral cortex. The deep layers contain the premotor neurons that project to the reticular formation near the ocular motor nuclei and to the reticular gaze control areas. Both the superficial and the deep layers of each colliculus contain a representation of roughly half the visual field by virtue of topographically arranged inputs from the corresponding hemiretinas (see Chapter 27). Thus, the superior colliculi seem ideally suited to the task of translating visual signals into motor commands. Somewhat counterintuitively, however, the superficial layers of the colliculi are apparently not a direct source of visual data to the deep layers. Instead, the superficial neurons send their axons to the lateral posterior–pulvinar complex of the thalamus, from which the visual data are forwarded to secondary visual areas of the parietal and occipital lobes. These visual cortices provide the data to the deep layers of the superior colliculus by way of corticotectal projections. It is this reprocessed information that is used by the deep layer neurons to compute saccade commands. Somewhat amazingly, the precise retinotopic map is not lost through this processing circuit; it arrives intact at the deep layers of the colliculus.

Importantly, *the deep layers of the colliculi also receive auditory and tactile input relevant to the spatial world that is in register with the visuotopic map* (Fig. 17.4). The auditory input arrives mainly by way of projections from the cerebral cortex and the inferior colliculi. The tactile input likewise comes in part from the cortex and in part from the ascending trigeminal

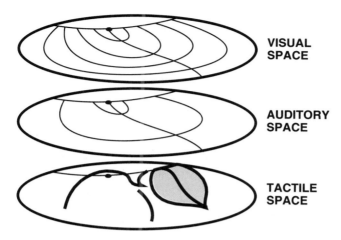

Figure 17.4 The representational maps of visual, auditory, and tactile space are in register in the deeper layers of the superior colliculus.

and spinal pathways. By convergence, these sensory data are conveyed to individual collicular neurons and have an additive effect on their responses.

Neurons in the deeper layers of the colliculus discharge massively prior to a saccadic eye movement. The population that fires lies in the region of the topographic map that registers the visual, auditory, or tactile target to be foveated. This discharge triggers the saccade by way of synaptic connections with the saccade control circuits of the brain stem, which in turn connect to the ocular motor nuclei to elicit the appropriate discharge pattern from the motoneurons. The foveal representation zone of the colliculus contains neurons that are inhibitory to the saccade-generating system and serve to halt the saccade upon foveation of the target stimulus. If a target is stationary on the foveas, we can keep it there simply by consciously suppressing any saccades—a behavior that depends in large part on the saccade inhibitory neurons of the foveal part of the superior colliculi. Of course, a visual stimulus that was held absolutely constant on any part of the retina would soon disappear due to repeated bleaching of the photopigments. This is prevented from happening by involuntary *microsaccades* of unknown generation that are ever-present and go unnoticed.

Remember that the deep layers of the colliculi also give rise to spinal projections that descend as the tectospinal tract to impinge on motoneurons of the upper cervical segments. *Through the tectospinal pathway, the colliculi can coordinate head and eye movements in gaze shifts.* Such coordination is essential when a target object is located in the far periphery of the visual field or beyond, since the eyes do not comfortably make saccades of more than about 60° of arc.

The Reticular Gaze Control Areas

The operational circuit of the horizontal gaze control center has been carefully studied. Four different cell types have been identified within the area that can be distinguished on the basis of their activity patterns (Fig. 17.5). Prominent among them are the so-called *burst* and *pause* neurons, which are important in the initiation of horizontal saccades. Pause neurons are tonically active and suppress the output of the burst cells by way of direct inhibitory synapses. When the pause neurons cease their activity due to inhibitory input from the superior colliculus or cerebral cortex, the burst cells are *released from this tonic inhibition,* and their discharge excites motoneurons in the abducens nuclei, producing a horizontal saccade. Although less is known about the midbrain vertical gaze control center, it may function in a similar fashion. These functionally identifiable areas of the reticular formation are often referred to as "centers," but it should be noted that anatomically they are not easily delimited, and when examined physiologically, they often overlap with other "centers" that influence different behaviors (e.g., swallowing, emesis, locomotion).

The brain stem mechanisms of saccade control that receive collicular signals are also under the influence of inputs from the cerebellum and a number of areas of the cerebral cortex. One such cortical area, called the **frontal eye field,** lies within Brodmann's area 8 and is particularly important for voluntary saccades (Fig. 17.2). The frontal eye field on each side is considered to be important for the initiation of *voluntary* conjugate

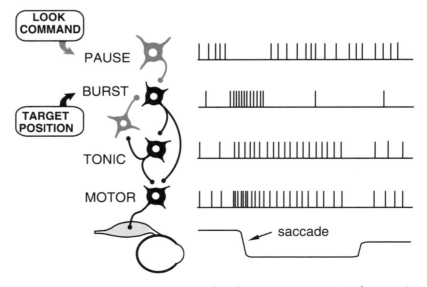

Figure 17.5 The responses of the local circuit neurons in the pontine horizontal gaze control center have been studied. Jet black neurons are excitatory and red are inhibitory.

eye movements to the contralateral visual field. Predictably, then, a lesion of the frontal eye field of one hemisphere (due to a stroke, for instance) causes a patient to have a sustained gaze to the side of the lesion. The patient gradually recovers from this impairment.

The brain stem saccade control mechanisms are subject to dysfunction in certain types of vascular accident or sometimes in cases of multiple sclerosis. For instance, damage to the pontine horizontal gaze control center on one side results in an inability to make horizontal eye movements to the side of the lesion.

The Smooth Pursuit System

If a visual target moves, as it often does in the real world, the smooth pursuit circuits *compute the direction and velocity of the retinal image and generate an ongoing command signal to the ocular motor complex.* Successful quail hunters and fighter pilots necessarily have well-exercised smooth pursuit circuits. Smooth pursuit, unlike saccadic movement, requires a retinal target. People cannot perform smooth eye movements in total darkness.

Although the brain stem circuits that control smooth pursuit movements are not well understood, it is clear that the data about the direction and velocity of the target must be processed through secondary visual areas of the cerebral cortex. One such area that is essential to visual tracking is called the **occipital eye field.** It is located rostral to the primary visual area in each cerebral hemisphere. If one or the other occipital eye field is damaged, smooth pursuit becomes impossible to perform. The brain stem circuits that mediate smooth pursuit have not been discovered.

Sometimes smooth pursuit movements and saccades occur together in an alternating pattern called *optokinetic nystagmus.* This reflexive eye movement is triggered when a visual scene with a repetitive pattern passes across the visual field. Riding as a passenger in a car traveling down a country road bordered by a picket fence will evoke the reflex. The eyes first foveate one of the fence posts, follow it smoothly until it slips beyond the maximum excursion of the eye in the orbit, then flick back in the opposite direction to foveate another post. Having locked onto the new post, the eyes again follow it smoothly. This alternating *slow* (smooth pursuit) and *fast* (saccadic) phase will persist until consciously overridden. Optokinetic nystagmus can be induced in a more controlled setting by placing a subject inside a large rotating drum painted with stripes. The drum will elicit the oscillatory eye movements. If the period of the stripes is incrementally reduced until the subject can no longer resolve them, the nystagmus will cease. Thus, the phenomenon can be used as a reliable test of visual acuity, since the subject cannot fool the tester as he might by memorizing the eye chart!

The Vestibulo-Oculomotor Reflex

Much of the time, we are required to maintain foveation of an object while our head moves through space. Imagine, for instance, a pass receiver in the National Football League who must keep his eyes on the football while he sprints across the field and is jostled about by a series of linebackers. He depends heavily on the integrity of his vestibulo-oculomotor reflexes to make the catch. This reflex serves to stabilize the eyes during head movements. The afferent signals that trigger the reflex are not visual, but originate instead in the membranous labyrinth of the inner ear, which detects displacements of the head. The *semicircular canals,* which detect angular displacements along the three axes of space, are especially important. The *utricle* and *saccule,* which detect linear accelerations of the head, though not irrelevant by any means, are more important for postural reflexes. The hair cells in each of the three semicircular canals on either side of the head transduce the angular acceleration of the head around a different axis and influence the ongoing baseline activity of the sensory fibers of the 8th cranial nerve (see Fig. 16.4). The transduction process is the same as that of the hair cells in the utricle and saccule described in the preceding chapter. The 8th nerve axons transmit the impulses to neurons in the four vestibular nuclei of each side, which translate the *pattern* of discharges to compute the exact direction and amplitude of head displacement. The *superior* and *medial vestibular nuclei are most essentially involved in the vestibulo-oculomotor reflexes* (Fig. 17.6). The neurons in these vestibular nuclei integrate the pattern of incoming signals and generate a correction signal that is sent directly to the ocular motor nuclei to move the eyes counter to the head

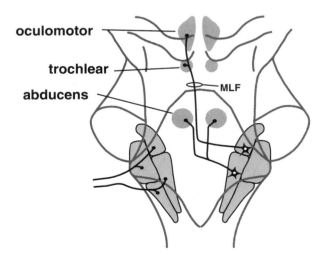

Figure 17.6 The vestibulo-oculomotor projections mediate reflexive eye movements. Abbreviation: MLF, medial longitudinal fasciculus.

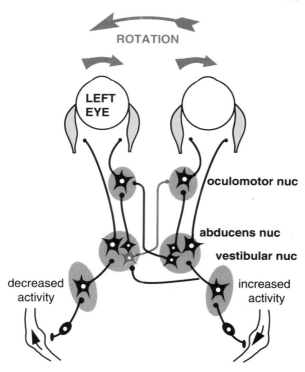

Figure 17.7 The diagram shows the basic circuit of vestibular nystagmus. The jet black neurons are excitatory; the red is inhibitory.

rotation and thus stabilize the image on the foveas. Both the medial and the superior nuclei send their ascending axons with the medial longitudinal fasciculi.

Remember that the medial vestibular nucleus also gives rise to axons that descend to cervical levels of the spinal cord, where they form monosynaptic connections on motoneurons that control neck muscles and, hence, head movements. This descending system is important for the compensatory head movements that commonly occur together with the vestibular-induced eye movements.

An interesting phenomenon associated with the vestibulo-oculomotor reflex is called *vestibular nystagmus*. Although the eye movements consist of the same slow and fast phases seen in optokinetic nystagmus, vestibular nystagmus is evoked by a phasic vestibular signal rather than a continuous visual signal. Vestibular nystagmus can be easily demonstrated by rotating a subject about the vertical axis on a spinning stool in the dark (Fig. 17.7). If the person is rotated to the left, her eyes undergo a slow conjugate deviation to the right and thereby tend to remain fixed on a particular point in space. However, when the rotation has proceeded to the degree at which the eyes reach the limit of their excursion, they make a saccade to the left. If the rotation is held at a constant velocity, the sensory signal will adapt and the eye movement will cease. If the rotation is suddenly stopped, it is the vestibular equivalent of producing a sudden acceleration to the right due to the inertia of the fluid in the semicircular canals. As a result, the eyes now undergo slow movements to the left alternating with fast movements to the right. This *postrotatory nystagmus* can be used to diagnose the functional state of the vestibular system in the neurological examination.

Vergence

The movements described so far are all conjugate eye movements. Let's conclude this chapter on gaze with a look at the disjunctive movements that occur during vergence of the eyes. To keep an object centered on both foveas when it moves toward or away from a viewer, the eyes must converge or diverge, respectively. This disjunctive movement sustains the fusion of a single image in depth—the function of *stereopsis.* This eye movement system works in concert with the pupil and lens control (visceromotor) systems that change the focal length of refraction according to the distance of the image from the eye. The visceromotor neurons that influence the ciliary ganglion neurons lie in the Edinger–Westphal nucleus, which is conveniently associated with the oculomotor complex. The constellation of vergence, lens, and iris movements is called the *accommodation reflex.* This reflex depends upon convergent projections from a part of the cerebral cortex in the occipital lobe (near the primary visual cortex) where depth perception is mediated and on some small nuclei that lie at the junction of the midbrain with the thalamus (the *pretectal area*) and receive axons from the retina.

Obviously, all of the eye movements described above are designed to maintain binocular vision (see Chapter 27). If the eyes are misaligned by abnormalities in the muscles, nerves, or central mechanisms, double vision (*diplopia*) occurs. The gaze control systems are vulnerable to damage at a number of sites. Most obviously, the relevant cranial nerves can be injured by trauma. The oculomotor nerve, because it passes between the posterior cerebral artery and the superior cerebellar artery, is easily compressed in the case of an extradural bleeding that forces the brain stem downward and stretches the two arteries taut across the nerve. Loss of 3rd nerve function can be readily ascertained even in an unconscious patient by shining a pen light into the eyes and looking for a pupillary response.

Detailed Reviews

Fuchs AF, Kaneko CRS, Scudder CA. 1985. Brainstem control of saccadic eye movements. *Annu Rev Neurosci* 8: 307.

Lisberger SG, Morris EJ, Tychsen L. 1987. Visual motion processing and sensory-motor integration for smooth pursuit eye movements. *Annu Rev Neurosci* 10: 97.

Sparks DL. 1986. Translation of sensory signals into commands for control of saccadic eye movements: role of primate superior colliculus. *Physiol Rev* 66: 118.

Wurtz RH, Komatsu H, Dürsteler MR, Yamasaki DS. 1990. Motion to movement: cerebral cortical visual processing for pursuit eye movements. In Edelman GM, Gall WE, Cowan WM (eds), *Signal and Sense: Local and Global Order in Perceptual Maps*. Wiley-Liss, New York.

18 Volitional Movement

Over a century ago, it was discovered that electrical stimulation of the cortex of the precentral gyrus of a dog could elicit movements of the animal's contralateral limbs. A short time after this observation, neuroscientists learned two related and significant facts from experimental transections of the pyramidal tracts in mammals. First, the neurons that undergo retrograde degeneration after such a lesion are the large pyramidal cells in layer V of the cerebral cortex of the precentral gyrus. Second, the animals exhibit severe movement disorders, mainly involving their extremities. Using more refined electrical stimulation techniques, Charles Sherrington later confirmed that movements are elicited with the lowest current intensities from the cortex of the precentral gyrus. Thus the notion was born that the M_I cortex is the main arbiter of the complex, discrete movements that characterize much of our voluntary behavior.

Today, we know that there are at least three distinct areas in the frontal lobes of either hemisphere that have a rather direct impact on the control of voluntary movements. In the **premotor cortex** (Brodmann's area 6) rostral to M_I are the **premotor area** and, dorsal to it, the **supplementary motor area** (Fig. 18.1), both of which contain corticospinal neurons and neurons that communicate with M_I. In addition to these cortical areas, the red nuclei of the midbrain figure prominently in the orchestration of goal-directed behavior. In fact, the red nuclei take their cue in large part from the motor cortices. The motor areas of the cerebral cortex depend on at least two other coprocessors of motor-related data: the cerebellum and the basal ganglia. We'll consider each of these coprocessors in turn in the following two chapters. For now, let's examine what is known about the computations performed by the motor cortices and see how they and the red nucleus cooperate to control behavior.

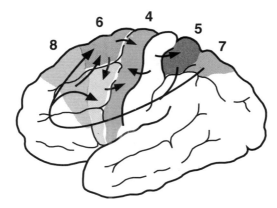

Figure 18.1 The primary motor and motor-related areas of the cortex are connected with one another by longer and shorter association axons.

The Primary Motor Cortex

The Canadian neurosurgeon Wilder Penfield mapped the primary motor cortex in humans and discovered that in each precentral gyrus *the muscles of the contralateral body and head are represented in an orderly, topographic fashion,* with the head muscles represented closest to the lateral fissure and progressively lower body parts represented on progressively higher parts of the precentral gyrus (Fig. 18.2). He also noted that the body parts were not proportionally represented in this cortical sheet. Instead, the *muscle groups used in movements that require finer control, such as those in the hand, are allotted disproportionately larger cortical fields.* You'll recognize that this principle is similar to that governing the representational maps of the primary somatosensory cortex. The proportionality differs from one mammalian species to the next, depending on which parts of the musculoskeletal system are most used for fine manipulative movements. For instance, a rat has only a minor forepaw representation but a relatively large snout representation in its M_I cortex. A raccoon, in contrast, has a forepaw representation comparable to the hand representation in a monkey or a human.

For a long time after Penfield's observation, a debate raged as to whether the representation in the primary motor cortex was one of particular movements or one of muscles. The debate was not resolved until 1967, when Hiroshi Asanuma used microstimulation techniques (tiny electrodes are used that activate only a few dozen neurons in a minute volume of cortex) in monkeys to show that *the cortex is arranged according to single muscles or small groups of muscles rather than particular movements.* The passage of a small current through the microelectrode reproducibly elicits a twitch in a particular muscle; it does not evoke a predictable movement. Moreover, *the sites where stimulation produces*

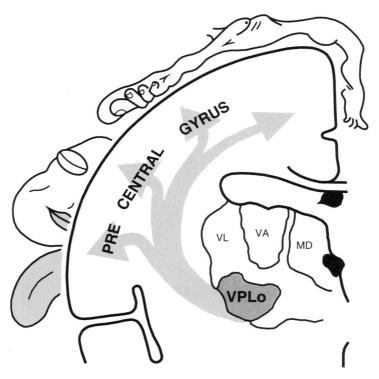

Figure 18.2 The primary motor strip on the precentral gyrus is topographically organized according to body musculature, as is its thalamic partner. Abbreviations MD, mediodorsal nucleus; VA, ventral anterior nucleus; VL, ventral lateral nucleus; VPL$_o$, oral part of the ventral posterolateral nucleus.

contractions within single muscles are arranged radially into columns akin to the columns in the S$_I$ cortex where all the cells are related to the same part of the body surface and the same somatosensory submodality. Some of the radial arrays produce excitation (contraction) of individual muscles, whereas other columns produce inhibition of the tonic activity (relaxation) of the muscles. Although the output of a single column is directed mainly at one muscle, the individual layer V pyramidal neurons within a single column can influence other muscles by way of axon collaterals that contact local circuit interneurons of the lower motor circuits. Asanuma also found that *individual muscles are represented by multiple columns,* especially the distal muscles. Thus, in the primary motor strip there is a redundancy of representation and, among the motor pools that innervate different muscles, a convergence of inputs from different cortical columns.

Edward Evarts and his collaborators recorded the activity changes in individual precentral corticospinal neurons in the arm–hand area of

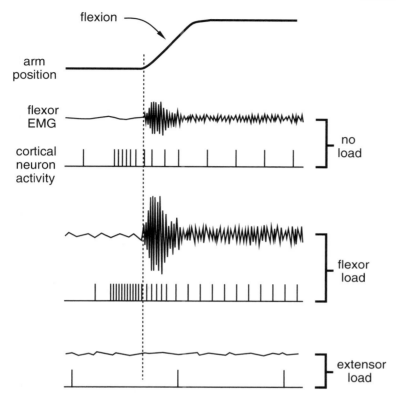

flexion

arm
position

flexor
EMG

cortical
neuron
activity

no
load

flexor
load

extensor
load

Figure 18.3 Edward Evarts's experiments showed that corticospinal neurons in the primary motor cortex of the monkey discharge prior to muscle contraction in a manner that indicates they dictate the force of contraction, not the direction of movement. Abbreviation: EMG, electromyogram.

awake monkeys that had been trained to perform specific tasks with their contralateral arm (Fig. 18.3). The monkey, when signaled, was to move a lever by wrist flexion to a fixed position in order to receive a juice reward. Weights could be added to the lever so that the monkey would have to generate more (loaded) or less (unloaded) force to accomplish the identical flexion movement. When more force was required (flexor load), the corticospinal neurons fired at a higher frequency than when less force was needed. Moreover, if no force was required, i.e., if the extensors were loaded and the monkey needed simply to relax them to let the lever move in the reward-producing direction, the corticospinal neurons to the flexor motoneurons did not increase their discharge at all, even though the direction of the movement was the same! Using this clever paradigm, the experimenters were able to demonstrate conclusively that *motor corticospinal neurons become active prior to the onset of*

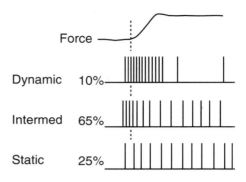

Figure 18.4 Claude Ghez identified three categories of motor cortex neurons by recording their activity changes during movements. The neurons encode different parameters of contraction force. The percentage of the total number of motor cortex neurons represented by each type is given.

voluntary movement and *their discharge frequency encodes the force of muscular contraction rather than the direction of limb movement.*

There are at least three types of cortical neuron in M$_I$ that encode different aspects of force generation (Fig. 18.4). *Dynamic* neurons encode for the rate of force development. They fire a burst prior to the change in force of the muscle contraction, then return to their baseline firing rate. Thus, the dynamic neurons influence the *speed* at which a movement is made. *Static* neurons encode the steady-state level of force used during the movement. They increase their activity prior to and throughout the duration of the movement. Finally, there are many neurons with *intermediate* properties that fire a modest burst prior to the onset of movement and sustain an elevated discharge throughout the movement. The intermediate neurons are in the majority in M$_I$. Incidentally, we have seen a similar range of properties in the premotor neurons that control saccadic eye movements, where they were called *burst* and *tonic* to describe the way in which their frequencies are modulated (Chapter 17).

Interestingly, the activity modulations of neurons of the red nucleus fall into the same three categories as those of the M$_I$ cortex, but it is the dynamic type that predominates (about 70%). That this should be the case seems entirely reasonable when one considers that the rubrospinal axons influence proximal limb muscles more than distal and play an essential role in generating gross arm movements like reaching (below). The loads encountered by the arm are typically greater than those encountered by the hand and fingers. The wielding of a tool or lifting of a heavy object may increase the need for the augmented force generation encoded by the dynamic type of rubrospinal neuron.

The Cortex and the Red Nucleus Have Complementary Roles in Arm Movements

If the medullary pyramids are surgically cut in a monkey, its finger and wrist movements are greatly disturbed. Gross arm movements fare considerably better. If the monkey attempts to retrieve a food pellet from

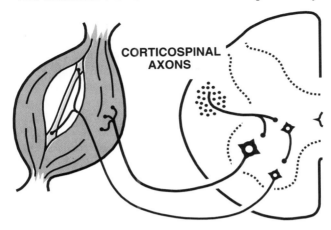

Figure 18.5 Descending axons coactivate α and γ motoneurons to maintain muscle tone and the integrity of the stretch reflex during muscle contractions.

a narrow well placed before it, it may extend its arm toward the well with reasonable accuracy and grace, but must struggle with its now spastic fingers to pluck the pellet from its niche. If the monkey is anesthetized and its M$_I$ cortex exposed, the monkey's proximal arm muscles can be made to contract upon electrical stimulation of the appropriate part of M$_I$, but stimulation of the hand area fails to elicit finger movements. These data indicate the existence of alternative, *parallel channels* from the motor cortex to the motor pools in the spinal cord that innervate the proximal arm muscles.

Although corticoreticular projections certainly play a role in conveying cortical commands to the spinal cord, the most important parallel route appears to be the projection from the primary motor strip to the red nucleus, whence the signals are relayed over the rubrospinal tract. These synaptic relationships and experimental results suggest that when we reach out to grasp an object, the primary motor cortex controls the gross movement (the reach) both directly, by way of its corticospinal projection, and indirectly, by way of its corticorubral projection. The final flexion of the fingers, however, seems to be decidedly under cortical control.

Although both the corticospinal and rubrospinal axons use glutamate at their synaptic terminals and therefore excite the cells that they contact, their ultimate effect on spinal motoneurons is for the most part mediated by local interneurons and is not uniform. Electrophysiological evidence indicates that in general, *corticospinal and rubrospinal axons facilitate motoneurons that supply flexor muscles and inhibit those that supply extensors.* This influence is similar on both α and γ motoneurons (Fig. 18.5). Individual corticospinal and rubrospinal axons that excite groups of motoneurons also produce inhibition in the motoneurons that innervate antagonist muscles. Much of the excitation and all of the inhibition of motoneurons by corticospinal axons is indirect, mediated by interneurons of the spinal cord intermediate zone. We learned in the preceding chapters that many movements can occur without voluntary control, and some occur even when we try to suppress them (one cannot hold one's breath indefinitely). Many spinal reflexes can function independently, as can rhythmic movements such as chewing and walking. Significantly, however, many of the interneurons contacted by corticospinal and rubrospinal axons to bring about volitional behaviors are the same ones that mediate the automatic reflexes. Thus, *the descending systems govern behavior by selectively engaging the movement patterns wired into the lower circuits.*

Motor Cortical Neurons Receive Sensory Feedback

Not only do primary motor cortex neurons participate in the initiation of voluntary movements, but they are also apprised of the consequences of the movements that are made. In other words, *they receive sensory feedback by way of association connections from the primary somatosensory cortex and by way of the thalamus.* The thalamic input to the primary motor cortex comes mainly from the oral (rostral) part of the ventral posterolateral nucleus (VPLo), which itself receives axons ascending with the medial lemniscus (Fig. 18.6). Indeed, neurons of the primary motor cortex evince sensory receptive fields like those in the S_I cortex. Examination of their receptive fields indicates a close relationship to the muscles they influence (Fig. 18.7). Individual motor cortex neurons receive either proprioceptive input from the muscle they control or tactile input from the region of skin related to the function of the muscle. Thus, in a sense, there is a long sensorimotor loop through the motor cortex, which resembles the segmental connections between muscle spindle afferents and motoneuron pools that occur at the level of the spinal cord. The type of sensory data fed back to the M_I neurons (cutaneous pressure and propriosensory, for instance) is highly relevant to the computations of contraction forces that are made by the neurons of the M_I cortex. Such sensory data may enable the M_I neurons to modulate their output according to the physical parameters (weight, texture, hardness) of objects that are manipulated.

The Contribution of the Supplementary Motor and Premotor Areas

In order to behave successfully (accomplish our intended goals), the brain must select a plan of action that specifies the *sequence* and *amplitude*

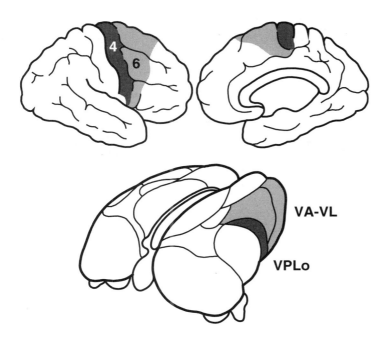

Figure 18.6 There are a number of motor-related nuclei in the thalamus that project to the different motor cortical areas. Abbreviations as in Figure 18.2.

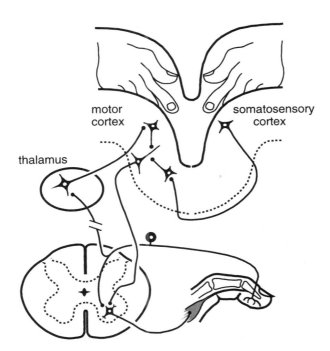

Figure 18.7 Hiroshi Asanuma showed that the tactile receptive fields of neurons in the primary motor cortex are related to the muscles that move the sensory surface.

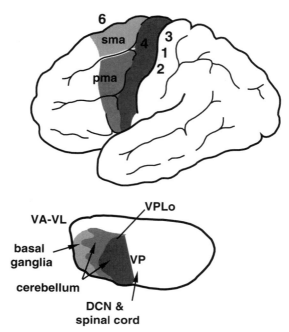

Figure 18.8 The ventral tier nuclei of the thalamus (bottom) project to particular motor-related areas of the cortex. Abbreviations: DCN, dorsal column nuclei; pma, premotor area; sma, supplementary motor area; VA, ventral anterior nucleus; VL, ventral lateral nucleus; VP, ventral posterior nucleus; VPL_o, oral part of the ventral posterolateral nucleus.

of contractions and relaxations in the different muscle groups to be used. This motor programming is accomplished not by the primary motor cortex, but by the supplementary motor and premotor areas of the frontal lobe (Fig. 18.1). It is noteworthy that both of these areas rely heavily on highly reprocessed sensory (perceptual) input that they receive from cortical areas 5 and 7 of the parietal lobe (below), and both produce their effects on movement through dense axonal projections to the primary motor cortex. One major difference between these two motor regions is in their thalamic dependencies; the supplementary motor area receives its major thalamic input from a part of the ventral anterior-ventral lateral (VA-VL) complex that is heavily influenced by the globus pallidus, whereas the premotor cortex receives its major thalamic input from a part of VA-VL influenced mainly by the cerebellum (Fig. 18.8).

The supplementary motor cortex is topographically arranged like the primary motor cortex, but stimulation of the supplementary motor cortex produces slowly developing, complex movement patterns rather than simple muscle contractions, and the movements are often bilateral. Moreover, there seems to be little correlation between the discharge of neurons in the supplementary motor area and the details (such as the

Box 18.1   Functional Imaging of the Living Human Brain

The four existing neuroimaging techniques are computed tomography (CT), magnetic resonance (MR), single photon emission computed tomography (SPECT), and positron emission tomography (PET). The term *tomography* comes from the Greek verb to cut and refers to the ability of the computer to reconstruct the images as though they were a series of "slices" through the brain. All four methods follow the same set of principles—a signal-emitting stimulus is applied that conveys information about the tissues, detectors monitor the signal, and a high-speed computer digitizes and analyzes the signal and converts it to a gray-scale image of the tissue. The signal for CT is an X-ray beam; for MR, it is the hydrogen protons within the tissue itself, which are oriented by a powerful magnetic field and then differentially reoriented by a radio-frequency signal. For SPECT and PET, radioactive tracers are injected and detected with, respectively, photon or γ-ray counters arrayed around the head. CT and MR provide high-resolution static images that are useful primarily for clinical purposes: localizing brain lesions, tumors, and vascular anomalies. Since SPECT and PET can be used for functional imaging, they have become powerful research tools.

The radioactive tracers used for SPECT and PET studies are differentially taken up by brain tissues according to their metabolic activity or regional cerebral blood flow changes. The concept underlying these techniques is based on the use of γ-ray detectors to monitor regional cerebral blood flow as indicated by the nondiffusible tracer, ^{133}xenon. Because the bolus of xenon travels rapidly through the circulation, it provides a dynamic and quantitative estimate of regional cerebral blood flow with a resolution of about 2 centimeters. SPECT can also be used to examine changes in the levels of receptors for neurotransmitter by injecting radiolabeled ligands that are specific for a particular receptor. It has so far been used to localize and quantify muscarinic ACh and D_2 dopamine receptors.

PET is the major technique for mapping brain activities. A positron-emitting label, such as [^{18}fluorine]deoxyglucose, is taken up by brain areas in proportion to their level of physiological activity. In this way, the activity of different regions can be monitored under experimental conditions in which the subject performs a specific activity like visual examination, reading, listening, or executing various movements. Use of the method has taken off in recent years. It promises to provide exciting new data about the way the human cerebral cortex organizes and performs its mysterious functions.

force) of the produced movement. Studies of regional cerebral blood flow using radioactive xenon and scalp detectors have provided compelling evidence that the supplementary motor area is important in motor planning. Regional increases in blood flow can be considered an indication of increased neural activity (Box 18.1). If a human is asked to perform

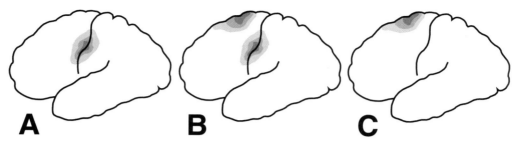

Figure 18.9 A human subject can be given an injection of radioactive xenon immediately before he is asked to perform three activities. The relative activity levels in his left cerebral cortex can then be indirectly monitored with γ-ray detectors mounted around his head. In these examples of patterns that could be expected to emerge, the increased activity is indicated by progressively darker red shading. The pattern shown in **A** results from the performance of a simple repetitive movement of the right index finger. The pattern in **B** unfolds when the subject performs a sequential movement using all the digits of the right hand. Finally, the pattern in **C** emerges when the patient is perfectly still but has been instructed to mentally rehearse the sequential movement he had performed previously.

a simple motor task such as compressing a spring between thumb and forefinger, only the hand representation areas of the contralateral primary motor and primary somatosensory cortices show an increased blood flow. However, if a subject is asked to perform a more complicated movement using all the digits in a coordinated fashion, blood flow is markedly increased bilaterally in the region of the supplementary motor area as well as in the primary motor and somatosensory cortex. Finally, when a subject is asked to mentally rehearse the movements without actually making them, blood flow is increased in the supplementary motor areas, but not at all in the primary motor cortex (Fig. 18.9).

The premotor area is less well understood than the other two motor areas. It seems to be important for the guidance of limb movements, and it influences motoneuron pools that innervate proximal limb muscles. Recordings have shown that neurons in the premotor cortex respond to sensory stimuli in a way that is contingent upon the *intent* of the animal to use the information to direct movements.

Spatial Information Comes from the Parietal Cortex

Both the supplementary motor and the premotor areas receive a dense input from the cortex of the superior parietal lobule (Fig. 18.1). This cortex, which consists of Brodmann's areas 5 and 7, is known to process the kind of sensory information necessary for the purposeful movements we make, that is, visual and tactile information about the physical qualities of objects in the environment and their spatial relationships to one another and to our own bodies. Destruction of the superior parietal lobule, in cases of stroke, for instance, often leads to a curious syndrome of sensory and motor *neglect,* in which the patient will fail to respond to stimuli on the contralateral side of the body or in the contralateral field of vision. The properties of neurons in areas 5 and 7 of the parietal cortex are very complex. For example, cells have been recorded in a monkey that fire only when the monkey reaches for a desired object within its immediate surroundings, but fail to fire when the same movement is made in the absence of the stimulus or when the stimulus is out of reach or no longer desired. Thus, it appears that *the parietal cortex supplies the motor planning areas with essential information about the quality of an object and its spatial relationship to the individual so that appropriate, goal-directed movements can be planned and generated.*

Detailed Reviews

Evarts EV. 1979. Brain mechanisms of movement. *Scientific Am* 241: 164.

Hepp-Reymond MC. 1988. Functional organization of motor cortex and its participation in voluntary movements. In Steklis HD, Irwin J (eds), *Comparative Primate Biology.* Alan R Liss, New York.

Kuypers HGJM. 1985. The anatomical and functional organization of the motor system. In Swash M, Kennard C (eds), *Scientific Basis of Clinical Neurology.* Churchill Livingstone, New York.

Wise SP. 1985. The primate premotor cortex: past, present and preparatory. *Annu Rev Neurosci* 8: 1.

Wise SP, Strick PL. 1984. Anatomical and physiological organization of the non-primary motor cortex. *Trends Neurosci* 7: 442.

19 The Cerebellum and Movement Precision

The cerebellum is essential for the full perfection of a wide variety of bodily movements. Once considered to be "the head ganglion of the propriosensory system," it is now apparent that the cerebellum contributes to all brain activities that are ultimately expressed in movement. To accomplish this role, the cerebellum receives sensory information of all kinds (except, perhaps, visceral, olfactory, and taste) from all parts of the body, information about motor commands from the descending systems, and a diversity of reprocessed information from higher brain structures, including, most prominently, the cerebral cortex. In turn, the cerebellum's output is sent to all of the motor control systems of the brain. The cerebellum is best known clinically for its contribution to the *coordination* of skeletomotor movement; clinical signs of cerebellar dysfunction are ataxia of limb and eye movements, postural disturbances, and decreased muscle tone. In this chapter, let's examine the cerebellar processing circuit and see how its design enables it to contribute precise timing to specific muscle contractions.

Gross Structure

Three major lobes of the cerebellum can be distinguished on phylogenetic grounds (Figs. 19.1 and 19.2). The oldest and smallest part, called the **flocculonodular lobe,** is closely related to the vestibular system. It is separated by the **posterolateral fissure** from the newer, larger **corpus cerebelli.** Of the latter, the part rostral to the **primary fissure** remains relatively constant in mammals and is called the **anterior lobe.** The larger part caudal to the primary fissure expands greatly in humans and is called the **posterior lobe.** Distinctive surface features of the cerebellum include the **vermis** lying longitudinally along the midline, the **lobules** (there are ten of them, numbered with roman numerals from rostral to

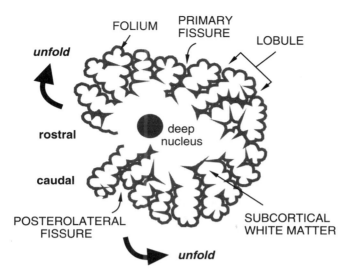

Figure 19.1 This drawing of a sagittal section through the cerebellum schematizes the superficial, folded cortex and the deep nuclear group embedded in the subcortical white matter. The fourth ventricle is to the left. Note that the rostral and caudal ends of the cortex recurve under the cerebellum until they are almost touching. For diagrammatic purposes, the cortex can be "unrolled," as indicated by the large arrows, and laid flat as in Figure 19.2.

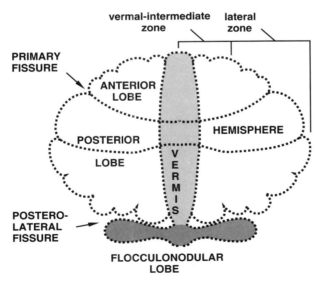

Figure 19.2 A schematic representation of the "unrolled" and "flattened" cerebellar cortex shows the regional subdivisions used for reference. The rostralmost edge of the cortex is at the top.

caudal), and the many folia. The lobules are larger convolutions separated by deep sulcuses, and the folia are the redundant mediolaterally oriented ripples. The cerebellum is connected to the brain stem by afferent and efferent axons that pass in three pairs of massive fiber bundles: the superior, middle, and inferior cerebellar peduncles (often called, respectively, the **brachium conjunctivum, brachium pontis,** and **restiform body; see Figs. B2–4).**

The largest, middle peduncles each contain the incoming axons from the contralateral pontine nuclei. The main outflow of the cerebellum travels in the superior peduncles, which carry axons to several brain stem structures, the most prominent of which are the red nuclei, and to the thalamus. The superior peduncles also contain a relatively small number of incoming axons. The inferior peduncles carry mainly incoming fibers from the spinal cord and lower brain stem structures, and outgoing axons to the lower brain stem.

From a surface perspective, one sees little of the complexity of the cerebellum. Buried beneath the surface are the layers of the cerebellar cortex, which cover a core of white matter consisting of axons passing to and from the cortex. Embedded deep within this white matter core are three pairs of deep cerebellar nuclei (see Figs. 19.5 and B3). On each side, the most medial of these nuclei is the small **medial** (or fastigial) nucleus, the most lateral is the large, ribbonlike **lateral** (or dentate) nucleus, and between them is the **interposed** nucleus (archaically referred to as two nuclei: emboliform and globose).

Cellular Organization of the Cerebellar Cortex

The cerebellar cortex is histologically uniform throughout. Nowhere else in the CNS is three-dimensional regularity so evident. Five morphologically unmistakable cell types are present, called the **granule, Purkinje, stellate, basket,** and **Golgi** cells, which are arranged in a precisely reiterated pattern in the three-layered cerebellar cortex (Fig. 19.3). Such regularity suggests that the information processing in the cerebellar cortex is everywhere the same, and that the functional differences within the cerebellar cortex are determined by regionally distinct input (see below).

The outermost layer of the cerebellar cortex is the **molecular layer.** It consists mainly of axons (parallel fibers and climbing fibers), the dendrites of Purkinje, stellate, and basket cells, and only a few cell bodies (basket and stellate). Subjacent to the molecular layer is the **Purkinje cell layer,** which consists of a **monolayer** of the large (about 80 micrometers in diameter) Purkinje cell bodies. The dendritic tree of each Purkinje cell is extensively branched in the molecular layer, but only in a plane perpendicular to the long axis of the folium. The dendrites of the Purkinje

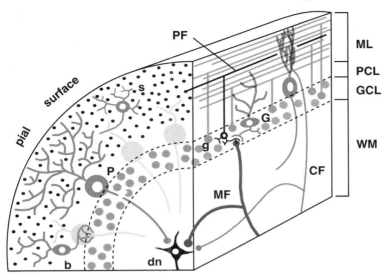

Figure 19.3 A wedge cut from a single folium is schematized to show the five cell types of the cerebellar cortex, their arrangement and orientation, and their synaptic relationships to one another. The red neurons use GABA and are inhibitory. Abbreviations: b, basket cell; CF, climbing fiber; dn, deep nucleus neurons; G, Golgi cell; g, granule cell; GCL, granule cell layer; MF, mossy fiber; ML, molecular layer; P, Purkinje cell; PCL, Purkinje cell layer; PF, parallel fiber; s, stellate cell; WM, subcortical white matter.

cells are covered by short, thick dendritic spines. The axons of the Purkinje cells carry the sole output of the cerebellar cortex, and they are directed at the cells of the deep cerebellar and the vestibular nuclei. The third, innermost layer is the **granule cell layer** which contains large numbers of densely packed granule cells. The granule cell axons ascend to the molecular layer and bifurcate into the **parallel fibers** that run, beamlike, parallel to the long axes of the folia. Such axons can travel several millimeters along a folium, passing through the planes of several Purkinje cell dendrites on which they form synaptic contacts with the dendritic spines. The granule cell layer also contains the large Golgi cells, which have their dendrites in the molecular layer and send out highly branching, short axons that synapse on many neighboring granule cells.

   The incoming fibers to the cerebellum send axon collaterals to the deep cerebellar nuclei and the cerebellar cortex (Fig. 19.3). Upon reaching the cerebellar cortex, the afferent axons are described as two general types: **climbing fibers,** which "climb" (like a grapevine on a trellis) and synapse repeatedly on the somas and dendritic trees of the Purkinje cells, and **mossy fibers,** which terminate on the granule cell dendrites. Climbing fibers originate only from the inferior olivary nuclei. Cerebellar afferent axons from all other sources appear as mossy fibers.

The synaptic relationship between the granule cell dendrites and the axonal endings of the mossy fibers and Golgi cells is a special one. The granule cells extend short, clawlike dendrites that seem to grasp the lobulated terminals of the mossy fibers, and these dendritic claws are in turn covered by the axonal endings of the Golgi cells. This peculiar synaptic configuration is referred to as a **glomerulus.**

Activity in the Cerebellar Circuit

Most of the incoming fibers to the cerebellum are excitatory to neurons in both the deep nuclei and the cortex (Fig. 19.5). *The tonic activity of the neurons in the deep cerebellar nuclei is high and is driven by these excitatory afferent fibers.* For now, let's disregard the climbing fibers and concentrate on the activity induced by the mossy fibers. Incoming activity along a circumscribed group of mossy fibers activates a local cluster of granule cells (Fig. 19.4) at their glomerular connections. The glutamatergic granule cells, in turn, sequentially activate the Purkinje cells in the long row that happens to be aligned with their parallel fibers. In addition to the sequential activation of a row of Purkinje cells, the parallel fibers also activate nearby basket and stellate cells in the molecular layer. The GABAergic axons of the basket cells wrap around the cell bodies of the Purkinje cells on either side of the activated row and potently inhibit them. *This mechanism of lateral inhibition sharpens the* spatial focus *of activity along the folium.* The GABAergic stellate cells activated by parallel fibers send their axons back to the Purkinje cells just activated by the parallel fibers and inhibit them. Like basket and stellate cells, the Golgi cells are also activated by granule cell axons. Since the synapses of the Golgi cell axons on the dendrites of the granule cells at the glomeruli release GABA and inhibit the granule cells, this part of the circuit represents another feedback pathway, which introduces a pulsatile qual-

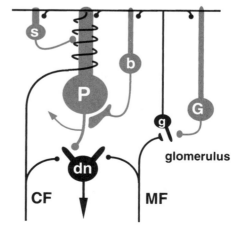

Figure 19.4 The basic cerebellar circuit depends on precisely timed excitatory and inhibitory transmission events. Abbreviations are as in Figure 19.3. Jet black neurons are glutamatergic and excitatory; the red are GABAergic and inhibitory.

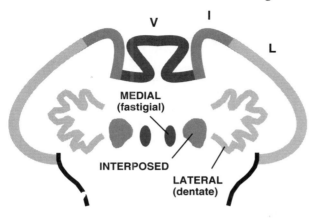

Figure 19.5 The projection of Purkinje cell axons to the deep nuclei of the cerebellum is topographically organized. Abbreviations: I, intermediate zone; L, lateral zone; V, vermal zone.

ity to the incoming activation of the Purkinje cells. Thus, *the inhibitory connections of the stellate and Golgi cells increase the temporal resolution of Purkinje cell activity.* The circuit of the cerebellar cortex, thus, serves as a time-chopper with a high degree of spatial resolution.

The output of the cerebellar cortex is conveyed exclusively by the axons of Purkinje cells to the deep cerebellar nuclei and, to a lesser extent, to the vestibular nuclei. The action of Purkinje cell axons on the postsynaptic cell is always inhibitory and is mediated by the neurotransmitter GABA. Thus, *the outflow of the cerebellar cortex carried by the Purkinje cell axons serves to modulate the high level of afferent-driven activity of the cells in the deep nuclei.*

This corticonuclear projection is topographically organized so that the **medial** (vermal) **zone** projects to the medial nuclei, the **intermediate** (paravermal) **zones** project to the interposed nuclei, and the **lateral** (hemispheric) **zones** project to the lateral nuclei (Fig. 19.5). The cells of the deep cerebellar nuclei give rise to the axons that convey information to other brain structures, as well as axons that return to the cerebellar cortex (nucleocortical fibers of obscure function).

The Functional Organization of the Cerebellar Cortex

It is most useful to identify three cerebellar subdivisions on the basis of afferent fiber connections (Fig. 19.6): the **vestibulocerebellum,** which corresponds to the flocculonodular lobe; the **spinocerebellum** (so called because of its heavy input from the spinal cord), which consists of the rostral and caudal parts of the vermis and adjacent portions of the

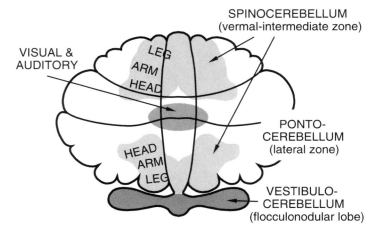

Figure 19.6 The functional areas of the cerebellar cortex are determined by dominant inputs from the spinal cord, vestibular system, and cerebral cortex via the pontine nuclei.

intermediate zone; and the **pontocerebellum** (or cerebrocerebellum), which consists of the middle portion of the vermis and the large, lateral zone of the cerebellar hemisphere and receives its major input from the cerebral cortex by way of the pontine nuclei. It is important to note, however, that there is extensive overlap of the terminal distribution of pontine fibers with those from the vestibular complex and spinal cord. These three main regions, vestibulo-, spino-, and pontocerebellum, distinguished on the basis of their inputs, exert their major influence on descending motor systems of the CNS by way of their efferent projections through the deep cerebellar nuclei.

The major inputs to the cerebellum are summarized in Figure 19.7. The vestibulocerebellum receives axons from the vestibular nuclear complex (secondary sensory axons) and, remarkably, from the vestibular ganglion itself (primary sensory axons). Both the primary and the secondary vestibular axons enter the cerebellum through the inferior cerebellar peduncle and appear as mossy fibers in the nodular and floccular cortex. Perhaps not surprisingly, some of the Purkinje cell axons that leave the vestibulocerebellar cortex bypass the deep cerebellar nuclei and instead terminate directly in the vestibular nuclei of the medulla to influence the vestibulospinal motor system. *This close alliance of the vestibular sense with the cerebellum is important for the maintenance of posture and balance in terrestrial animals.* Why is the vestibulomotor system given special access to and from the processing circuit in the cerebellar cortex, bypassing the deep nuclei? Remember that vestibulospinal postural reflexes must work quickly if they are to be effective. Compensation for a false step on a narrow mountain trail must be well coordinated and occur with minimal delay if it is to prevent a plummet into the Stygian valley.

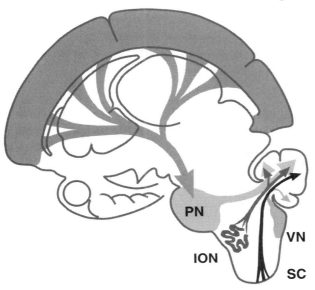

Figure 19.7 The cerebellum receives four major inputs. Abbreviations: ION, inferior olivary nucleus; PN, pontine nuclei; SC, spinal cord; VN, vestibular nuclei.

The spinocerebellum receives axons from the spinal cord by way of two tracts, the dorsal and ventral spinocerebellar tracts (Fig. 19.8). The **dorsal spinocerebellar tract** conveys uncrossed axons from the **dorsal nucleus** (of Clarke), which is located in the spinal cord (from segments T_1 to L_2). You'll recall that the dorsal nucleus receives primary sensory

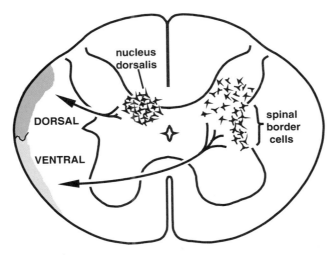

Figure 19.8 The two spinocerebellar tracts (shaded red) arise from functionally distinct populations of spinal cord neurons.

fibers from the periphery that convey propriosensory information. The dorsal spinocerebellar axons enter the cerebellum through the inferior peduncles and appear as mossy fibers in the cortex, where they terminate in a somatotopic fashion representing the lower limb.

The **cuneocerebellar tract** originates in the **external** (or lateral) **cuneate nucleus** of the medulla and is functionally the upper limb equivalent of the dorsal spinocerebellar tract. The **trigeminocerebellar tract** arises from the trigeminal complex, enters via the inferior peduncle, and provides the face representation to the **spinocerebellar** part of the cortex. The nucleus dorsalis, external cuneate nucleus, and spinal trigeminal nucleus, all convey information about limb position and movement from transducers in the periphery to the spinocerebellum. These inputs impose two separate somatotopic representations on the cortex of each cerebellar hemisphere, one in the anterior lobe and one in the posterior lobe (Fig. 19.6).

The **ventral spinocerebellar tract** consists of crossed and uncrossed axons mainly from the base of the dorsal horn. Some of these neurons are called **spinal border cells** because of their location at the lateral edge of the spinal cord gray matter (Fig. 19.8). The spinal border cells, unlike those in Clarke's column, are not dependent on peripheral sensory information. Instead, they receive synaptic input from descending axons and thus intercept a copy of the motor command signals just as they are being delivered to the lower motor circuits. The axons of the ventral spinocerebellar tract ascend all the way to the superior peduncle to enter the cerebellum and terminate in the spinocerebellar area. They provide the spinocerebellar cortex with information about the movement commands.

The pontocerebellum is at its fullest development in the human. Its expansion corresponds to the increased development of the cerebral cortex and pontine nuclei. The pontine nuclei receive a topographically organized projection from all parts of the cerebral cortex and project to the contralateral cerebellar hemisphere (as mossy fibers) by way of the middle peduncles. The motor and premotor cortex contribute especially to the corticopontine projections, as do the parietal sensory regions. By way of this corticopontocerebellar connection, the cerebral cortex can access the services of the cerebellum for the control of voluntary movements.

Other Cerebellar Inputs

A number of nuclei scattered through the lower brain stem send axons to the cerebellum. They are collectively referred to as *precerebellar* nuclei. The **lateral reticular nucleus** is an example of a rather prominent precerebellar nucleus located in the ventrolateral part of the medulla. Its functional impact on the cerebellum is unknown, as are those of several smaller nuclei embedded in the reticular formation that project selectively to the cerebellum. The cerebellum also receives monoaminergic axons from the locus coeruleus (norepinephrin) and the raphé nuclei (serotonin) that terminate in a diffuse manner and appear to have a widespread modulatory action on the cerebellar cortex.

Certainly the most conspicuous of the precerebellar nuclei is the inferior olivary nucleus (Fig. B2). The axons from this nucleus are crossed, enter the cerebellum through the inferior peduncle, and appear as the climbing fibers in the cerebellar cortex. The inputs to the inferior olivary nuclei come from the cerebral cortex, spinal cord, and several brain stem structures. A single climbing fiber may contact anywhere from one to ten Purkinje cells, but each Purkinje cell is contacted by only one climbing fiber. Remarkably, each climbing fiber directly and powerfully excites the Purkinje cells it contacts by multiple synapses on their dendritic trees. A single action potential in a climbing fiber can cause a prolonged train of action potentials in the Purkinje cells it innervates. The initial action potential in the train is larger than those that follow. This is because the ionic basis of this initial action potential is unlike any other in the CNS—it is mediated by an influx of extracellular Ca^{2+} rather than Na^+ and is aptly referred to as a Ca^{2+} spike. Interestingly, the climbing fiber system is thought *not* to play a direct role in ongoing movement control, but instead to be involved in the adaptation of skilled movements due to experience—a *learning* process that will be discussed in Chapter 29.

Outputs of the Deep Cerebellar Nuclei

In general, the outputs of the functional zones of the cerebellar cortex are directed by way of the deep nuclei to the brain regions that provide them with input (Fig. 19.9). The fastigial nuclei receive their cortical input from the vermis and the flocculonodular lobe. They in turn send axons to the vestibular nuclei and reticular formation by way of the inferior peduncle and to more rostral brain stem regions via a smaller number of axons that pass through the superior peduncle. Thus, the fastigial nuclei exert their major influence on the medial descending motor systems (vestibulo- and reticulospinal). The interposed and dentate nuclei send axons through the superior peduncle to the contralateral red nucleus, the premotor layers of the superior colliculus, the motor-related nuclei of the thalamus (VA-VL complex and VPLo), and, to a lesser degree, the pontine nuclei, inferior olive, and ocular motor nuclei (follow the superior peduncle from the deep nuclei to the red nucleus in Figs. B3–7). The interposed nuclei receive much of their input from the spinocerebellum, and they influence descending motor systems of both the medial and the lateral systems. The dentate nuclei receive their input from the pontocerebellum and influence mainly the descending motor systems of the lateral group (rubro- and corticospinal), which affect mainly limb and hand movements. The cerebellar output to the VPLo influences the

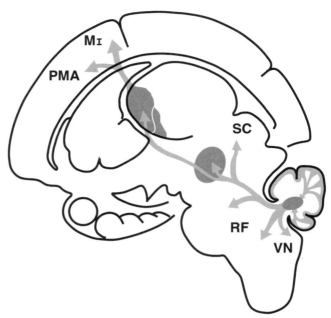

Figure 19.9   The outputs of the cerebellum are directed to motor control nuclei. Abbreviations: M$_I$, primary motor cortex, PMA, premotor area; RF, reticular formation; RN, red nucleus; SC, superior colliculus; VA-VL, ventral anterior-ventral lateral nuclei; VN, vestibular nuclei; VPL$_o$, oral part of the ventral posterolateral nucleus.

primary motor cortex, and that to the VA-VL complex contributes to the computations performed in the premotor area (see Fig. 18.8).

The Role of the Cerebellum in Movement Control

It is clear that the cerebellum has a close association with the vestibular complex, forcefully influencing all four vestibular nuclei, and with the reticular formation. Through the descending systems that originate from these motor control regions, the cerebellum can influence postural reflexes. By way of its connections with the red nucleus and the premotor and motor cortex, the cerebellum impacts on voluntary movements as well. What is the contribution that the cerebellum makes to each of these motor control systems? The topographic arrangements of inputs and outputs of the cerebellar cortex and the sequential engagement of a series of Purkinje cells by parallel fibers suggest that the various motor systems use the cerebellum to *govern the timing of muscular contractions* so as to smooth out individual bodily movements and blend the individual movements smoothly together during sequential movements. In this way, the cerebellum provides the major part of the *precision* so typical of our skeletal movements.

Consistent with this notion of the cerebellum as a timing device is the long-standing clinical observation that the most striking features of cerebellar dysfunction are disturbances of muscle coordination. As you might surmise, damage to the intermediate region of the cerebellar cortex, which you'll recall harbors the spinocerebellar terminals, produces impairments in axial (trunk) muscle control. An obvious sign of spinocerebellar damage is an inability to balance the trunk over the legs while walking, a condition often described as "drunken sailor's gait." To cause extremity ataxia of the ipsilateral arm, a lesion, necessarily large, must be placed more laterally in the pontocerebellum. A special kind of tremor results. It's called *intention tremor* because it appears only when a goal-directed arm movement is made and, in fact, intensifies as the hand approaches the target. In any case, the deficit caused by damage to the cerebellum appears to be a *temporal decomposition of the way muscles cooperate*. It is worth keeping in mind that a remarkably large loss of cerebellar cortex is necessary to produce severe signs, but even minor damage to the deep cerebellar nuclei is extremely debilitating.

Detailed Reviews

Gilman S. 1985. The cerebellum: its role in posture and movement. In Swash M, Kennard C (eds), *Scientific Basis of Clinical Neurology*. Churchill Livingstone, New York.

Llinás R. 1981. Electrophysiology of the cerebellar networks. In Brooks VB (ed), *Handbook of Physiology*. American Physiological Society, Washington, DC.

Palay SL, Chan-Palay V. 1974. *Cerebellar Cortex: Cytology and Organization*. Springer-Verlag, New York.

20 The Basal Ganglia System and Movement Sequencing

Like the cerebellum, the basal ganglia system is known from clinical and experimental evidence to constitute an important integrative mechanism of higher motor control. Comparison with the cerebellum, however, stops there. Unlike the unitary cerebellum with its stereotypic circuit, the basal ganglia system includes several large, dissimilar nuclei distributed in the fore- and midbrain, each with a complex microcircuitry of its own. And the synaptic interrelations *among* the components of the basal ganglia defy any simple logic; collectively, they use most of the classic neurotransmitters and multiple peptide cotransmitters, each of which interacts with an array of receptor subtypes. Whereas the cerebellum receives relatively raw sensory data, the basal ganglia system is synaptically removed from the sensory periphery; most sensory data presented to it arrive only after substantial reprocessing through the cerebral cortex. Likewise, the output of the basal ganglia system affects the final common pathway mainly indirectly by way of the corticospinal system, whereas that of the cerebellum impacts directly on all of the sources of descending motor control. This complexity, together with a relative inaccessibility to experimental manipulation, has hampered studies of the connections between the components of the basal ganglia and of the computations they perform; mechanistic knowledge is still sketchy. What is well known, however, is that lesions within the basal ganglia system due to vascular accident or inherited disease give rise to movement disorders (akinesia, tremor, involuntary movements, and postural disturbances) that are commonly encountered by the neurologist. In fact, the most common of the adult-onset brain diseases, Parkinson's disease and Alzheimer's disease, have been linked to basal ganglia dysfunction. Even though the functional role of the basal ganglia system remains poorly understood, it is nonetheless imperative that future physicians become familiar with what is known about them. The subject of basal ganglia dysfunction is destined to concern them greatly as our society continues to gray.

Component Nuclei

As originally used, the term *basal ganglia* referred to all of the subcortical nuclei of the telencephalon. Over the years, however, the meaning shifted from one of purely anatomical location to one of functional relationships. Today, the term has evolved to exclude some of the original telencephalic members (amygdala and septal nuclei) and to include several larger and smaller cell masses buried deep within the telencephalon, diencephalon, and even the midbrain. Since *basal ganglia* refers to these cell groups in the plural, an appropriate collective term that embraces them all, their interconnections, and their synaptic relationships to other parts of the brain is the basal ganglia *system* (Figs. 20.1 and 20.2). The largest of the cell masses is the corpus striatum, which itself is divided into the caudate nucleus and the putamen (together called simply the **striatum**) and the **external** and **internal** segments of the globus pallidus—together, the **pallidum** (Figs. B8–10). Incidentally, the term *lentiform nucleus* was used by early anatomists to refer collectively to the putamen and globus pallidus, which lie side by side and, in transverse sections, appear as a lens-or wedge-shaped mass (Fig. B9). Neuroradiologists still use this anatomical designation.

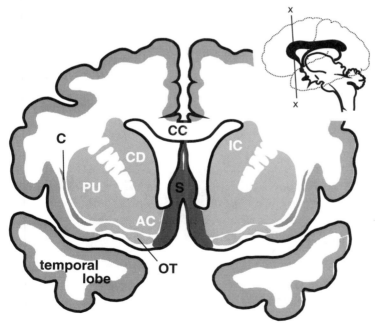

Figure 20.1 The main nuclei of the striatum (red shaded)—the caudate nucleus (CD), putamen (PU), nucleus accumbens (AC), and olfactory tubercle (OT)—are buried deep in the cerebral hemisphere (compare to Fig. B10). Abbreviations: C, claustrum; CC, corpus callosum; IC, internal capsule; S, septum; V, lateral ventricle.

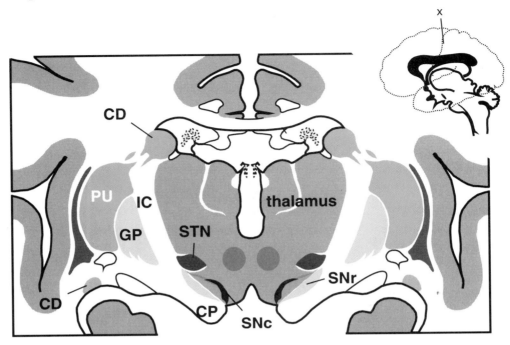

Figure 20.2 Several component nuclei of the basal ganglia can be seen in this oblique section (compare to Figs. B8 and B9). Like the caudate nucleus (CD) and putamen (PU), the globus pallidus (GP) and pars reticulata of the substantia nigra (SNr)—both shaded a lighter red—are separated during development by the growing axons of the internal capsule (IC) as they descend to form the cerebral peduncle (CP). The dorsal surface of SNr is capped by the pars compacta (SNc). The subthalamic nucleus (STN) lies near the pallidal structures and is an important participant in basal ganglia circuitry.

Because of recent experimental work, we now include the **nucleus accumbens** as part of the striatum. The **olfactory tubercle,** which lies subjacent to the accumbens, is considered to contain both striatumlike and pallidiumlike regions. The nucleus accumbens and the striatal part of the olfactory tubercle are often described as the "ventral striatum," whereas the pallidal part of the olfactory tubercle is viewed as the "ventral pallidum." This ventral part of the basal ganglia system is related closely to brain structures involved with emotions (see below).

The **pars reticulata** of the substantia nigra, which lies nested in the cerebral peduncle of the midbrain, can be considered yet another subdivision of the pallidum on embryological, histological, and connectional grounds (Fig. B7). Two structures so closely related to the corpus striatum that they are considered parts of the basal ganglia system are the **pars compacta** of the substantia nigra in the midbrain and a nucleus rostral to it in the diencephalon, the **subthalamic** nucleus. Now let's examine the interconnections and the flow of information between these component nuclei of the basal ganglia system.

Striatal Inputs

The primary input–output organization and internal circuitry of the basal ganglia system are summarized in Figure 20.3. The striatum constitutes the *input side* of the system. *The major input comes in the form of an excitatory glutamatergic projection from all parts of the cerebral cortex.* Corticostriatal axons come from pyramidal cells mainly (though not exclusively) in layer V. The projection is topographically organized and is especially heavy from association areas. On the basis of its cortical input, the striatum can be divided into at least three broad functional domains (Fig. 20.4). The sensorimotor (M$_I$ and S$_I$) cortices project to the putamen; the association cortices, especially the prefrontal areas, project mainly to the caudate nucleus; the olfactory and hippocampal (limbic) cortex, along with a related but noncortical structure, the amygdala, project to the ventral striatum (nucleus accumbens and olfactory tubercle). Thus, the striatal cells take their cues in large measure from the cerebral cortex, and the striatum as a whole serves all areas of the cortex, regardless of functional affiliation.

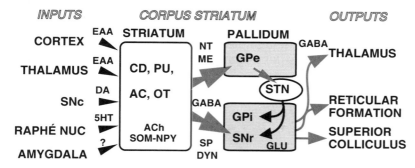

Figure 20.3 Information flow through the basal ganglia is mostly unidirectional. The striatopallidal axons (large red arrows) exert a profound GABA-mediated inhibition of the pallidal neurons. The pallidal output (red arrows) to motor-related cell groups is also inhibitory and mediated by GABA. Abbreviations: AC, nucleus accumbens; ACh, acetylcholine; CD, caudate nucleus; DA, dopamine; DYN, dynorphin; EAA, excitatory amino acids; GABA, γ-aminobutyric acid; GLU, glutamate; GPe, external segment of the globus pallidus; GPi, internal segment; 5HT, 5-hydroxytryptamine; ME, met-enkephalin; NPY, neuropeptide tyrosine; NT, neurotensin; OT, olfactory tubercle; PU, putamen; SNc, pars compacta of the substantia nigra; SNr, pars reticulata; SOM, somatostatin; SP, substance P; STN, subthalamic nucleus.

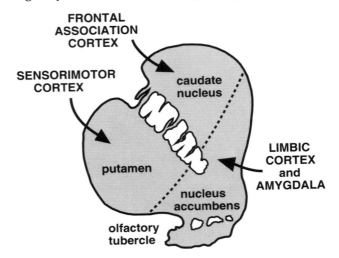

Figure 20.4 This schematic of a frontal section through the striatum shows three regions associated with different functional domains on the basis of their major inputs. Medial is to the viewer's right and dorsal is at the top of the image.

The electrophysiological studies of Mahlon DeLong have shown that the somatotopic organization of the sensorimotor cortex is maintained in its projection to the putamen. Microelectrode recordings reveal a dorsal region in which the neurons respond to leg movements, a ventral region responsive to orofacial movements, and a region in between responsive to arm and hand movements (Fig. 20.5). This somatotopy is maintained in the putamen's projection to the globus pallidus. Moreover, groups of neurons in the putamen appear to be organized according to movements around particular joints. Although individual striatal and pallidal cells respond more vigorously to active than to passive movements, the fact that they respond to both indicates that they receive sensory feedback from the movements as well as information about the motor commands issued by the cortex.

The second densest input to the striatum comes from the thalamus. The intralaminar nuclei, especially the centromedian nuclei, are the major origins of the thalamostriatal projection. Thalamostriatal axons, like corticostriatal, use excitatory amino acids (glutamate or aspartate) as neurotransmitters. Since the intralaminar nuclei receive synaptic input from several sources, most prominently the brain stem reticular formation, they convey multimodal information about somatic and vestibular sensations and eye movement-related data to the striatum. This thalamic input, together with the cortical input, selectively and intermittently drives the activity of striatal neurons.

The pars compacta of the substantia nigra and some paranigral cell clusters contain the neurons that *synthesize the neurotransmitter dopamine.*

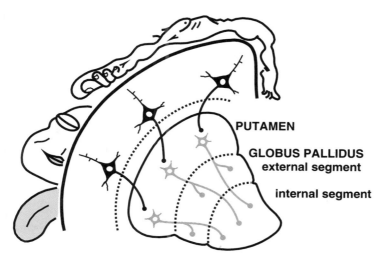

Figure 20.5 Mahlon DeLong showed that the axonal projection system from the sensorimotor areas of the cortex to the corpus striatum is somatotopically organized. The jet black neurons are excitatory, glutamatergic neurons; the red neurons are GABAergic and inhibitory.

Many of these dopaminergic neurons send their axons to the striatum (Fig. 20.6). Another set of monoamine axons that synapse in the striatum, these containing serotonin, originates in the midbrain from the raphé nuclei. Although the bulk of evidence suggests that serotonin inhibits striatal cells, the action of dopamine on striatal cells depends on the receptor subtypes with which it interacts (Fig. 20.7). Further, the subtype appears to vary from one subpopulation of striatal neuron to another. For instance, striatal neurons that project to the external segment of the globus pallidus express mainly D_2 receptors, which are associated with the inhibition of adenylyl cyclase, whereas striatal cells that project to the internal segment and the substantia nigra express mainly D_1 receptors, which activate the cyclase. Other dopamine receptor subtypes (called D_3, D_4, and D_5) related to these two have been cloned recently and are also differentially distributed among striatal neurons.

Like its impact on striatal computations, the role of the dopaminergic nigrostriatal projection in behavior remains mysterious. Conveniently, dopaminergic neurons can be selectively destroyed with either of two neurotoxins—6-hydroxydopamine or 1-methyl-4-phenyl-1,2,3,6-tetrahydropyridine (MPTP)—and the behavioral consequences monitored in experimental animals. Monkeys with such lesions assume a fetal posture and are unable to generate any skeletal movement at all (akinesia). Two young men who were inadvertently dosed with MPTP as a contaminant in a synthetic recreational drug suffered the same "frozen" condition as the monkeys. This observation of akinesia after nearly total depletion of

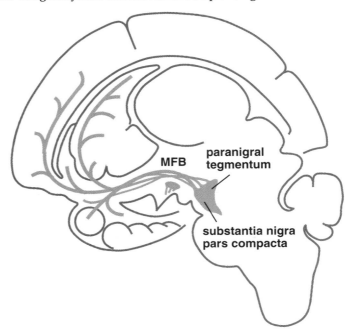

Figure 20.6 The dopaminergic projections to the forebrain arise from neurons in the pars compacta of the substantia nigra and paranigral cell groups of the ventral tegmentum. The axons ascend as part of the medial forebrain bundle (MFB).

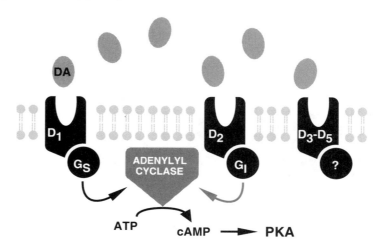

Figure 20.7 There are multiple dopamine (DA) receptor subtypes (D_1–D_5) expressed differentially by striatal and other forebrain neurons. Some have been linked to the stimulation of cAMP production, whereas other subtypes inhibit adenylyl cyclase. Abbreviations: G_S, G_I, stimulatory and inhibitory GTP-binding proteins; PKA, cAMP-dependent protein kinase.

nigrostriatal dopamine surprised no one, however, since it had already been known for some time that the primary pathology in Parkinson's disease is the loss of pigmented (dopaminergic) neurons in the substantia nigra pars compacta. The most prominent features of Parkinson's disease are progressively worsening hypokinesia, skeletomuscular rigidity, and resting tremor.

The cause of Parkinson's disease is unknown. It occurs in one in a thousand of the general population, making it one of the two most common adult-onset brain disorders (the other is Alzheimer's disease). The therapy of choice is systemic treatment with L-DOPA, the immediate precursor of dopamine (see Chapter 3). The disease typically shows a progression to the point where drug treatment is no longer effective, presumably due to progressive loss of the dopaminergic neurons that would otherwise convert the L-DOPA to dopamine. An exciting new therapeutic approach to Parkinson's disease has been attempted in recent years. It involves the transplantation of fetal dopaminergic neurons directly into the striatum of afflicted patients. If enough of the fetal cells take up residence and begin to squirt dopamine, an amelioration of the akinesia results. That the L-DOPA and transplant therapies work at all suggests that the dopaminergic signaling that goes on in the striatum may have a more relaxed requirement for intricate spatiotemporal discharge patterns than do the majority of synaptic connections in the brain. Perhaps bathing striatal cells in dopamine somehow readies them to perform their duties when called upon by the cortex or thalamus.

The Striatum Is Organized into Compartments

The striatum is nearly featureless when stained with nonspecific dyes. Predictably so, considering that more than 95% of its cells are of a common shape and size, and that the other 5%, though distinctly larger, are evenly dispersed among their smaller cohorts. Nevertheless, when Captain Ahab intoned to the reticent Starbuck that *"All visible objects, man, are but as pasteboard masks,"* he might just as well have been remarking on the nature of the striatum. For, beginning with the histochemical observations of Ann Graybiel, many recent histochemical and physiological studies have exposed a recondite mosaic behind its nondescript facade (Fig. 20.8). Immunocytochemistry reveals clusters of cells of a common peptide cotransmitter phenotype; the clusters are separated from one another and from cell clusters of different cotransmitter phenotypes. Receptor-binding studies indicate heterogeneous densities of neurotransmitter receptors, such as the D_1 and D_2 types of dopamine receptor. Axon-tracing studies show that the striatal afferent axons distribute in discontinuous terminal patches of higher and lower density. This mosaic appearance has been described in terms of *patches* or *islands*

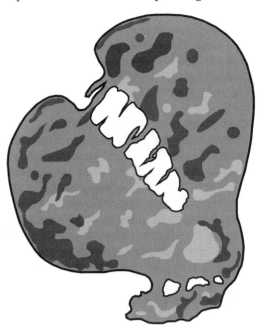

Figure 20.8 When stained with any one of several histochemical procedures, a striatal tissue section often has a mottled appearance of high- and low-intensity staining against a more homogeneously stained background. Observations of this kind have led to the notion that the striatum is anatomically and functionally organized into compartments.

or *striosomes* ("striatal bodies") embedded in an evenly textured *matrix*. Since, in some cases, the striatal mosaics identified by different staining techniques appear to be in register with one another, the notion has emerged that the striatum is organized into subsets of *modules* that perform a basic striatal computation. In other cases, however, the modular hypothesis is difficult to reconcile with mosaic patterns that show little, if any, obvious correspondence. Electrophysiological data are encouraging with respect to this concept of modular organization. Recordings in the sensorimotor zone of the monkey striatum have shown that cells related to movements around particular joints are segregated into clusters that are separated from one another by clusters related to other joints. Correspondence between any of the histochemically revealed compartments and this joint-related clustering has not been conclusively demonstrated to date. Ahab again: "*there, some unknown but still reasoning thing puts forth the mouldings of its features from behind the unreasoning mask.*"

Most striatal neurons participate in an extensive local circuitry that covers domains of modest dimension. The smaller striatal cells whose axons project to pallidal targets also elaborate intrastriatal axon branches. While highly branched nearby, the local collaterals do not stray far from their parent cells. For instance, there are no connections from the caudate to the putamen or vice versa; the head of the caudate does not appear to communicate with the tail. Since these local connections are GABAergic, they tend to cast an inhibitory pall over the striatum. The larger striatal neurons, which are either cholinergic or use the peptide cotransmitters somatostatin and neuropeptide tyrosine, also figure prominently in the local striatal circuitry. Moreover, these larger neurons, unlike their smaller, more abundant neighbors, do not send their axons out of the striatum at all. Although detailed observations on striatal synaptology have been published recently, the computation performed by the striatum remains cryptic. The striatal cells seem to act something like a reconnaissance squad on a moonless night in a deep jungle acrawl with enemy troops; with a restraining hand on a shoulder, a finger to the lips, the troops remind one another to maintain strict silence while listening intently for the metallic click of a thumbed safety that will elicit their swift and violent response. So, too, does an eerie silence prevail in the striatum probed with a recording electrode. The cells are not languid, however, but fully alert and waiting to be set off by the signal sent down from the cortex.

The Striatum Projects to the Pallidum and Nigra

The axonal output of the striatum arises from the smaller cells and is directed to both segments of the globus pallidus and to both the pars compacta and the pars reticulata of the substantia nigra (Fig. 20.9). Although all the output neurons of the striatum are GABAergic, there appear to be at least two populations of output neuron based on their peptide cotransmitter: one that uses the inhibitory cotransmitter enkephalin and another that uses the excitatory cotransmitter substance P (and maybe dynorphin; Fig. 20.3). These two populations appear also to express different types of dopamine receptor, the substance P-containing group expresses D_1 receptors, whereas the enkephalin group expresses mainly D_2 receptors. Curiously, the two populations project differentially to the pallidum. The substance P axons distribute mainly in the internal segment of the globus pallidus and in the pars reticulata of the substantia nigra, whereas the enkephalin axons abound in the external segment of the globus pallidus. Although this dual peptidergic organization in the striatopallidal projection may relate to the distinct connections and functions of the two pallidal components (below), no one knows yet what it signifies for basal ganglia processing. What is known with certainty is that *the main effect of striatopallidal transmission is to inhibit pallidal neurons.*

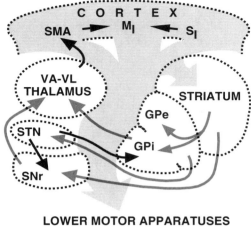

Figure 20.9 This diagram shows the motor-related circuit of the basal ganglia. The red arrows indicate inhibitory connections. Abbreviations: M$_I$, primary motor cortex; S$_I$, primary somatosensory cortex; SMA, supplementary motor area; other abbreviations as in Figure 20.3.

Outputs of the Basal Ganglia System

As the striatum is the input side of the corpus striatum, so the pallidum is the output side (Fig. 20.3). The major outflow originates from the GABAergic cells of the internal segment of the globus pallidus and is directed primarily at the thalamus. These pallidothalamic axons terminate in the ventral anterior-ventral lateral (VA-VL) nuclear complex and in the intralaminar nuclei (especially the centromedian nucleus). You'll recall from Chapter 18 that the thalamocortical projection of the VA-VL complex is directed at the supplementary motor area of the frontal lobe, where it can influence the planning of complex movement sequences and the corticospinal (pyramidal) motor system (Fig. 20.9). The GABAergic cells of the pars reticulata of the substantia nigra also receive striatal axons and, like the internal pallidal cells, project to the VA-VL complex. Remember that the cerebellothalamic axons also distribute in the VA-VL complex, but they do not contact the same population of neurons as the pallido- and nigrothalamic axons. Instead, the cerebellum contacts neurons that project to the premotor area. Since the pallidal projections to the thalamus are inhibitory, *the basal ganglia output serves to suppress the thalamic excitation of the supplementary motor area* (Fig. 20.9).

Another output of the corpus striatum that originates from cells in both the internal pallidal segment and the pars reticulata of the substantia nigra leads to a circumscribed region of the midbrain reticular formation called the **pedunculopontine nucleus** (Fig. 20.3). The efferent connections of this nucleus are not fully known, but it may provide a link with descending reticulospinal systems that would enable the corpus striatum to influence the spinal and brain stem motor circuits.

A third important output of the basal ganglia system is the GABAergic projection from the pars reticulata of the substantia nigra (but *not* from any part of the globus pallidus) to the deeper layers of the superior colliculus (Fig. 20.3). As you learned in Chapter 17, the superior colliculus is the source of signals that are essential for saccadic eye movements and eye–head coordination in gaze control. Recently, Robert Wurtz has shown that the nigrotectal neurons reduce their activity prior to and during the execution of saccadic eye movements. This release of the premotor tectal cells from inhibition may represent a mechanism by which the basal ganglia system can influence gaze.

The Subthalamic Sidetrack

The external segment of the globus pallidus projects mainly to the subthalamic nucleus, which, in turn, sends its axons back to both segments of the pallidum and to the pars reticulata of the substantia nigra (Fig. 20.10). Since the internal pallidum and pars reticulata are the sources of the corpus striatal outflow, it can be appreciated that *the subthalamic nucleus sits in a strategic position to modulate that output.* Significantly, the subthalamic nucleus also receives a direct input from the M_I cortex.

A clue to the role of the subthalamic nucleus comes from observations of the movement anomalies that ensue when the nucleus is damaged unilaterally by hemorrhage. The major anomaly, *hemiballismus,* is an uncontrollable flailing of the contralateral limbs (most commonly the arm). This clinical sign, together with the knowledge about its synaptic relationships, initially led to the hypothesis that the subthalamic nucleus *suppresses* the activity of the internal pallidum and pars reticulata, and that this activity is releas*ed* when the subthalamic nucleus is destroyed. This hypothesis received further support from the observation that

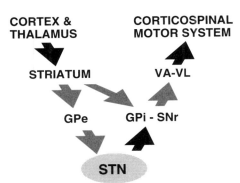

Figure 20.10 The circuit through the subthalamic nucleus (STN) modulates the inhibition of the corticospinal motor system. Jet black arrows represent glutamatergic, excitatory projections; red arrows represent GABAergic, inhibitory projections. Abbreviations: VA-VL, ventral anterior–ventral lateral nucleus; other abbreviatiuons as in Figure 20.3.

neurosurgical ablation of the internal pallidal segment of the VA-VL complex will cause the motor signs of hemiballismus to abate. However, the recent discovery that glutamate is the neurotransmitter used by the efferent axons of the subthalamic nucleus and that its action on pallidal neurons is excitatory has caused a re-evaluation of the suppression hypothesis. It now seems that *the excitatory transmission from the subthalamic nucleus to the output neurons of the pallidum serves to amplify their inhibitory influence over the thalamocortical neurons.* Thus, the subthalamic output serves indirectly to depress the thalamic excitation of the motor-related cortex.

The Role of the Basal Ganglia System in Movement

Although the contribution of the basal ganglia system to behavior remains very uncertain, its circuit properties and the dysfunctions that occur when it is damaged suggest that *it is important for the translation of higher cortical activity—perception, thought, and emotion—into volitional behavior.* What is the nature of the translation? Some have surmised that the basal ganglia system tonically suppresses movement and phasically releases specific movements when deemed appropriate by the cerebral cortex. Indeed, this has been an attractive notion to basal ganglia theoreticians for many years. In this scenario, the intent to act—a cognitive event—would generate a burst of activity in certain areas of the cerebral cortex that concern themselves with spatial perception and strategic movement planning. Following along in Figure 20.10, we see that this activity would be conveyed, in part, to the striatum, where it would phasically excite the ordinarily quiescent striatal neurons, which in turn would directly inhibit the pallidal and nigral neurons that tonically suppress the corticospinal system, thus "releasing" movement. With but a few more synaptic delays, the same burst of striatal activity would restore the tonic inhibition of the thalamus by way of the subthalamic sidetrack (striatal inhibition of the internal pallidum releases the subthalamic neurons to excite the inhibitors of the internal pallidal segment and the pars reticulata).

One argument that has been leveled against this theory is based on the experimental observation that striatal neurons related to particular limb movements do not modulate their activity until *after* a movement has begun. This finding, however, does not preclude the likely possibility that neurons in other (unrecorded) parts of the striatum, such as those in the head of the caudate, had already come on line *prior to* the initiation of the movement. These neurons receive their cortical input from the prefrontal areas that may be more concerned with elaborating complex behavioral strategies (see Chapter 28). Extending this notion, it can be imagined that the basal ganglia system may contribute to the necessary *sequencing* of individual movements that unfold during complex behaviors. After all, sequencing can be viewed as simply the preferential *initiation* of one movement after another.

A clinical clue that is consistent with this theory of the role of the basal ganglia system in behavior comes from clinicopathological observations in cases of *Huntington's disease* (made famous by one of its victims, the 1930s folk singer, Woody Guthrie). This disease is inherited as an autosomal dominant with complete penetrance. It is associated with a progressive degeneration of the smaller striatal neurons that constitute more than 95% of all striatal cells and are the neurons that give rise to striatal output. Prominent among the behavioral symptoms are excessive, spontaneous, involuntary movements of the hands and head. Because Huntington's disease is a progressive degenerative condition that takes decades to unfold, the total pathology involves degeneration of neurons in the cerebral cortex as well as all components of the basal ganglia. It is difficult to determine which lesions are primary and which are secondary. Moreover, because the damage is usually so extensive by the time a postmortem inspection is performed, it is difficult to correlate particular dysfunctions with specific lesions. However, it is notable that striatal degeneration from causes other than Huntington's disease has been described in which the dyskinesia consists of similar involuntary movements of the limbs and face. It seems, therefore, that the motor disturbances stem from the derailment of inhibitory striatal activity. The disease has an extremely poor prognosis and its symptoms are difficult to treat. Fortunately, carriers of the Huntington's gene can now be identified before the onset of dysfunction and, more importantly, before reproductive age by a screening technique that involves the mapping of restriction fragment length polymorphisms.

Evidence has also begun to accumulate that would implicate the basal ganglia in the *guidance of ongoing movements on the basis of perceptual feedback about their accuracy in achieving a desired outcome.* In this context, the close connection of the prefrontal association cortex with the caudate nucleus again seems significant; the prefrontal cortex exercises preferential control over a rather large volume of the striatum. One of the functional correlates of the prefrontal cortex, as we shall further explore in Chapter 28, is concerned with the planning of complex behavioral strategies and with *anticipation* of their potential outcomes. It could be that the command to execute a particular act originates in the far frontal regions of the cortex (after all, it has to originate somewhere) and activates a particular subterritory of the striatum that in turn releases a more caudal part of the cortex, the premotor cortex, from the inhibitory influence of the globus pallidus. Such a reverberating exchange between cortex and basal ganglia could assist with the planning of behavioral details by the premotor cortex and serve the ongoing guidance of skilled behavior by finely modulating motor cortical activity as the movement sequences unfold. In any case, it seems clear that the basal ganglia system provides a corollary processing circuit that serves cortical computations and enables the cortex to translate ideas into action and bring about successful behaviors.

Detailed Reviews

Albin RL, Young AB, Penney JB. 1989. The functional anatomy of the basal ganglia. *Trends Neurosci* 12: 366.

Alexander GE, Crutcher MD. 1990. Functional architecture of basal ganglia circuits: neural substrates of parallel processing. *Trends Neurosci* 13: 266.

DeLong MR. 1990. Primate models of movement disorders of basal ganglia origin. *Trends Neurosci* 13: 281.

VISCERAL SENSORIMOTOR CONTROL SYSTEMS

IV

Metabolism in the cells, tissues, and organ systems of the body requires a steady supply of oxygen and other nutrients, removal of waste, and the maintenance of ionic, osmotic, and temperature conditions within a relatively narrow physiological range. The CNS systems that ensure the integrity of these internal conditions are analogous to the systems that control skeletomotor behavior in the sense that they are organized hierarchically and depend on sensory feedback. In Chapters 21 through 24, we will analyze the organization of the visceral control systems. We'll begin, as with the skeletomotor systems, by identifying and following the relevant sensory data through the CNS. Next we'll examine some of the visceromotor reflex circuits that are organized at the level of the spinal cord and brain stem. Finally, we'll explore the higher-level control systems—the hypothalamus and the components of the limbic system—that regulate these lower visceromotor apparatuses. Along the way, it should become obvious that the visceromotor and skeletomotor systems are far from independent of one another. They work in concert most of the time.

Smell, Taste, and 21
Viscerosensation

The neural pathways that process the sensations of smell, taste, and viscerosensation are, at least initially, anatomically separate from one another. Nevertheless, each of these three senses ultimately provides essential data to the skeletomotor and visceromotor systems that maintain the internal environment (survival of the individual) and control reproduction (perpetuation of the species). In fact, the three usually work in concert to guide the acquisition of metabolic fuels as well as successful mating behavior.

Smell and Taste Depend on Chemosensors

Olfactory transducer cells are bipolar neurons that reside in the epithelium of the nasal mucosa (Fig. 21.1). Each transducer cell extends one of its two processes to the mucosal surface, where it ends as a knob from which springs a tentaclelike array of immobile cilia that are bathed in an ever-present film of mucus. The ciliary membrane is studded with metabotropic receptor proteins that associate with a unique form of GTP-binding protein called G_{olf} (Fig. 21.2) The receptors interact stereo-specifically with odorants dissolved in the mucus. When odorant molecules bind to such receptors, an intracellular signal transduction cascade, usually involving an olfactory-unique form of adenylyl cyclase, is initiated, leading to a transient opening of Na^+ channels. The depolarizing Na^+ flux that ensues is a generator potential, which upon exceeding threshold triggers impulses in the other process of the bipolar neuron, an axon that passes with others of its kind through one of several perforations in the thin cribriform plate of the ethmoid bone to enter and synapse in the olfactory bulb (Fig. 21.1).

The varied receptors for odorants are products of a large family of genes selectively expressed by the neurons in the olfactory epithelium.

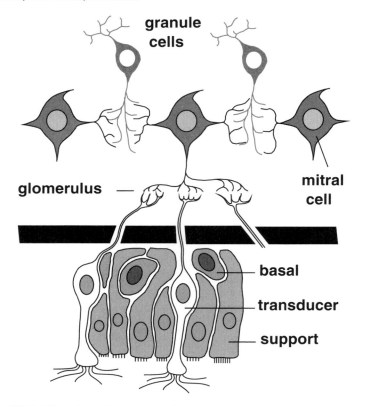

Figure 21.1   This drawing shows the structures involved in the earliest stages of olfactory processing. The horizontal black bar represents the ethmoid bone. Below it is the nasal epithelium, above it is the olfactory bulb. Axons of the bipolar transducer neurons in the epithelium pass through tiny openings in the bone to enter and synapse in the bulb. The tufted and periglomerular cells have been omitted for clarity.

Since single transducer neurons can respond to more than one (sometimes many) odorant molecule, it seems probable that each expresses some limited subset of receptor subtypes. Interestingly, olfactory bipolar neurons die after a life span of some sixty days and are replaced by new transducers that differentiate from nearby basal cells of the epithelium. Necessarily, the central synapses with mitral cells in the olfactory bulb are regularly lost and replaced. Moreover, there is no guarantee that the new transducers will express exactly the same set of odorant receptors as their predecessors. Paradoxically, this constant turnover of receptor subtypes and central connections comes at no apparent expense to the recognition of smells or to the often vivid memories they tend to evoke.

Taste transducer cells are also chemosensitive, but they are not neurons. They are specialized epithelial cells that occur in the tongue, palate, pharynx, epiglottis, and upper esophagus, where they cluster with one

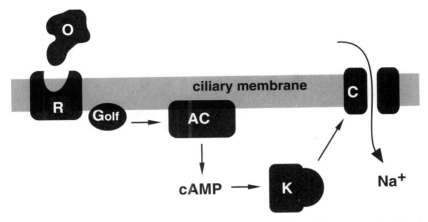

Figure 21.2 Olfactory receptors couple to GTP-binding protein (G_{olf}) and influence ion channels through second messenger pathways. Abbreviations: AC, adenylyl cyclase; C, ion channel; K, protein kinase; O, odorant molecule; R, odorant receptor protein.

another and with other types of epithelial cells in organs called **taste buds** (Fig. 21.3). The taste buds are found most often in larger organs of variable size and shape called **papillae.** Microvilli extend from the apical ends of the taste transducer cells into a shallow depression of the epithelial surface. The transducers respond to stimulus molecules that are dissolved in the saliva. Each transducer appears to respond preferentially to one of four primary classes of stimulus molecule that are

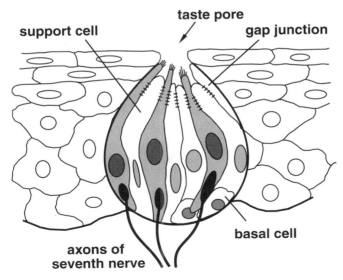

Figure 21.3 Taste-sensitive cells are contained in the taste buds that are embedded in the oral epithelium.

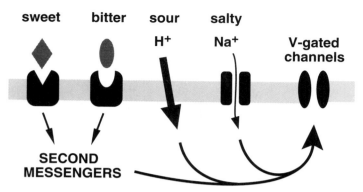

Figure 21.4 Different tastes are transduced by different molecular mechanisms.

perceived as *sweet, bitter, salty,* or *sour.* The transducer cells that respond to sweet and bitter stimuli express receptors like those of the olfactory transducers (Fig. 21.4). That is, the sapid molecules bind to receptors that are linked to second messenger systems that result in the opening of cationic channels causing correlated changes in membrane potential. Acid molecules, which induce sour tastes, and ions (especially Na^+), which induce salty tastes, do not act through such receptors. Instead, the acids readily diffuse through the membrane of the transducer cells and alter ion channel conduction; ions can directly alter membrane potentials by entering the transducer cells through voltage-independent (leaky) channels.

Neural fibers enter the taste buds to lie in contact with the base of the receptor cells; there is membrane juxtaposition, and the presynaptic element (the transducer cell) contains synaptic vesicles much like the hair cells of the vestibular transduction apparatus. Several receptor cells are innervated by single, highly branching nerve fibers so that the activity in a taste axon is the product of transduction by several receptor cells. Moreover, single taste fibers respond to different taste qualities, but each fiber has a "preferred" taste quality to which it responds most readily. That is, the fibers do not show absolute selectivity, but they do discriminate among the different taste qualities. This mode of sensory processing is in principle no different from that used by the other sensory systems. Central processing stations in the taste system must *compare* the activities in the array of active afferent axons in order to code the wide range of flavors we can discriminate.

Central Olfactory Pathways

Upon entering the cortex of the olfactory bulb, the small, nonmyelinated axons of the olfactory nerve synapse on the dendrites of large **mitral cells** and small **tufted** and **periglomerular cells** in an elaborate configuration called a **glomerulus** (Fig. 21.1) The periglomerular cells and a fourth cell type in the bulb, the **granule cells,** form a dense local meshwork of inhibitory connections and maintain a potent tonic inhibition of the output neurons of the bulb. The granule cells accomplish this inhibition through bidirectional dendrodendritic synapses with the mitral cells; the mitral cell dendrite releases an excitatory neurotransmitter that triggers the release of an inhibitory transmitter from the granule cell dendrite. Impulses arriving over the olfactory nerve fleetingly penetrate this veil of inhibition to trigger impulses conveyed by the axons of mitral and tufted cells, which project through the **olfactory tract** to several distant targets, largely cortical in structure, in the base of the forebrain (Fig. 21.5).

One such target is the **anterior olfactory nucleus,** which lies near the point of attachment of either olfactory tract to the base of the frontal lobe. Neurons in this nucleus send axons to the contralateral olfactory bulb by way of the anterior commissure. The bulk of the axons of the olfactory tract pass beyond the anterior olfactory nucleus and fan out to terminate in a small patch of cortex called the **piriform cortex** (Fig. 21.5). This primary olfactory cortex forms a continuous sheet extending from the base of the frontal lobe onto the rostromedial aspect of the temporal lobe (the uncus). Axons from the olfactory bulb also terminate in the olfactory tubercle, the corticomedial amygdaloid nucleus, and parts of the parahippocampal gyrus caudal to the uncus (the entorhinal cortex).

The connections of these secondary olfactory processing stations are diverse and of obscure functional significance. Many, such as the amygdala, parahippocampal cortex, and olfactory tubercle, are closely associated with the hypothalamus. Such a relationship is intuitively appealing, since the hypothalamus plays a crucial role in the orchestration of visceromotor and endocrine activities (see Chapter 23). The piriform cortex and amygdala

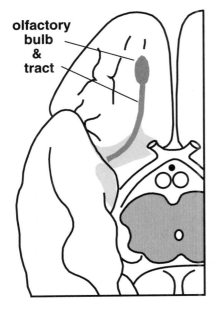

olfactory bulb & tract

Figure 21.5 In this schematic of a ventral view of the right cerebral hemisphere, the olfactory or piriform cortex (red shaded area) lies partly on the orbital surface of the frontal lobe (anterior perforated substance) and partly on the rostromedial surface of the temporal lobe (uncus).

send axons to the mediodorsal nucleus of the thalamus, which in turn projects to the prefrontal association cortex. Anecdotal clinical observations and less than definitive experimental data suggest that this cortex is involved with the elaboration of complex behavioral strategies, on the one hand, and social behaviors, on the other (see Chapter 28).

The Central Taste and Viscerosensory Pathways

The nerve fibers that innervate taste transducers travel in cranial nerves 7, 9, and 10, and their cell bodies are located in the sensory ganglia associated with these nerves (Fig. 21.6). Such axons form their first

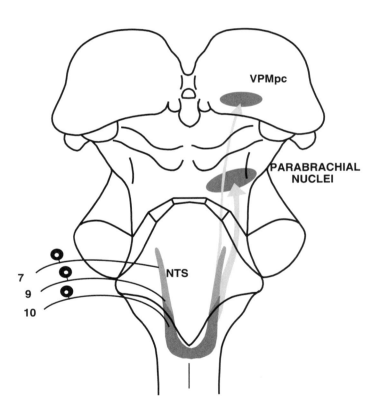

Figure 21.6 The nucleus of the solitary tract (NTS) receives primary axons of the 7th, 9th, and 10th cranial nerves (left) and projects ipsilaterally to the parabrachial nuclei and the parvicellular part of the ventral posteromedial nucleus (VPMpc) of the thalamus (right). The rostral NTS (lighter shading) is dominated by taste input, whereas the caudal part (darker shading) is dominated by viscerosensory information.

central synapse in the elongated **nucleus of the solitary tract** or solitary nucleus, which extends through the entire medulla and into the caudal pons (Fig. B2). Axons conveying taste information synapse at rostral levels of the solitary nucleus, a part reasonably called the *gustatory nucleus*. Many fibers conveying taste information travel in the *chorda tympani*, a branch of the facial (7th) nerve, from transducers on the rostral two-thirds of the tongue. Taste receptors of the palate are also innervated by fibers of the facial nerve, its *greater superficial petrosal branch.* The cell bodies of these facial nerve axons lie in the **geniculate ganglion.** The *lingual branch* of the glossopharyngeal nerve carries the taste fibers from the caudal one-third of the tongue. The **superior laryngeal nerve,** a branch of the vagus, conveys taste information from the epiglottis and the upper part of the esophagus.

After taste information reaches the rostral part of the nucleus of the solitary tract, it is sent over an ipsilateral, ascending pathway to a special parvicellular part of the ventral posteromedial nucleus (VPMpc) of the thalamus, and from there to the primary gustatory cortex (Fig. 21.7). The gustatory cortex is located rostrally adjacent to the oral somatosensory area at the base of the precentral gyrus (in Brodmann's area 3) and

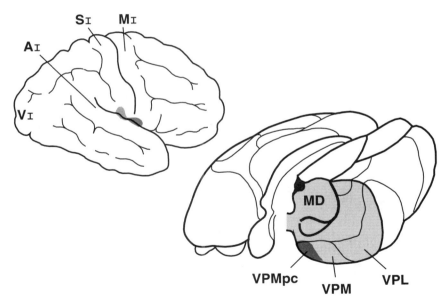

Figure 21.7 The primary taste cortex (red) receives axons from the parvicellular part (VPM$_{pc}$) of the ventral posteromedial nucleus (VPM) (also red), the thalamic target of axons ascending from the rostral part of the nucleus of the solitary tract. Abbreviations: A$_I$, primary auditory cortex; M$_I$, primary motor cortex; MD, mediodorsal nucleus; S$_I$, primary somatosensory cortex; V$_I$, primary visual cortex; VPL, ventral postero-lateral nucleus.

extends as a continuous sheet onto the upper bank of the lateral fissure and then onto the superior part of the insular cortex. Although responses can be evoked in the gustatory cortex by application of sapid stimuli to the tongue, very little is known about how this cortex is organized or about the transfer of taste information beyond the gustatory cortex. Some axons from the gustatory part of the solitary nucleus make local, reflex connections in the medulla, mainly by way of cells in the reticular formation.

More caudal levels of the solitary nucleus receive viscerosensory input from transducers in the aortic arch, the carotid body, the wall of the right atrium, the lungs, the gastrointestinal tract, and the liver. Axons conveying this information travel in the 9th and 10th cranial nerves. Additionally, much of the sensory innervation of the viscera is mediated by afferent axons that synapse in the spinal cord. Although there is a high probability that many spinal cord visceral afferents influence the visceromotor neurons, there are few conclusive data about the nature or the transmission of visceral information in the spinal cord. Unfortunately, this is only slightly less true of the brain stem viscerosensory circuits. The secondary viscerosensory neurons in the caudal part of the solitary nucleus project to two major target areas in the brain stem, the parvicellular reticular formation (especially the part located in the ventrolateral medulla) and the **parabrachial nuclei** of the pontine tegmentum (Figs. 21.6 and B5). The projections to the reticular formation are especially important in a number of vital blood pressure and heart rate reflexes involving the medullary and spinal cord visceromotor nuclei (see Chapter 22). The parabrachial cells probably also play a role in such reflexes by way of descending projections to the visceromotor nuclei, but they also convey viscerosensory data to higher levels of the brain, especially structures associated with the limbic system and hypothalamus (Fig. 21.8), which may influence emotions, drives, and the higher integration of visceromotor behavior (see Chapters 23 and 24). It does not appear that raw viscerosensory signals reach the cerebral cortex, which perhaps explains our failure to consciously appreciate such internal states as blood pressure or oxygen tension.

Smell, Taste, and Viscerosensations Guide Appetitive, Consummatory, and Reproductive Behaviors

Anyone who has seen the family dog adrool over the aroma of breakfast bacon appreciates that smell exercises forceful control over feeding and drinking behavior. Both smell and taste enable the selection of nutrients for consumption and the rejection of harmful substances (usually bitter-tasting) before ingestion. Incidentally, much of what we perceive as taste in fact depends on olfactory stimulation. Recall how bland even the most

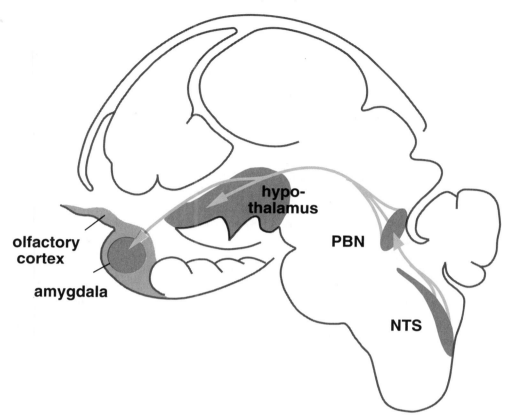

Figure 21.8 Viscerosensory and olfactory sensation have relatively immediate access to the hypothalamus and to limbic circuits by way of the amygdala. Abbreviations: NTS, nucleus of the solitary tract; PBN, parabrachial nuclei.

piquant stew tastes when a head cold has caused occlusion of the nasal passages and olfaction is lost (*anosmia*). Smell and taste also serve to trigger gastrointestinal (visceral) reflexes requisite to ingestion (salivation) and digestion (enzyme secretion, gut motility) of food. Viscerosensory signals that arise in the gut and liver contribute to the reflex secretion of digestive enzymes, gut motility, and shunting of blood to the gastrointestinal tract. Viscerosensory signals also play a role in triggering feeding and drinking as well as in curtailing these behaviors when sufficient food and fluid has been ingested.

Many taste preferences appear to be inborn, whereas others are clearly learned. Animals have repeatedly been shown to have the capacity to compensate for dietary deficiencies by selecting foods that contain the missing nutrients. Although most taste preference in adult humans is based on long experience, young children have been observed to display *specific hungers.* For example, sodium-deficient children (made so by the absence of the adrenal cortex and a resultant inability to maintain salt balance) show a craving for salt when allowed free access to a variety of foods. The mechanism for this phenomenon of specific hunger is unknown. Experiments in rats suggest that salt deprivation results in a decreased responsivity to sodium in primary afferent axons, even though their threshold sensitivity is not changed; higher concentrations are required for detection. Learned taste preferences are easily demonstrated experimentally. When delivery of a food is paired with a tone and nausea is subsequently induced by mildly poisoning the food, rats will learn to avoid eating the food in future trials but show no aversion to the tone. Thus, the rats seem to instinctively associate gastrointestinal distress with the taste (and smell) of food but not with auditory stimuli; other foods are not avoided when paired with the tone.

The sense of smell is essential to mating behavior in animals. In many insects, the individuals, usually the females, secrete odorant molecules called **pheromones** that attract potential mates. In estrous mammals, vaginal aromas function like pheromones; they signal the stage of the estrous cycle, informing a potential suitor whether the female is likely to welcome or rebuff his sexual overtures. The right fragrance can arouse the human libido as well. Clearly, the perfume industry thrives not only because their products satisfy the compulsion of civilized men and women to mask their bodily effluvia, but also because they promise potent erotic allure.

Detailed Reviews

Finger TE, Silver WL (eds). 1987. *Neurobiology of Taste and Smell.* Wiley, New York.
Loewy AD. 1990. Central autonomic pathways. In Loewy AD, Spyer KM (eds), *Central Regulation of Autonomic Functions.* Oxford University Press, New York.
Norgren R. 1984. Central neural mechanisms of taste. In Darian-Smith I (ed), *Handbook of Physiology, Sec 1: The Nervous System, Vol III: Sensory Processes, Part 2.* American Physiological Society, Bethesda, MD.

Visceromotor Reflexes 22

When we play basketball, our cardiac output is increased, we sweat, kidney filtration is reduced, and digestion is all but suspended. In contrast, after a heavy meal, gastric motility and the secretion of digestive enzymes are increased and blood flow is selectively diverted to the gut. The visceromotor systems of the CNS regulate these internal responses through direct actions on the contraction of cardiac and smooth muscle and glandular secretions. These effector apparatuses, in turn, adjust blood flow, breathing, digestion, waste elimination, and temperature. The visceromotor systems are also essential for the appropriate responses of the sex organs during mating behavior. Because of their critical involvement in these general functional domains, the visceromotor systems are said to control the bodily responses during the "three Fs"—*fighting, fleeing,* and. . . .

The sensory triggers and feedback that enable the visceromotor systems to function properly are primarily those described in the preceding chapter. In the present chapter, we'll see more specific examples of how viscerosensory signals influence the visceromotor output of the CNS. In the two chapters that follow this one, we'll explore the higher brain regions, such as the hypothalamus, that exert a higher control on the visceromotor systems. For now, let's explore the functional organization of the lower visceromotor apparatuses as we did earlier for the skeletomotor system.

Visceromotor Neurons

Central visceromotor neurons differ from skeletomotor neurons in important ways (Fig. 22.1). They are smaller than most α motoneurons, and they are segregated from α and γ motoneurons into separate cell groups in the spinal cord and brain stem (below). Like other motoneurons, they use the neurotransmitter ACh at their peripheral synapses, but they also release one or more of a large number of peptide cotransmitters and other types of molecules that may serve a signaling function (e.g., ATP).

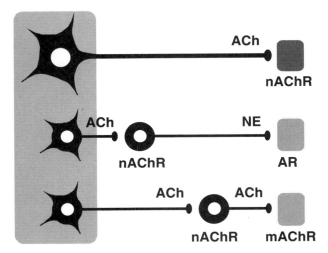

Figure 22.1 The diagram summarizes the fundamental similarities and differences between skeletomotor (top) and visceromotor (middle and bottom) neurons, and between the sympathetic (middle) and parasympathetic (bottom) divisions of the peripheral visceromotor neurons. The shaded area to the left is the CNS, and the shaded boxes to the right are the peripheral effector tissues. Abbreviations: ACh, acetylcholine; AR, adrenergic receptors; nAChR, nicotine acetylcholine receptors; NE, norepinephrin; mAChR, muscarinic acetylcholine receptor.

Whereas α and γ motoneurons innervate striated skeletal muscle, *the visceromotor efferent axons terminate only on neural crest-derived neurons that* reside in assorted ganglia (except for the chromaffin cells of the adrenal medulla, which are also derived from neural crest). The ACh receptors that are expressed by the ganglionic neurons are of both the *nicotinic* (nACh-R) and the *muscarinic* (mACh-R) subtypes. The ganglionic neurons also express peptide receptors. Although *the predominant response of the ganglionic neurons to input from the visceromotor neurons is a rapid EPSP mediated by the nACh-R,* they also show slower-onset, complex EPSP-IPSP responses mediated by the mACh-R and peptide receptors. *The axons of the ganglionic neurons innervate the effector tissues* (smooth and cardiac muscle and secretory glands) of the body. And, unlike the invariably excitatory neuromuscular junctions, the *ganglionic axons can either excite or inhibit the effector tissues,* depending on the receptor subtypes expressed by the target cells.

Divisions of the Peripheral Visceromotor System

The peripheral visceromotor system is often referred to as the *autonomic nervous system* because it is largely independent of voluntary control. The

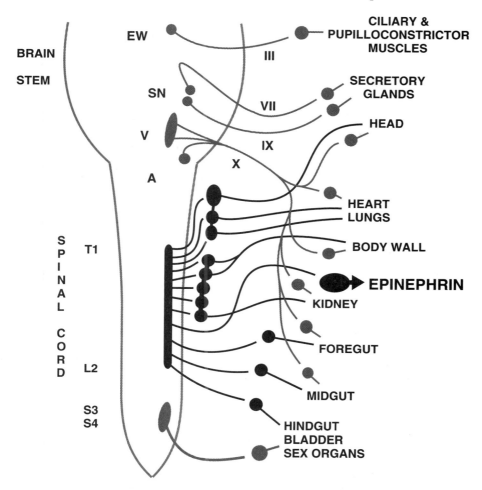

Figure 22.2 Shown in the diagram are the central locations and general peripheral domains of the sympathetic (jet black) and parasympathetic (gray) divisions of the visceromotor system. Abbreviations: A, nucleus ambiguus; EW, Edinger–Westphal nucleus; SN, salivatory nuclei; V, vagal motor nucleus.

autonomic ganglia are functionally divisible into separate **sympathetic, parasympathetic,** and **enteric** groups (Fig. 22.2). The sympathetic ganglia constitute the paravertebral chain, the cervical and the prevertebral ganglia, whereas the parasympathetic ganglia are more numerous and lie in or near the tissues they innervate. The enteric ganglia are diffuse, netlike collections of neuronal cell bodies and fibers that occur in the gut wall (see below). All of the ganglionic sympathetic neurons use no-repinephrin as a neurotransmitter, with the single exception of those that innervate sweat glands, which use ACh. Norepinephrin interacts with both α and β adrenergic receptors expressed by effector tissues. The neurotransmitter of all ganglionic parasympathetic neurons is ACh, which most often interacts with mACh-Rs expressed by the innervated target tissues. In addition to these main neurotransmitters, ganglionic neurons of both the sympathetic and parasympathetic groups and the enteric neurons release peptide cotransmitters.

Since the central visceromotor neurons that innervate, respectively, the sympathetic and parasympathetic ganglia are segregated into separate nuclei, they also can be classified as sympathetic and parasympathetic. The intermediolateral nucleus appears throughout the thoracic and upper lumbar segments of the spinal cord as a small, distinct spur of the central gray matter (Fig. 22.1). All of the visceromotor neurons that innervate the *sympathetic* ganglia are contained in this column. Some of these sympathetic visceromotor neurons innervate the chromaffin cells of the adrenal medulla rather than ganglionic neurons. Like ganglion neurons, adrenal chromaffin cells are derived from the neural crest. Importantly, the chromaffin cells are also biochemically similar to ganglionic sympathetic neurons in that they synthesize catecholamines. *The chromaffin cells secrete epinephrin* (adrenalin) *into the systemic circulation; this epinephrin activates adrenergic receptors and augments the sympathetic responses mediated by the tissue-specific release of norepinephrin from ganglionic sympathetic axons.*

In sacral segments of the cord, in a position analogous to the intermediolateral column (but not readily visible as a discrete spur), are the *parasympathetic* visceromotor neurons. The axons of these neurons, like the visceromotor axons that emanate from the parasympathetic nuclei of the brain stem, are generally longer than sympathetic visceromotor axons because they must reach more distantly located ganglia. Because of the anatomical distribution of the central sympathetic and parasympathetic visceromotor neurons, they are often collectively called the *thoracolumbar* and *craniosacral* components of the visceromotor system. In either case, the central visceromotor neurons integrate a great many inputs from other parts of the brain, mostly by way of nearby interneurons, and are thus the final common pathway of the visceral effectors.

Visceromotor Reflexes

Both the sympathetic and the parasympathetic systems work largely on the basis of short- and long-loop reflex connections and require no attention or volitional input. In general, the sympathetic system acts more globally than the parasympathetic. It is highly responsive to changes in the external environment that might be harmful or life-threatening. It also mediates the constellation of responses to increased internal demands for blood flow, gas exchange, and heat dissipation during strenuous activity. In contrast, the parasympathetic system augments

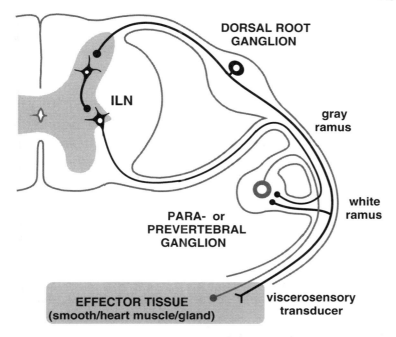

Figure 22.3 The diagram shows the basic organization of lower sympathetic reflex circuits. The ganglionic neuron is shown in gray for contrast. Abbreviation: ILN, interomedial nucleus.

vegetative activities such as gastric secretion and motility, and restores heart rate and blood pressure to resting levels. Because of this general dissimilarity in function, it has often been said that the sympathetic nervous system mediates the "fight or flight" responses, whereas the parasympathetic mediates the "rest and digest" responses.

Most (but not all) effector tissues are innervated by both the sympathetic and parasympathetic divisions and are thus coordinately regulated by the two. In some instances, such as the baroreflex described below, the two divisions appear to exert *opposing* effects on the effector system. In other cases, such as the reflex control of salivation, their actions are *synergistic*. In still other cases, such as bladder control and urination, they are closely related to a coordinate output of α motoneurons that innervate striated muscle of the urethral sphincters.

The basic lower circuit of sympathetic reflexes is shown in Figure 22.3. It is notable that the peripheral neurons of the sympathetic ganglia often receive direct synaptic input from collateral branches of primary viscerosensory axons. This very short circuit, which does not occur in the skeletomotor control systems, indicates again the highly *primal* and *peripheral* nature of the life-sustaining visceromotor systems. It is reminiscent of the primitive two-neuron nervous system of the simple jellyfishes described in Chapter 1. The central portion of the sympathetic

reflexes is most often mediated by local circuit interneurons of the intermediate zone that are interposed between the primary viscerosensory axons and the visceromotor neurons.

The Bladder Reflex

A good illustration of a lower visceromotor reflex is the control of the urinary bladder. As the bladder fills, the sensory feedback from the stretch transducers in the bladder wall triggers an excitation of the sacral visceromotor neurons. The discharge of these parasympathetic motoneurons activates the ganglionic neurons that innervate the smooth muscle of the bladder, producing contraction. At the same time, the α motoneurons (skeletomotor) that control the external urethral sphincter are inhibited so that the sphincter tone is relaxed. A vigorous urination occurs. In most of us beyond the age of two or three, this reflex is brought under the control of descending systems that block its activation until we can get to the rest room or find a concealing bush. In paraplegic patients who have suffered a spinal lesion above the sacral level, the bladder is released from descending control. A condition known as *automatic bladder* ensues in which the voiding reflex is triggered by the sensory input from bladder distention. Automatic bladder is distinct from the effects of direct damage to the parasympathetic motoneurons in the sacral cord or transection of the pelvic splanchnic nerve, which innervates the bladder. The latter lesions will produce a condition called *atonic bladder* in which the bladder fills excessively due to the absence of parasympathetic activation of the voiding reflex. The patient must be catheterized.

The Baroreflex

Blood pressure is normally regulated by a negative-feedback reflex circuit (Fig. 22.5). This baroreflex is a good illustration of a coordinated visceromotor response and is important for medical students to understand, since the medicinal treatment of heart disease is often based on pharmacological manipulation of this reflex circuit. The sensory impulses that trigger the reflex originate from nerve endings that monitor the degree of stretch in the walls of the carotid sinus and aorta. These mechanosensory fibers, which travel in the glossopharyngeal and vagal nerves, are tonically active and modulate their activity up or down according to changes in blood pressure that manifest as distention of the vessel walls (Fig. 22.4). The barotransducer afferent fibers terminate on neurons in the caudal end of the nucleus of the solitary tract. Solitary nucleus neurons, in turn, make both direct and indirect synaptic contacts on parasympathetic visceromotor neurons that reside in close association with the nucleus ambiguus. The indirect connections, which are in

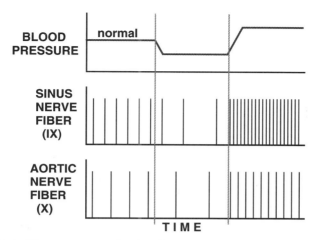

Figure 22.4 The response of individual nerve fibers to changes in blood pressure can be recorded in the glossopharyngeal (IX) and vagal (X) nerves.

the majority, are mediated by neurons in the parvicellular reticular formation of the medulla. The activation of periambigual visceromotor neurons decreases both heart rate and force of contraction via postganglionic cholinergic effects on the mACh-Rs expressed by cardiocytes.

Concurrent with the increase of parasympathetic tone, the secondary sensory neurons of the solitary nucleus reduce sympathetic output over a more complex circuit that involves cells in a special *ventrolateral part of the medullary reticular formation* (Fig. 22.5). The ventrolateral medulla has been known for several years to be critically involved in cardiovascular regulation. If the caudal part of this region is destroyed in rats, arterial pressure is permanently increased. Conversely, local injection of L-glutamate causes depressor effects. This caudal part of the ventrolateral medulla sends inhibitory GABAergic axons to a slightly more rostral part, where several neurons are found that exhibit pacemaker activity and project to the sympathetic visceromotor neurons of the intermediolateral nucleus; the neurons of the rostral ventrolateral medulla augment sympathetic tone. The activation of the inhibitory neurons in the caudal part of the ventrolateral medulla abolishes the pacemaker activity of the neurons in the rostral part and thus decreases the excitatory sympathetic output to the heart. Sympathetic stimulation of adrenal chromaffin cells, which tends to accelerate the heart through the systemic release of epinephrin, is also depressed. Finally, the tone of arterial smooth muscle, which is innervated by noradrenergic sympathetic ganglionic neurons but not by cholinergic parasympathetic neurons, is decreased, leading to vasodilation and decreased vascular resistance. Thus, the net effect of increasing blood pressure is to trigger a negative feedback circuit that decreases heart output and vascular resistance.

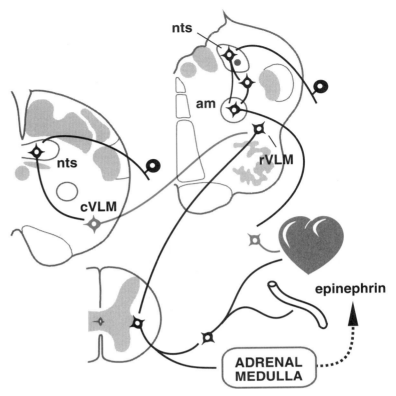

Figure 22.5 The basic circuit of the baroreflex is diagrammed. Stretch in the walls of the carotid sinus or aorta activates mechanotransducers that send an excitatory afferent volley to the nucleus of the solitary tract (nts) via the vagus and glossopharyngeal nerves. The red neurons are inhibitory. Abbreviations: am, nucleus ambiguus; cVLM, caudal ventrolateral medulla; rVLM, rostral ventrolateral medulla.

The Enteric Nervous System

The enteric division of the autonomic nervous system is composed of two concentric sheets of neurons within the wall of the gut (Fig. 22.6). The outer sheet lies between the layer of longitudinal smooth muscle and the layer of circular smooth muscle. It's called the **myenteric plexus.** The inner sheet lies subjacent to the mucosal epithelium; call it the **submucosal** plexus. Within each sheet, there are clusters of neuronal cell bodies amidst a network of processes that enable communication between the clusters. It's a design not unlike the primitive nerve net of the jellyfishes. The enteric nervous system functions autonomously; all of the basic gut functions occur in the absence of extrinsic autonomic innervation.

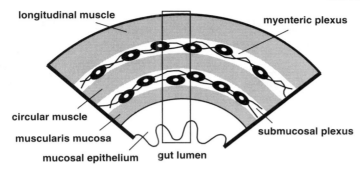

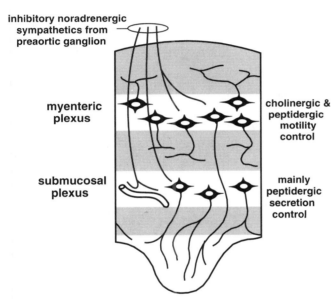

Figure 22.6 The top drawing shows a wedge of the intestine with its concentric layers of muscle (shaded) and neural plexuses surrounding the mucosal epithelium of the gut lumen. The portion contained within the rectangle is shown at higher resolution in the lower drawing.

The myenteric plexus is primarily responsible for gut motility. Its neurons send processes into both the longitudinal and the circular muscle layers, where they release ACh and excite contraction. The submucosal plexus influences motility indirectly by way of axons that extend to the neurons of the myenteric plexus. But this motility function is secondary to their major role in the control of epithelial functions. *The submucosal plexus activates water and ion transport across the epithelial cells and secretion of digestive juices from crypt cells.* Although some of the neurons in the submucosal plexus are cholinergic, they all *release one or more of a large variety of gut peptides* that control the epithelial functions in complex and subtle ways.

Both the myenteric and the submucosal plexuses are innervated by axons that arise from the sympathetic and parasympathetic ganglia. The sympathetic axons also end on blood vessels in the gut wall and regulate local circulation. Sympathetic noradrenergic transmission inhibits enteric neurons and thus decreases gut peristalsis and secretion. The cholinergic parasympathetic innervation augments both gut functions.

Detailed Reviews

Appenzeller O. 1990. *The Autonomic Nervous System*. Elsevier, Amsterdam.
Furness JB, Costa M. 1987. *The Enteric Nervous System*. Churchill Livingstone, New York.
Loewy AD, Spyer KM. 1990. *Central Regulation of Autonomic Functions*. Oxford University Press, New York.

Hypothalamic Regulation of the Internal Environment 23

Metabolism in the cells, tissues, and organ systems of the body requires a steady supply of oxygen and nutrients, waste removal, and the maintenance of ionic, osmotic, and temperature conditions within a relatively narrow physiological range. The hypothalamus performs many ongoing computations that are essential to the maintenance (homeostasis) of these internal conditions. It regulates circulation, respiration, digestion, and excretion, as well as body temperature and the relative concentrations of extracellular fluid components. It also regulates primal appetitive behaviors, such as eating and drinking, that have direct bearing on the internal environment. During development, the hypothalamus influences growth. Finally, the hypothalamus is essential to reproductive functions, such as gamete production, mating, parturition, and mothering, which perpetuate not the individual but the species. To accomplish these varied functions, the hypothalamus receives appropriate sensory input (visceral, somatic, olfactory, gustatory, visual) and expresses its output through both *synaptic* and *hormonal* routes. The synaptic routes lead, on the one hand, to the preganglionic efferent neurons of the lower visceromotor systems and, on the other, to parts of the brain that control skeletomotor behavior. The *hormonal* route is a direct regulation of the secretions of the pituitary gland. Most often, the hypothalamus exerts its influence over all three of these avenues simultaneously in a carefully choreographed output.

Structural Organization

The hypothalamus is the smaller part of the diencephalon, and like the thalamus above, it lies on either side of the narrow third ventricle (Figs. 23.1, B8, and B9). Its borders are ill-defined, but roughly speaking, the hypothalamus extends rostrocaudally from the lamina terminalis to

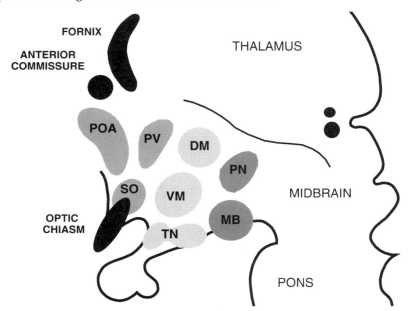

Figure 23.1 This schematic midsagittal section through the hypothalamus shows the rostrocaudal zones and some of the major nuclei. Abbreviations: DM, dorsomedial nucleus; MB, mammillary body; PN, posterior nucleus; POA, preoptic area; PV, paraventricular nucleus; SO, supraoptic nucleus; TN, tuberal nucleus; VM, ventromedial nucleus.

the caudal edge of the **mammillary bodies** (Figs. 23.9 and B8), which are themselves hypothalamic nuclei. The region immediately rostral to the optic chiasm, extending dorsalward to the anterior commissure, is called the **preoptic area.** This rostralmost part of the hypothalamus merges indistinctly with the ventral forebrain. At its caudal end, the hypothalamus likewise merges insensibly with the central gray substance and the reticular formation of the midbrain tegmentum. From its middle levels caudalward, the lateral border of the hypothalamus on either side is more or less defined by the medial edge of the internal capsule (Fig. 23.2). Rostrally, its lateral extensions blend into the ventral forebrain (Figs. 23.5 and B9). Dorsally, the hypothalamus fuses with the subthalamic area (Fig. 23.9).

The hypothalamus can be divided both anatomically and functionally into *medial* and *lateral* parts, which are separated by a sagittal plane passing through the descending fornix (Figs. 23.2 and B9). The medial hypothalamus consists of several nuclei and more diffusely arranged neurons and is the location of the bulk of the neurons that control pituitary secretions. The lateral hypothalamic area has few, if any, discrete nuclei. Instead, it is a diffuse collection of cells scattered along the **medial forebrain bundle** and serves as part of a multineuronal system

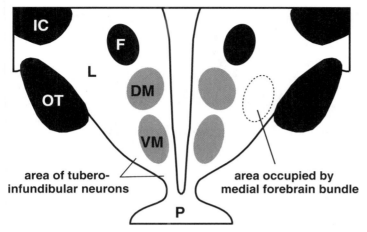

Figure 23.2 This sketch of a frontal section through the hypothalamus shows some of the major nuclei that occupy its middle zone. Abbreviations: DM, dorsomedial nucleus; F, fornix; IC, internal capsule; L, lateral hypothalamic area; OT, optic tract; P, pituitary; VM, ventromedial nucleus.

linking the basal forebrain with the reticular formation of the midbrain tegmentum (see Chapter 24). The lateral hypothalamus is, therefore, in a position to receive information from both ends of what has been referred to as the *septohypothalamomesencephalic continuum,* and can relay impulses to more medial hypothalamic nuclei by way of short intra-hypothalamic axons. Thus, *the lateral hypothalamus resembles the brain stem reticular formation in several anatomical and functional respects.* The circuitry in which the hypothalamus is involved is extremely complex; even the architecturally distinct nuclei of the hypothalamus are heterogeneous in function, histochemical composition, and connectivity. The medial hypothalamus, with its more discrete nuclear arrangements, can be divided into *four* rostrocaudal zones (Fig. 23.1)—*preoptic, rostral, middle,* and *caudal*—that are to some extent associated with distinct functions (see below).

Major Tracts and Connections

The four major axonal inroads to the hypothalamus are the fornix, the **stria terminalis,** the **ventral amygdalofugal pathway** and the medial forebrain bundle. These pathways connect different parts of the hypothalamus with other nuclei of the fore- and midbrain (Fig. 23.3). The fornices carry fibers from cells of the hippocampal cortex, especially a part called the **subiculum.** As the columns of the fornix penetrate the ventral forebrain, they split into pre- and postcommissural portions,

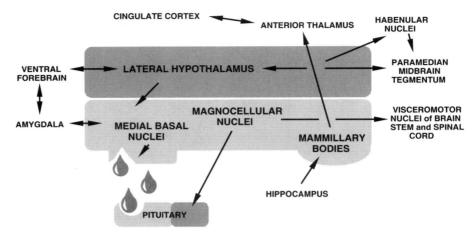

Figure 23.3   This diagram summarizes the major parts of the hypothalamus and their synaptic relations to nuclei in the fore- and midbrain.

which pass, respectively, rostrally and caudally to the anterior commissure (see Fig. 24.3). The precommissural portion distributes to the septal nuclei in the ventral forebrain. As one follows the larger postcommissural bundles of the fornix along their course, they seem to be directed preferentially at the mammillary nuclei, where indeed a large number of their axons make synaptic contacts. However, many of the fornix fibers distribute throughout the preoptic area and rostral and middle levels of the hypothalamus. Some fibers of the fornix turn dorsalward early in this main trajectory to reach the anterior nuclei of the thalamus. Moreover, other axons of the fornix extend caudalward beyond the mammillary bodies to the paramedian part of the midbrain tegmentum.

The stria terminalis carries fibers from the amygdala mainly to the medial preoptic area and more caudal levels of the medial hypothalamus. Like the fornix, the stria terminalis takes a long arching course, following the curve of the lateral ventricle on either side. The amygdala also sends axons to the lateral hypothalamus by way of the ventral amygdalofugal pathway, which takes a less circuitous route than the stria terminalis, passing medially across the ventral forebrain (see Fig. 24.6).

The medial forebrain bundle carries descending fibers from olfactory structures, the septal nuclei, the nucleus accumbens, and the cerebral cortex, and ascending fibers from the midbrain tegmentum. It also carries many axons of hypothalamic origin that enter and exit all along its course through the lateral hypothalamus. This bundle is far less discrete than the fornix and stria terminalis both in its compactness and in the diversity of axons that travel within it.

The major outflow of the hypothalamus travels in the **mammillothalamic tract, stria medullaris, mammillotegmental tract,** and medial forebrain bundle of either side. The mammillothalamic tract carries axons from the mammillary complex to the anterior nuclear group of the thalamus, in part continuing the flow of information conveyed by the fornix from the hippocampus (Fig. B9). The stria medullaris carries axons from the lateral preoptic area and lateral hypothalamus to the **habenular nuclei,** which are themselves involved in connections to other parts of the brain stem. The mammillotegmental tract carries axons from the mammillary complex and other parts of the hypothalamus to the paramedian tegmentum of the midbrain. Through these efferent synaptic routes, the hypothalamus can influence the lower visceromotor apparatuses and the higher brain activities that guide motivated behavior.

Temperature Regulation

The monitoring of internal conditions and the elaboration of a choreographed visceromotor, endocrine, and behavioral response by the hypothalamus are well illustrated by its role in the regulation of body temperature. Electrical stimulation of the preoptic area in a rat elicits vasodilation, water retention, panting, and, if possible, relocation to a cooler part of the cage. Lesion of the preoptic area produces chronic hyperthermia. In contrast, if the caudal zone of the hypothalamus is electrically stimulated, the rat undergoes peripheral vasoconstriction, shivers, and burrows under the wood shavings. Thus, the preoptic area and parts of the caudal zone of the hypothalamus are essential for the maintenance of optimal physiological temperature (Fig. 23.4). Accordingly, preoptic neurons respond to changes in body temperature that are signaled to them over multisynaptic ascending somatosensory path-

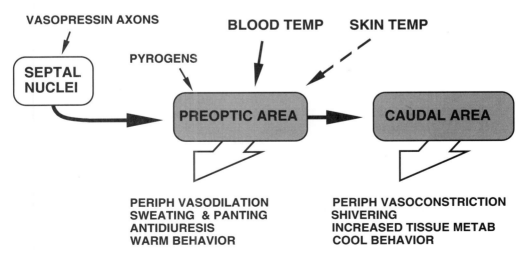

Figure 23.4 Different regions of the hypothalamus interact to regulate body temperature.

ways. Somewhat surprisingly, however, the temperature-related neurons of the preoptic area can also *directly transduce temperature fluctuations of the blood flowing through nearby capillaries.* Some preoptic neurons are warm-sensitive whereas others are cold-sensitive. Preoptic area neurons also respond to circulating pyrogens, such as the macrophage product *interleukin-1,* by resetting their temperature sensitivity. The neurons of the preoptic area are also subject to strong synaptic influences from the septal nuclei. The septal area is in part an antipyretic area; when activated, it limits the magnitude of the fever response. Thus, it appears that *most of the computations relevant to temperature regulation occur in the preoptic area,* whereas the temperature-related cells of the caudal hypothalamus simply respond to inputs from the preoptic cells.

The Preoptic Area Plays a Pivotal Role in Reproductive Processes

If the preoptic area is destroyed in a female rat, she will cease cycling through estrus, fail to mate, and if possessed of a litter, fail to nurse and nurture the pups. These internal reproductive processes and related behaviors are under the direct control of the gonadotropic hormones, follicle-stimulating hormone (FSH) and luteinizing hormone (LH), which are released phasically from the anterior pituitary. Not surprisingly, the hypothalamic neurons that secrete the neurohormonal factors that govern the synthesis and release of the gonadotropins are located in the preoptic area. Their axons extend to the infundibulum, where they secrete the releasing hormones into the pituitary portal circulation (see below). The activity of such preoptic neurons exhibits a pulsatile discharge pattern such that the bursts of spike activity correlate with the intermittent elevations in plasma LH and FSH.

Neuroanatomists have noted that in many mammalian species, including humans, there are a number of small nuclei in the preoptic area that are sexually dimorphic. Since some of them influence penile erection, they are more fully developed in males. Interestingly, it has recently been reported that some of the sexually dimorphic nuclei in male homosexuals are similar in size to their counterparts in females. Such an observation has led many people to speculate that these preoptic nuclei may influence masculine and feminine forms of behavior and sexual preferences in humans.

The Hypothalamus Contains the Biological Clock

The suprachiasmatic nuclei are located in the rostral hypothalamic zone, one on each side of the midline cradled by the optic chiasm. The relationship is no coincidence; the suprachiasmatic nuclei receive a direct input from the retina by way of retinal ganglion cell axons that enter them from the underlying chiasm. Together, the suprachiasmatic nuclei function as a biological "clock" producing circadian and perhaps seasonal rhythms of brain activity based on information about the light–dark cycle conveyed by the retinal axons. They appear to influence the sleep–waking cycle (see Chapter 13) and daily cycles of certain hormones. They may also control hibernation in some mammalian species.

Eating and Drinking Are Influenced by the Hypothalamus

If the ventromedial nucleus of the rat hypothalamus is ablated bilaterally, the rat will eat incessantly to the point of obesity (Fig. 23.2). Electrical stimulation of the nucleus produces the opposite effect: a hungry rat will stop eating. If the lateral hypothalamic area (lateral to the ventromedial nucleus) is destroyed bilaterally, the rat will dehydrate and starve itself to death. Although these early observations led some investigators to postulate that the medial and lateral hypothalamic areas function as opposing "centers" that control eating and drinking, it soon became apparent that these regions are but a part of an extended brain circuitry that regulates consummatory behavior. The lesions and stimulations respectively impair or activate an extensive sensorimotor and motivational circuitry and have endocrine consequences as well, any or all of which may contribute to the ultimate behavioral effects. For example, there are many neurons in the middle zone of the hypothalamus that respond to the sight or taste of food. There is evidence that taste perceptions are altered in rats that have received lesions. Also, neurons in this region respond directly to circulating glucose levels. Thus, as for other functions of the hypothalamus, the spectrum of stimulating inputs and the output circuits are manifold and complex.

Hypothalamic Relations to the Pituitary

The most recognizable nuclei in the rostral hypothalamic zone are the **paraventricular** and **supraoptic nuclei,** together called the **magnocellular secretory nuclei** because of the large size of their cell bodies and the fact that they project axons to the posterior pituitary to release oxytocin and vasopressin (Figs. 23.5 and 23.6A). Incidentally, the direct hypothalamic projections to visceromotor cell groups in the brain stem and the spinal cord (e.g., the dorsal motor nucleus of the vagus and the intermediolateral nucleus) originate from the neurons of the paraventricular nucleus (Fig. 23.7). Interestingly, these neurons use the same peptide hormones, oxytocin and vasopressin, as peptide cotransmitters at their

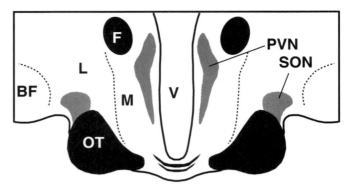

Figure 23.5 This drawing of a frontal section through the rostral hypothalamus shows the magnocellular secretory cell groups. Abbreviations: BF, basal forebrain; F, fornix; L, lateral hypothalamic area; M, medial hypothalamic area; OT, optic tract; PVN, paraventricular nucleus; SON, supraoptic nucleus; V, third ventricle.

axon terminals. *This descending system probably serves to integrate lower visceral reflexes with endocrine reflexes and emotional behavior.* It should be kept in mind that this direct outflow of hypothalamic signals to preganglionic efferent neurons is quantitatively minor compared to other descending systems. *Hypothalamic influences on visceromotor structures are mediated, for the most part, by multisynaptic reticulospinal pathways.*

Magnocellular neurons that secrete vasopressin and thereby increase water resorption in kidney tubules are directly responsive to osmolality or to changes in sodium levels in the blood of nearby brain capillaries. Like the temperature-sensitive neurons in the preoptic area, the magnocellular hypothalamic neurons can respond to conditions of the internal environment independently of synaptic input arriving from peripheral transducers.

The middle hypothalamic zone contains the **parvicellular secretory nuclei** (e.g., **arcuate nucleus** and **tuberal nuclei**) clustered mainly around the median eminence and infundibulum (Fig. 23.2). Axons of these peptide-containing (e.g., somatostatin, thyrotropin-releasing hormone) and dopamine-containing (prolactin inhibitory factor) cells form the so-called **tuberoinfundibular tract.** The neurons in these nuclei produce and secrete releasing factors into the pituitary portal circulation to regulate the synthesis and release of hormones by anterior pituitary cells (Fig. 23.6B).

Secretory neurons of the hypothalamus are not radically different from any other CNS neurons; they have similar morphology, conduct action potentials along their axons, and release chemical messenger substances. However, secretory neurons only rarely synapse on other neurons; instead, they form close contacts with capillaries called **neural–hemal organs.** Moreover, the signal molecules they release are not neurotrans-

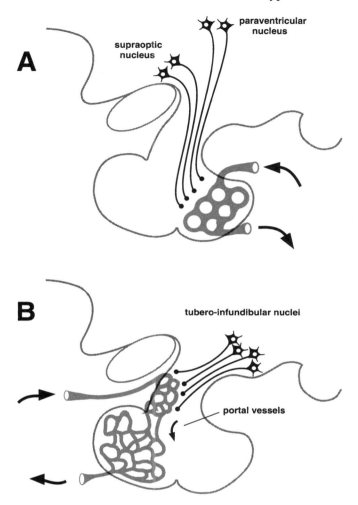

Figure 23.6 The magnocellular secretory nuclei send their axons to the posterior pituitary, where they release their peptide hormones into the systemic circulation (**A**). The parvicellular secretory nuclei send their axons to the median eminence and the infundibulum, where they release their hormones near fenestrated capillaries of the pituitary portal system (**B**).

mitters, but *neurohormones* that are released into the perivascular space to be carried by the blood to target organs. Neural-hemal endings are located in the posterior pituitary and the median eminence–infundibular region. Stated otherwise, such neurons function as *neuroendocrine transducers that receive synaptic signals from other neurons and transmit signals via hormones.*

The *milk-ejection reflex* in rats, which involves the magnocellular neurosecretory neurons, illustrates the role of hypothalamic neurons in

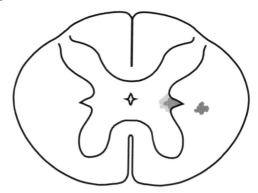

Figure 23.7 The location of the hypothalamospinal tract and its terminal distribution are shown on a cross section of the spinal cord.

converting a synaptic input to a hormonal output, a process aptly called *neuroendocrine transduction*. Suckling is the natural stimulus for the reflex. The sensations elicited by suckling are conveyed over the somatosensory system to ultimately trigger a burst of spikes in the oxytocin-synthesizing neurons of the anterior hypothalamus. This burst of activity causes the release of oxytocin from axonal endings in the posterior pituitary. The hormone, in turn, travels in the systemic circulation to the mammary glands, where it interacts with receptors expressed by the myoepithelial cells, causing them to contract and eject milk.

The interaction of hypothalamic secretory neurons with other endocrine tissues of the body is not unidirectional; hypothalamic neurons are themselves influenced by circulating hormones secreted by endocrine glands of the body. Receptors for several circulating hormones have been demonstrated in hypothalamic neurons by injecting animals with hormones labeled with radioactive isotopes. The radiolabeled hormones bind to the specific receptors and can be detected by subsequent autoradiographic processing of the hypothalamic tissues. Hormone-concentrating neurons have a widespread distribution in the hypothalamus and several structures that have close synaptic relationships with the hypothalamus. Such hormone receptors are considered to mediate the long-loop feedback regulation of neurosecretion by these brain cells.

It is easily appreciated that hypothalamic cells that function as neuroendocrine transducers can respond to synaptic inputs, to circulating hormones, and even to local conditions in the hypothalamus. In turn, they can influence the effector systems of the body through synaptic contacts on neurons in other parts of the brain and by secreting hormones that act on distant receptors in target organs. From the general description of hypothalamic circuitry given above, it is clear that the neurosecretory cells lie amidst a matrix of diverse signal traffic. Just as numerous systems converge on the motor neurons of the brain stem and spinal cord, these neurosecretory cells, which are the final common pathway for hormonal regulation by the brain, must integrate many inputs to perform effectively.

The Hypothalamus Is Involved in Emotion

From the above discussion, it can be appreciated that the neural transformations of the hypothalamus are ultimately expressed by the visceromotor, endocrine, and behavioral effector systems of the body (Fig. 23.8). Mainly on the basis of relatively sparse clinical data, we suspect that the hypothalamus also influences the mental experience of emotion. Hans-Lukas Teuber, one of the pioneers of modern neuropsychology, used to describe one of the brain-damaged patients he studied to illustrate the role of the hypothalamus in emotion. During the Second World War, the patient was wounded by a small piece of shrapnel that penetrated his skull and brain and came to rest in the third ventricle. The fragment could not be surgically removed because of the life-threatening risk of damaging the hypothalamus. Since the time of his wounding, the patient complained that he would suddenly and unaccountably experience "feelings," often of sadness so extreme that he would sob uncontrollably. As the story goes, the man eventually noticed that his mood swings would occur whenever he had reason to shake his head vigorously. X-rays showed that the tiny metal fragment was moving about in the lower part of the ventricle, irritating the walls of the hypothalamus!

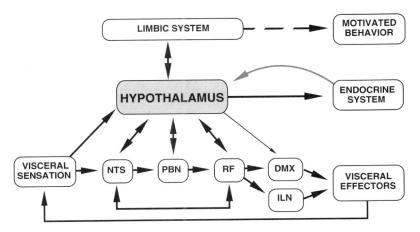

Figure 23.8 The diagram summarizes the synaptic relationships of the hypothalamus with other parts of the brain. Abbreviations: DMX, dorsal motor nucleus of the vagus; ILN, intermediolateral nucleus; NTS, nucleus of the solitary tract; PBN, parabrachial nuclei; RF, reticular formation of the brain stem.

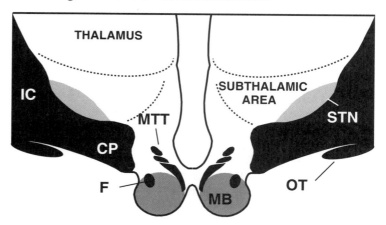

Figure 23.9 A sketch of a frontal section through the caudal hypothalamus shows the mammillary bodies (MB) and their related fiber bundles. Abbreviations: CP, cerebral peduncle; F, fornix; IC, internal capsule; MTT, mammillothalamic tract; OT, optic tract; STN, subthalamic nucleus.

Exactly how the hypothalamus contributes to emotion remains a mystery. Much of the hypothalamus is synaptically closely linked to a set of forebrain structures (e.g., hippocampus, amygdala, septal nuclei) collectively referred to as the "limbic system," which itself has been associated with emotional expression. As already noted, the hypothalamus receives the bulk of the fornix fibers (Fig. 23.9). These relations of the hypothalamus to the limbic system will be taken up in the next chapter.

Clinicopathological Observations Reveal the Full Spectrum of Hypothalamic Functions

One of the more common ways in which the hypothalamus is bilaterally damaged is by the dorsal expansion of a pituitary tumor. In addition to the severe hormonal abnormalities due to the pituitary dysfunction (for example, *hypothyroidism* from decreased release of thyrotropin from the anterior pituitary and *diabetes insipidus* from decreased release of vasopressin by axon terminals in the posterior pituitary), patients with an invasive tumor can be expected to display one or more visceral, emotional, or behavioral disturbances. For instance, cold or hot "flashes" may occur due to dysfunction of circuits in the preoptic area. Persistent hunger or thirst is commonly reported due to disruption of activity in the middle hypothalamic zone. The patient may experience bouts of sadness or euphoria with no apparent cause. A decrease in libido or impotence is common. Incidentally, the tumor usually also compresses the fibers that travel through the posterior part of the optic chiasm, leading to bilaterally diminished visual acuity in the temporal visual fields (see Chapter 26). In fact, it can be expected that bilateral hypothalamic damage will produce a greater diversity of overt abnormalities than any other brain lesion.

Detailed Reviews

Haymaker W, Anderson E, Nauta WJH (eds). 1969. *The Hypothalamus.* Charles C Thomas, Springfield, IL.

Morgane PJ, Panksepp J. 1979. *Anatomy of the Hypothalamus.* Marcel Dekker, New York.

24 Motivation and Emotion

How the brain encodes motivation and emotion is both a compelling and a frustrating issue for many neuroscientists. This is in no small part because both of these mental phenomena are ill-defined psychological constructs that cannot be directly studied. Instead, we are forced to make inferences about them from observations of behavior in experimental circumstances or from verbal reports of "feelings" by human subjects. Unfortunately, the same behavior can be executed for any number of reasons, and aside from the inherent ambiguity of language, verbal reports, ironically, are subject to emotions and hidden motives. Even despite these impediments, progress has been made on the basis of experimental work and careful observations in clinical cases of brain damage. As a result, several circuits have been implicated as contributors to either motivation, emotion, or both. Some neuroscientists find it useful to lump such circuits together into a hypothetical computational unit called the *limbic system*.

The Limbic System

The term *limbic system* is a vague one that originates from the French term *le grand lobe limbique* introduced more than one hundred years ago by Paul Broca. Broca coined the term to indicate the prominent ring of cortex along the medial, free edge (*limbus*) of the cerebral mantle, composed of the cingulate and parahippocampal gyruses and the hippocampus proper (**CA fields** and **dentate gyrus**). Later, the ring came to be called the *rhinencephalon* (nose-brain) because it was discovered that in lower mammals the olfactory cortex was directly linked to the parahippocampal gyrus. The amygdala became included with the rhinencephalon because of its similarly close association with the olfactory cortex. When this functional notion of a "nose-brain" proved too restrictive (it was emphasized by James Papez that the overwhelming input to the ring comes from association areas of the neocortex rather than the olfactory

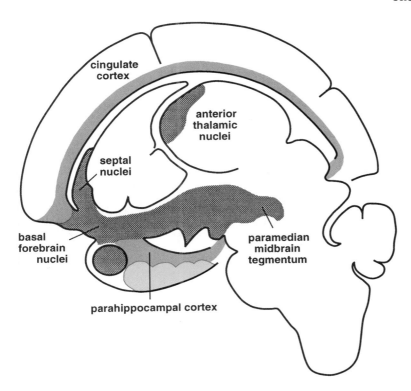

Figure 24.1 One version (and there are many) of the limbic system includes the closely interconnected cortical and subcortical cell groups that have been shaded in this diagram.

cortex), Paul McLean suggested reversion to the term *limbic system* to designate this diversity of neural structures.

Why bother at all with a collective term for neural structures so separate and structurally different? There are two reasons. First is the circumstance that *all components of the limbic system are synaptically associated with the subcortical neural continuum that extends from the septal region through the hypothalamus into a paramedian zone of the midbrain tegmentum* (see Chapter 23). A second feature common to limbic system structures is that they have a *low threshold for the inducement of seizure activity.* Before considering the latest functional concepts of the limbic system, let us consider in greater detail its components and their relationship to the septohypothalamomesencephalic continuum (Fig. 24.1).

The Entorhinal Area

The neocortex that is contiguous with the hippocampus and lies on the medialmost aspect of the parahippocampal gyrus (on the medial side of the temporal lobe) is called the **entorhinal area** because of its proximity to the rhinal sulcus. *This cortex is the major information inroad to the hippocampal formation.* It receives information from many brain structures, including direct projections from the olfactory cortex, amygdala, septal nuclei, some small midline and medially located thalamic nuclei, the paranigral dopaminergic cell groups in the ventral tegmental area of the midbrain, the serotonergic raphé nuclei, and the noradrenergic locus coeruleus. Its most significant input comes indirectly from all cortical areas by way of a multisynaptic transcortical progression through association areas to the temporal lobe. In other words, *virtually all perceptual, behavioral, and cognitive computations performed by the cerebral cortex are delivered to the hippocampal processing circuit by way of the entorhinal cortex.*

The Hippocampus

The hippocampal complex is entirely cortical in structure and includes the CA fields (for *cornu ammonis* or Ammon's horn, which the hippocampus resembles on gross inspection), the dentate gyrus, and the subiculum (Fig. 24.2). The subicular cortex is transitional between the three-layered hippocampal cortex and the six-layered cortex of the entorhinal area. The three hippocampal layers are the molecular layer, pyramidal cell layer, and polymorphic layer; in the dentate gyrus, the granule cell layer replaces the pyramidal cell layer. The major source of input to the hippocampus comes by way of the axons that arise from pyramidal neurons in the entorhinal area. Such axons cross the hippocampal fissure (which is actually an adhesion of the pial surfaces of the entorhinal cortex and dentate gyrus) as a diffuse bundle called the **perforant path.** The granule cells of the dentate gyrus relay the information they receive to field CA3; field CA3 projects in turn to field CA1 and the subiculum. Finally, the outflow from the hippocampal formation (the axons of the fornix) arises mainly from the pyramidal neurons of CA1 and the subiculum. The fornix fibers from CA1 terminate mainly on neurons of the septal nuclei in the basal forebrain. Those from the subiculum project to the hypothalamus (Fig. 24.3). The septal nuclei reciprocate the projections they receive from the hippocampal formation.

Papez Circuit

The axons that emerge from the hippocampal complex convene as the fornix and travel together to terminate along the septohypothalamomesencephalic continuum. Their most obvious target in the continuum is the mammillary bodies. By way of their axons in the mammillothalamic tracts, the mammillary nuclei forward the signals borne to them by the fornix to the **anterior nuclei** of the thalamus. The anterior nuclei provide the thalamic input to the cortex along the **cingulate gyrus** on the medial

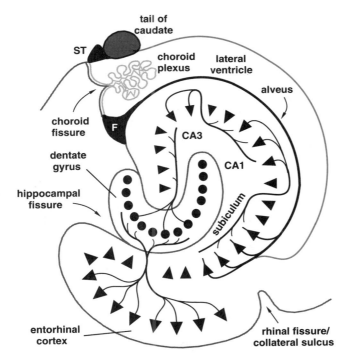

Figure 24.2 This schematic of a frontal section through the temporal lobe (lateral is to the viewer's right) shows the basic hippocampal processing circuit. Abbreviations: CA1, CA3, zones of the cornu ammonis; F, fornix; ST, stria terminalis.

surface of the hemisphere, itself a part of Broca's limbic lobe (Fig. 24.4). From the cingulate cortex, signals are delivered to the parahippocampal gyrus via the cingulum bundle and from there gain re-entry to the hippocampal formation. This anatomical gyre was described in detail by James Papez and was reverently dubbed the *Papez circuit* by his contemporaries (Fig. 24.5). The signals that circumnavigate the gyre are subject to reprocessing at each synaptic station, where the loop is opened to outside influences. Although it is suspected to somehow mediate emotions, the service performed by the gyre remains obscure, as does the nature of the reprocessing that occurs in each of its segments.

The Amygdaloid Complex

The amygdala is a telencephalic mass that rests beneath the cortex near the rostral end of the temporal lobe (Fig. B8). It is divisible into *three* main

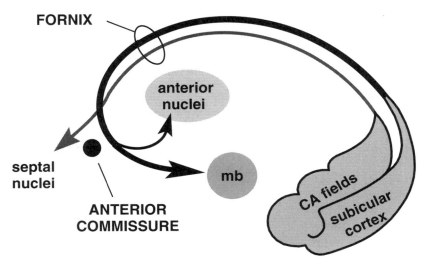

Figure 24.3 The origin, arc, and distribution of the two components of the fornix are schematized in this Daliesque image. The larger number of axons (thick black line and arrow) arise from the subiculum and pass caudal to the anterior commissure (black dot). The fornix fibers from the CA fields (especially CA1) of the hippocampus proper are fewer in number and pass mainly rostral to the anterior commissure to reach the septal nuclei. Abbreviations: atn, anterior thalamic nuclei; F, fornix; mb, mammillary bodies.

nuclear groups: a **corticomedial** group, a **basolateral** group, and the **central nucleus** (Fig. 24.6). The best-established input to the amygdala is that which comes from the olfactory bulb to the corticomedial group, and from the olfactory cortex to the basolateral group. The central nucleus receives taste and viscerosensory fibers from the brain stem and dopamine fibers from the ventral tegmental area of the midbrain. Other afferent axons come from thalamic nuclei, the hypothalamus, and the frontal and temporal association cortices.

The major outputs of the amygdala are directed to the ventral forebrain and hypothalamus via the stria terminalis and ventral amygdalofugal pathway. Thus, like the hippocampus, the amygdala expresses itself in large measure over the continuous mass of cells and neuropil lying along the medial forebrain bundle (see Fig. 23.3). The continuum not only receives amygdaloid fibers, but also sends fibers back to it. You'll recall from Chapter 20 that the amygdala also projects to the ventral striatum, where it may influence goal-directed movement.

The Functional Role of the Limbic System

We know already that the septohypothalamomidbrain continuum over which the limbic system expresses itself is centrally involved in the

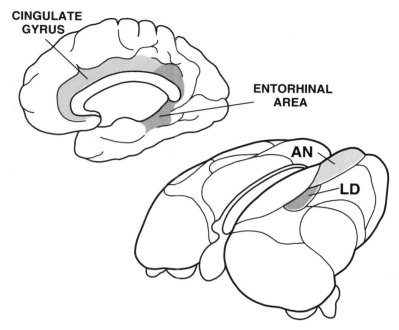

Figure 24.4 The drawings indicate the anterior nuclear group (AN) shown in red on the thalamus diagram and the anterior part of the cingulate cortex (also red) to which they send axons. The laterodorsal nucleus (LD) may be closely associated with the anterior group and projects to a more caudal part of the cingulate gyrus that leads around the splenium of the corpus callosum into the entorhinal area (gray).

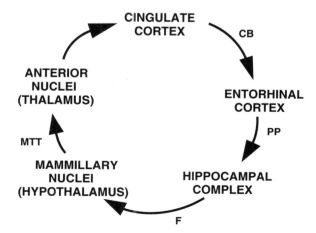

Figure 24.5 In 1937, the neuroanatomist James Papez suggested a closed cortical circuit that mediated emotions. Abbreviations: CB, cingulate bundle; F, fornix; MTT, mammillothalamic tract; PP, perforant path.

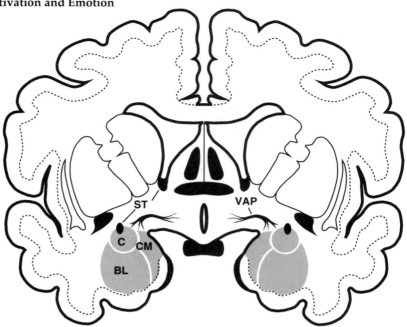

Figure 24.6 This drawing of a frontal section through the forebrain shows the three nuclear groups (BL, basolateral; C, central; CM, corti-comedial) of the amygdaloid complex (shaded in red). Efferent axons of the amygdala are sent to the hypothalamus by way of two pathways: the stria terminalis (ST) and the ventral amygdalofugal pathway (VAP).

regulation of endocrine and visceral effector mechanisms (see Fig. 23.6). But we have overlooked a large body of data that indicate still a third function of this extended hypothalamic circuitry—emotional experience. Because so many emotional disorders may be due in part to degenerative or biochemical diseases that affect components of the limbic system, it is the psychiatrist's favorite part of the brain. For instance, extreme hostility and aggressive reactions sometimes occur in patients with temporal lobe tumors that compromise the amygdala or its connections. It is well known that a tumor of the hypothalamus in humans is often accompanied by spontaneous fits of anger or deep sadness or mirth. Much of the time, however, there is no overt organic cause for the affective disorders displayed by patients. In such cases, neurotransmitter imbalances become the primary suspect, and treatments are based on drugs that modify neurotransmitter levels or interact with specific receptors.

In animals, a lesion along the hypothalamic continuum, depending on its location, can elicit a wide spectrum of emotional responses that range from agitation and aggression to all-consuming pleasure. That the continuum mediates experiences of pleasure can be inferred from experiments in which rats are allowed to self-stimulate their rostral hypothalamus through a chronically implanted electrode. Rats will obsessively press a lever that sends a mild current through the electrode. So rewarding is the electrical stimulation that the rats will neglect to groom or even feed themselves. They will also willingly negotiate forbidding obstacles, even walk across an electrified grid, to reach the lever! Thus, it would seem that the hypothalamic continuum is centrally involved in states of emotion and motivation. Emotion can be defined as an elicited *feeling*, whereas motivation indicates a *drive* that strongly directs behavior toward a goal that will satisfy a particular need. Hunger and thirst are examples of motivations. Needless to say, higher emotions like hate and love are some of the most forceful motivators of all for humans.

What kind of information is required by a system that mediates emotion and motivation? One major input to the hypothalamus and amygdala comes from the nucleus of the solitary tract, which you'll recall is the first central terminus for taste and viscerosensory signals. The sense of smell also has privileged access to the amygdala and entorhinal cortex. Obviously, information about blood sugar and electrolyte levels is relevant to hunger and thirst, and taste and smell prove useful in the goal-directed behavior necessary to relieve these conditions (Box 24.1). But as is often pointed out, man does not live by bread alone. We are motivated by more noble desires than hunger and thirst, and we feel abstract emotions such as love and hate. Perhaps for these *higher* emotions and motivations to be expressed through the hypothalamus, a more elaborate input from the cerebral cortex is required, after appropriate recoding for the purpose by the limbic system.

Association of the amygdala with emotions is readily discovered by experimental manipulations of the nucleus in animals. If the amygdala of a well-fed, lounging cat is electrically stimulated through a chronically implanted electrode, the animal will suddenly leap to its feet, arch its back, unsheath its claws, and hiss as though highly enraged. It might pounce on a previously tolerated rat roaming nearby. If the amygdala is bilaterally destroyed in a monkey that previously enjoyed high status in his troop pecking order due to his traits of boldness and aggression, the monkey becomes overly fearful and immediately falls from his position of authority. His former supplicants now lord it over him.

The *psychomotor epilepsy* so readily elicited in limbic system structures may aptly reflect the functions of the system in humans. Such seizures typically begin with an *aura* made up of either olfactory or gustatory hallucinations—a foul smell or metallic taste. These illusions are followed by a mood change—feelings of anxiety or severe loneliness—and a lapse into a dreamy state. Finally, a motoric phase ensues that consists of a coordinated series of complex behaviors, often seemingly goal-directed. When recovered from the seizure, the patient cannot recall the experience at all, as though memory formation had been temporarily shut down.

Indeed, the limbic system has been considered to mediate some forms of memory formation. Alcoholics who neglect their nutrition (as they

Box 24.1   Smell and Taste Are Primally Linked to Emotions, Motivations, and Predictable Behaviors

Emotions and motivations are often hard to define. It may be useful to think of them as aids to survival and reproduction stemming from early evolution. The table below catalogs some of the smells and tastes that seem to be closely linked to visceral responses that accompany emotions and motivations and the goal-directed behaviors they tend to trigger.

Odorant	Emotion/Motivation	Behavior
Mom	Satiety/hunger	Suckle
Prey (food)	Bloodlust/hunger	Stalk/attack/eat
Predator	Fear/escape	Flee
Pheromone	Love/libido	Court (and get lucky)
Competitor	Rage/defend	Attack
Noxious	Revulsion/avoid	Retreat

Taste	Emotion/Motivation	Behavior
Sweet (nutritious)	Pleasure/hunger	Consume
Salty and sour		
Mild	Pleasure/hunger	Consume
Excessive	Aversion/avoid	Spit out
Bitter (toxic)	Disgust/avoid	Spit out

often do) become thiamine- (vitamin B_1) deficient. Thiamine deficiency leads to behavioral changes such as confusion and loss of short-term memory. *Postmortem* examination of the brain frequently reveals lesions of the mammillary bodies and adjacent parts of the posterior hypothalamus, the anterior nuclei of the thalamus, and the dentate gyrus. This clinical condition is called the *Wernicke–Korsakoff syndrome* after its initial describers. In some cases of otherwise intractable epilepsy, patients have undergone surgical excision of the temporal lobe cortex, including the hippocampus. The result is hippocampal *amnesia,* which is characterized by an inability to form new memories, although presurgical memories remain intact.

Consider for a moment that those things we tend to remember are things that are meaningful to us. Insignificant things, if they reach consciousness at all, are immediately forgotten. Meaningful things are things to which we attach some affective quality—they give us pleasure or pain. Could the association of events with affective qualities take place in the limbic system? Is the inability to form memories in the hippocampally damaged patient or the postictal amnesia of the epileptic a consequence of the brain's inability to elevate salient stimuli to an emotionally meaningful level? Although we remain for the time being unable to answer these intriguing questions, we have begun to make inroads to understanding the molecular and cellular mechanisms and the circuitry of learning and memory (see Chapter 29). *Dum spiro, spero!*

Detailed Reviews

Ben-Ari Y. 1981. *The Amygdaloid Complex.* Elsevier, Amsterdam.

Isaacson RL. 1982. *The Limbic System.* Plenum Press, New York.

Van Hoesen GW. 1982. The parahippocampal gyrus: new observations regarding its cortical connections in the monkey. *Trends Neurosci* 5: 345.

SPECIAL SENSES AND HIGHER BRAIN FUNCTIONS

V

In this book, the sensory systems are arranged according to an unconventional scheme. The somatosensory and vestibular systems were presented in such a way as to set the stage for analysis of the movement control systems. The perceptual roles of somatic sensation were somewhat neglected. Likewise, the smell, taste, and viscerosensory systems were presented in the context of visceromotor control. The two sensoria that we have left to consider, hearing and vision, can be considered "special" in the sense that humans depend very heavily on them as the primary conveyors of data used by the brain for complex behavioral and emotional guidance, for perception and memory, for much of creative thought, and for language. Not surprisingly, deafness and blindness are clearly the two most debilitating sensory deficits suffered by humans, especially so in our modern technology-based society.

Because of the special dependence of higher brain functions on hearing and vision, analysis of these sensory systems leads naturally into a discussion of the higher brain functions they serve so well. Therefore, we'll first examine the sensory transducers, anatomical pathways and central data processing in the auditory and visual systems. Incidentally, analysis of these systems is particularly useful for the illustration of many processing principles fundamental to all of the sensory systems. For instance, we'll see how convergence and divergence and parallel and serial processing lead to feature extraction and the construction of increasingly complex "images" of the world around us. After you have a firm understanding of such principles, we'll see how they serve higher functions mediated by different areas of the cerebral cortex. Finally, we'll take a brief look at the way the CNS learns and remembers.

Hearing 25

Although hearing is valuable to most animals—the ability to detect the footfall of a predator in the underbrush may be the difference between life and death—it is particularly useful to human beings, whose daily communication relies heavily on spoken language. Actually, the inability to *hear* speech is less an impediment in modern society than is the inability to *speak* understandably. Deafness per se is little more than an inconvenience, since the deaf can learn to read lips or sign language. Ironically, the greater handicap for a deaf person is her own inevitably poor enunciation. Even the speech of the most eloquent orator deteriorates rapidly if not self-monitored. This dependence of speech—clearly a motor function—on the sense of hearing is yet another dramatic example of the critical role of sensory feedback in behavioral control.

Although our modern technological society ensures that deafness is less likely to lead to personal demise, the extremely loud noises produced by that same technology—jet engines and electronically amplified music, for instance—damage our auditory apparatus almost daily. Hearing impairment has become a widespread complaint in the modern world that the medical neuroscientist must often confront. Because the auditory system is so important and so vulnerable to damage at several levels, it is essential that the student have a firm grasp of how the system is structured and how it encodes and processes sound information.

What Is Sound?

The energy that we sense as sound is a mechanical distortion of contiguous matter in the form of pressure waves (Fig. 25.1). Although we can hear the muffled warble of an outboard engine while submerged in the ocean or the rumble of an approaching freight train by placing an ear to the rail, the medium that most often conveys pressure waves to our ears

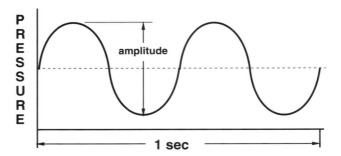

Figure 25.1 A vibrating source produces a pressure wave with two parameters important for hearing: frequency (in this case 2 hertz) and amplitude (measured in decibels referred to the sound pressure level necessary to just detect a 1-kilohertz tone).

is the air. The *frequency* of the pressure waves that reach our ears determines the **pitch** of the sound and is measured in cycles per second (hertz). Humans can detect frequencies over a wide range—about 20 hertz to 20 kilohertz—but are most sensitive between 1 and 3 kilohertz. The peak-to-peak *amplitude* of the pressure wave is measured in terms of the *decibel* sound pressure level scale, a logarithmic scale that correlates with the **loudness** of sound. Calm human speech is in the neighborhood of 65 decibels. Damage to the ear starts to occur at about 100 decibels, depending on the pitch of the sound and the length of exposure; 120 decibels is downright painful to most of us.

Sound Conduction to the Transducer Organ

The external auditory apparatuses are responsible for delivering pressure differences carried in the medium of air to the organ of Corti, where the pressure differences are converted into neural signals. First in line is the pinna, which collects pressure waves traveling through the air and funnels them to the ear canal (external auditory meatus; Fig. 25.2A). Its complex ridges facilitate the localization of a sound source. The ear canal ends blindly at a thin elastic membrane called the eardrum (tympanic membrane). Like the cone on your woofer, it is easily pushed inward by the high-pressure phase of the airborne sound wave and snaps back during the low-pressure phase. The eardrum is attached to a process of the first of the three middle ear bones, the **malleus.** The malleus articulates with the second bone, the **incus,** which, in turn, articulates with the third bone, the **stapes.** Because these three tiny bones (the smallest in the human body) weigh very little, they can resonate in accordance with the oscillations of the eardrum and efficiently transfer sound waves at high frequency.

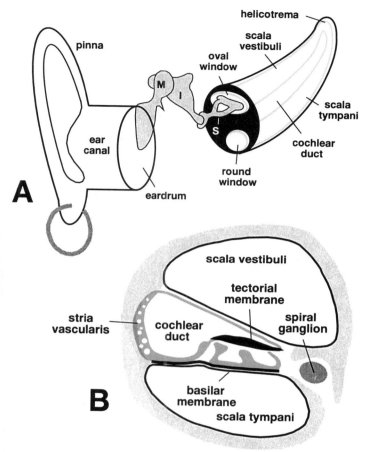

Figure 25.2 The sound transduction apparatus is diagrammed in sketch **A.** The cochlea has been unrolled and foreshortened for the sake of clarity. Sketch **B** shows a radial section through the cochlea. Abbreviations: I, incus; M, malleus; S, stapes.

Two tiny muscles in the middle ear can influence the movements of the malleus and the stapes: the **tensor tympani** and the **stapedius.** The tensor tympani is innervated by the trigeminal motor nerve and is activated by sudden contact in the facial area. This reflex may stabilize the middle ear bones during head impact. The stapedius is reflexively activated by motor axons of the facial nerve in response to sound. Stabilization of the stapes can protect the ear from damage due to loud sounds of low frequency. Since it is a sound-triggered *neural* reflex, it cannot protect against impulse sounds like gunshots. Impulse sounds and prolonged loud sounds can damage the mechanical apparatuses leading to the cochlea and cause a decrease in hearing or even deafness.

Hearing loss due to damage of the middle ear apparatuses is called *conductive* deafness to distinguish it from the deafness that results from damage to the 8th nerve. Conductive disorders usually cause a ringing in the ears called *tinnitus*. Conductive deafness that is caused by infection—*otitis externa* (swimmer's ear) or *otitis media*—is treated with antibiotics or Eustachian tube surgery and is usually transient.

The stapes is attached to a flexible membrane stretched across an opening in the bony labyrinth called the **oval window** (Fig. 25.2A). The oval window seals one end of a long, spiral, membrane-lined chamber of the cochlea. The chamber is filled with perilymph and is continuous around the apex of the cochlea (the continuity is called the **helicotrema**) where it spirals back in parallel with itself to end at another flexible membrane stretched across an opening called the **round window.** The part of the chamber leading from the oval window to the helicotrema is called the **scala vestibuli,** and the return leg is called the **scala tympani.** Spiraling the length of the cochlea between these two scalas is a second membrane-enclosed chamber called the **cochlear duct** (or scala media). *The cochlear duct is filled with endolymph* manufactured by an epithelial tissue called the **stria vascularis** that lies along the eccentric wall of the cochlear duct (Fig. 25.2B). Thus, the cochlea resembles a tightly coiled snail shell when viewed from without and internally consists of a triple helix of fluid-filled chambers.

Sound pressure actuates the oval window, which communicates the pressure waves to the perilymph of the scala vestibuli. Since fluid is not compressible, the compensatory flexing of the round window membrane absorbs the pressure waves at the other end of the chamber. *The oscillating pressure within the cochlear fluid sets in motion a wave that travels along the* **basilar membrane** *from its base to its apex.* The basilar membrane sits at the interface of the scala tympani and the cochlear duct and supports the organ of Corti, which contains the auditory transducer cells (see Fig. 25.4). *The passive mechanical properties of the basilar membrane vary along its length so that the traveling wave created by a given frequency of sound maximally activates a particular region of the membrane* (Fig. 25.3). In other words, it is tuned along its length to a continuum of frequencies. The end of the basilar membrane nearer the oval window (basal end) responds maximally to high frequencies; the end nearer the helicotrema (apical end) is actuated by lower frequencies. This passive mechanical property, first described by the Nobel laureate Georg von Békèsy, contributes to the optimal frequency–amplitude response of the hair cells that are arrayed along the length of the basilar membrane. But this first-order tuning is not the sole determinant of the frequency tuning of the hair cells. In a dead animal or one in which the *inner* hair cells are destroyed, hair cell frequency–amplitude tuning degenerates to a broader, lower-amplitude response profile (see Fig. 25.6). Thus, the outer hair cells serve as a "cochlear amplifier" that boosts and sharpens the frequency sensitivity of the basilar membrane (see below).

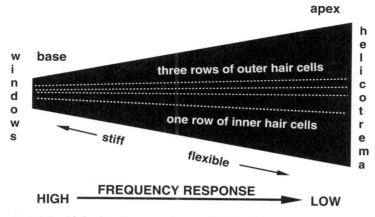

Figure 25.3 If the basilar membrane (jet black) were uncoiled and laid flat, it would be easier to appreciate that its width (and stiffness) changes from one end to the other; it is about five times wider at its apical end than at its basal end. This physical feature influences its intrinsic resonance properties.

Transduction by Hair Cells in the Organ of Corti

The organ of Corti contains the hair cells that transduce the mechanical distortion of the basilar membrane into neural signals (Fig. 25.4). There is one row of hair cells that lies proximally eccentric to the axis of the cochlear spiral (aptly called the *inner* hair cells) and three parallel rows that lie still more eccentrically (the *outer* hair cells). The bundle of variably stiff stereocilia atop each inner hair cell projects into the matrix of the overlying **tectorial membrane.** The apical ends of the hair cells are immersed in the high-K^+ endolymph of the cochlear duct. Thus, when the tectorial and basilar membranes move with respect to one another, the stereocilia are bent, triggering a molecular response mechanism that is essentially the same as that described for the hair cells of the vestibular apparatus (see Chapter 16, especially Fig. 16.4). Briefly, a shearing of the stereocilia results in an opening of mechanically gated channels, allowing an inward K^+ current that depolarizes the hair cell membrane; depolarization, in turn, leads to the Ca^{2+}-dependent release of a neurotransmitter from the base of the inner hair cell onto patiently waiting nerve fibers. *It is mainly the row of inner hair cells that transduce mechanical distortion of the basilar membrane into electrochemical energy for central auditory processing. The hair cells of the outer rows are contractile cells, innervated mainly by motor axons that emerge from the brain stem, and function primarily as effector cells that alter the response properties (tuning) of the organ of Corti.*

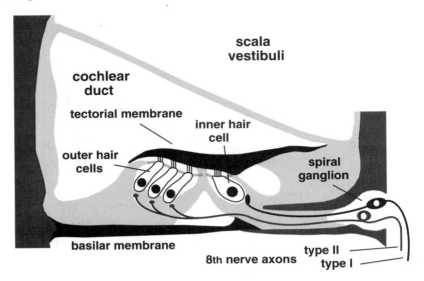

Figure 25.4 A section through the cochlear duct shows the organ of Corti astride the basilar membrane. Bony tissue is shaded dark gray and soft membranous tissue light red. The axis of the cochlear coil is to the viewer's right.

Hair Cells Are Tuned

The hair cells arrayed along the basilar membrane are morphologically dissimilar. Those nearer the basal end are squat and have shorter, stiffer stereocilia; they resonate at higher frequencies. The hair cells at the apical end are taller, with longer, more flexible stereocilia, and are low-frequency resonators. These inherent *mechanical* properties of hair cell stereocilia match those of the region of basilar membrane on which they are located and serve to sharpen their frequency tuning.

Microelectrode recordings from resting hair cells in turtles, frogs, and birds (mammalian inner hair cells don't hold up well under recording conditions) reveal that they also have an inherent *electrical* resonance. They oscillate between depolarization and hyperpolarization at a characteristic frequency. These oscillations in membrane potential are mediated by three types of ion channel located in different parts of the hair cell membrane (Fig. 25.5). Spontaneously active Ca^{2+} channels near the apex of the hair cells periodically allow a Ca^{2+} influx that depolarizes the membrane. The increase in cytosolic Ca^{2+} activates Ca^{2+}-sensitive K^+ channels in the membrane of the sides and bottom of the hair cell membrane. Since this part of the hair cells is bathed in low-K^+ perilymph, there is an outward K^+ current that repolarizes the membrane. Voltage-activated, delayed K^+ channels then hyperpolarize the membrane. The

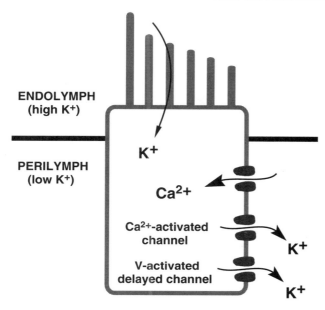

Figure 25.5 Ion channels determine the electrical resonance of hair cells. For explanation, see text.

system is reset by pumps in organelle membranes and the plasma membrane that remove Ca^{2+} from the cytosol. This electrical resonance varies continuously along the basilar membrane, from highest near the oval window to lowest near the helicotrema. Mechanical frequencies along the basilar membrane that match this electrical resonance will amplify the oscillations of the hair cells; mechanical frequencies that do not match will interfere with them. Thus, the physical properties of the basilar membrane combine with the mechanical and electrical properties of the hair cells to sharpen the frequency response tuning of the hair cells (Fig. 25.6). It has yet to be determined whether or not mammalian hair cells exhibit such electrical resonance.

Hair Cells Are Innervated by 8th Nerve Axons

Most of the sensory fibers of the cochlear portion of the 8th nerve (that is, those whose bipolar cell bodies lie in the **spiral ganglion**) innervate the inner hair cells. These *type I* sensory fibers innervate only one inner hair cell, but each inner hair cell receives up to twenty such fibers. Only about 10% of the sensory fibers innervate the outer hair cells (*type II*), and those that do branch profusely to many outer hair cells. The inner hair cells are the ones primarily responsible for the transmission of sound

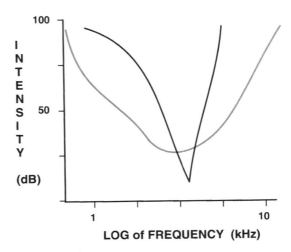

Figure 25.6 The isoresponse (threshold) curves as a function of frequency of hair cells at a particular place along the basilar membrane indicate that the frequency responses of a living hair cell (black curve) are sharper than the tuning curve produced solely by its location on the basilar membrane. If the animal is dead or missing its outer hair cells, the tuning curve is broader (red). The nerve fiber that innervates the hair cell would exhibit a nearly identical tuning curve.

information to the brain. The outer hair cells function as effector cells that respond to sound by changing their length. Since the outer hair cells have their tallest cilia embedded firmly in the tectorial membrane the shortening of their cell bodies tightens their grip on the tectorial membrane, rendering it less mobile. This in turn reduces the effect of sound waves on the cilia of the inner hair cells and thus decreases their responsiveness. The outer hair cells are directly innervated by *motor* axons arising from neurons in the brain stem near the **superior olivary nucleus** (see below) that also influence their contractility and hence the sensitivity of the organ of Corti. Thus, the outer hair cells modulate the sensitivity of the inner hair cells. *These outer hair cell mechanisms reduce the sensitivity of the organ of Corti* when sound amplitude is destructively high and, by way of central influence, can focus the sensitivity of the organ on particular frequencies of interest.

The Coding of Sound Frequency and Amplitude

Since the sound-induced movements of the basilar and tectorial membranes are oscillatory, the inner hair cells are alternately depolarized and hyperpolarized, causing a pulsatile release of neurotransmitter onto the type I sensory endings. Since type I axons can fire spikes up to about 500 hertz, they can only match the lowest of the sound frequencies we commonly hear. Moreover, we are sensitive to amplitude as well as frequency, and you'll recall that the preferred mechanism for coding amplitude in the somatosensory system is spike frequency. How then does the auditory nerve encode sound frequency and amplitude? In general, higher-frequency sounds are encoded by the topography of the innervation of inner hair cells along the length of the organ of Corti—a place code. *The discharge of a given fiber is interpreted as a particular frequency by the brain because of this* tonotopic *organization of the auditory nerve fibers.* Amplitude is coded mainly by the recruitment of additional sensory fibers—a population code. *The ten to twenty axons that innervate each inner hair cell exhibit a spectrum of thresholds for spike discharge.* As amplitude increases, more axons are triggered to discharge.

The Central Auditory Pathways

The sensory axons of the auditory portion of the 8th cranial nerve terminate in the ipsilateral **ventral** and **dorsal cochlear nuclei** (Figs. 25.7 and B2). Since the primary sensory axons terminate in a topographic

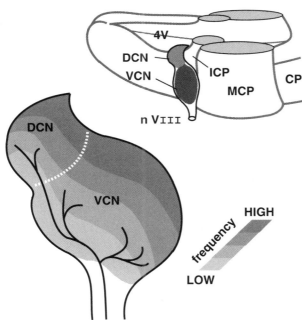

Figure 25.7 The position of the right cochlear nucleus with respect to the brain stem is shown in the top drawing. An enlargement of the cochlear nucleus in the bottom image shows its tonotopic organization. Each incoming primary axon branches to several subnuclei made up of specific cell types. Abbreviations: CP, cerebral peduncle; DCN, dorsal cochlear nucleus; ICP, inferior cerebellar peduncle; MCP, middle cerebellar peduncle; VCN, ventral cochlear nucleus.

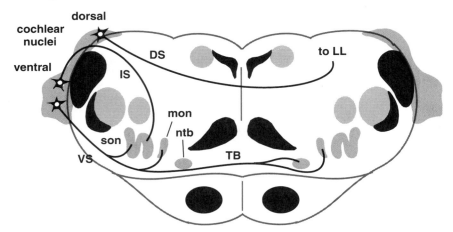

Figure 25.8 A cross section through a caudal pontine level shows the initial targets (shaded in red) of secondary sensory axons arising from the cochlear nuclei. Abbreviations: DS, dorsal stria; IS, intermediate stria; LL, lateral lemniscus; mon, medial olivary nucleus; ntb, nucleus of the trapezoid body; son, superior olivary nucleus; TB, trapezoid body; VS, ventral stria.

pattern, the cells of the cochlear nuclei are tonotopically organized. Each incoming axon branches to synapse in a number of subnuclei that are made up of secondary sensory neurons of a particular shape that have distinct physiological properties. For example, stellate (*chopper*) cells generate bursts of spikes at regular intervals and help to signal the different frequency components of a sound. Bushy cells, in contrast, generate one or two spikes in response to primary input and thus signal the onset of a sound stimulus. In this way, the extraction of particular stimulus features begins already at the level of the cochlear nuclei.

The secondary sensory axons that emerge from the cochlear nuclei follow three paths called the dorsal, intermediate, and ventral acoustic stria (Fig. 25.8). The dorsal stria contains axons that arise from the dorsal cochlear nucleus and ascend as part of the contralateral **lateral lemniscus.** Axons of the intermediate and ventral strias arise from the ventral cochlear nucleus and project *bilaterally* to the **superior olivary complex** in the pontine tegmentum (Fig. B3). The crossing fibers form a conspicuous bundle beneath the medial lemniscus called the **trapezoid body.**

The medial superior olivary nucleus contains spindle-shaped neurons with a medially and a laterally extended dendritic trunk. Each dendrite receives input from either the ipsilateral or the contralateral cochlear nucleus, and the cell is thus able to compare interaural time differences—acoustic data that are useful in sound localization. The neurons of the main superior olivary nucleus process differences in sound intensities.

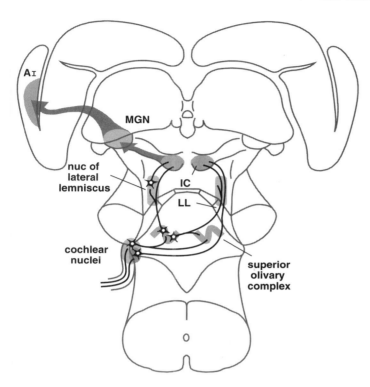

Figure 25.9 The diagram shows a simplified version of the major ascending auditory pathways. Fibers from the cochlear nuclei and the superior olivary nuclei form the lateral lemniscus (LL), which ascends to the inferior colliculus (IC). The shaded arrow leading from the IC to the medial geniculate nucleus (MGN) represents the brachium of the inferior colliculus.

Fibers of the superior olivary complex of each side also cross in the trapezoid body to reach their contralateral counterparts or to ascend with the contralateral lateral lemniscus. The lateral lemniscus on each side of the brain stem is the main ascending auditory pathway (Figs. 25.9, B5, and B6). Each lemniscus carries axons from the contralateral dorsal cochlear nucleus and the superior olivary complex of both sides to the **nucleus of the lateral lemniscus** and the inferior colliculus (Fig. B6). Thus, extensive bilateral connections occur early in the auditory processing circuits. For this reason, lesions above the level of the cochlear nuclei do not cause monaural deficiencies. The cells of the inferior colliculus are arranged tonotopically (Fig. 25.10). They project via the brachium of the inferior colliculus to the **medial geniculate nucleus** (MGN; Fig. B7) of the thalamus, which in turn projects to the primary auditory cortex (A$_I$) located on the long gyruses of the temporal plane (Brodmann's area 41). The inferior colliculus functions to relay data about the tonal quality and

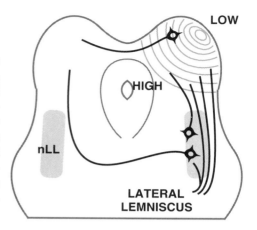

Figure 25.10 The tonotopic organization of the inferior colliculus is represented by concentric, curved layers of cells as indicated by the red shaded lines. It is imposed by the fibers ascending as the lateral lemniscus, some of which relay in the nucleus of the lateral lemniscus (nLL). The crossed axons emanating from cells of nLL and the inferior colliculus indicate that communication between the auditory structures on the two sides is a common occurrence along the ascending auditory pathways.

the location of sound to the cerebral cortex by way of the thalamus. It also contributes to reflexive orientation to sound sources by way of connections to the deeper layers of the superior colliculus (see Chapter 17). Both the MGN and the inferior colliculus receive feedback projections from the auditory cortex that can gate and modulate the ascending sound data processed at these way-stations.

The Functional Organization of A$_I$

Like the somatosensory (Chapter 12) and visual cortices (Chapter 27), A$_I$ is organized into functional columns that extend from the pial surface to the subcortical white matter. Most cells within a given column exhibit similar binaural properties. The most obvious columns are associated with tone. Positron emission tomography studies of the human A$_I$ cortex indicate a tonotopic organization that is arranged from high to low tones represented progressively along a fronto-occipital axis. Orthogonal to the tonal columns are columns of binaural cells related to sound localization. There are two types of such columns: *summation columns*, in which the response is greater to binaural than to monaural input, and *suppression columns*, in which the response is greater to monaural than to binaural input. Since the A$_I$ uses interaural time and intensity differences to localize sound, the summation and suppression columns represent a collective map of acoustic space. The perception of the complex sound stimuli that are constantly encountered—imagine the temporal, pitch, and intensity modulation that must be deconstructed in order to understand speech—requires that the computations of A$_I$ be routed to several secondary auditory areas. Area A$_{II}$ lies, beltlike, around the perimeter of A$_I$ and receives a dense axonal input from it. Little is certain about the integration of sound data that occurs in A$_{II}$. The perception of complex sounds and their integration with data supplied over the other sensory pathways takes place in the multimodal areas of the association cortex.

We will tackle the issues of higher cortical processing after we examine primary visual processing in the next chapter.

Detailed Reviews

Altschuler RA, Bobbin RP, Clopton BM, Hoffman DW. 1991. *Neurobiology of Hearing: The Central Auditory System.* Raven Press, New York.

Harrison RV. 1988. *The Biology of Hearing and Deafness.* Charles C Thomas, Springfield, IL.

26 Visual Transduction

Blindness is such a catastrophic disability that humans have invested monumental effort over the ages in combating deficits of eyesight. Corrective lenses, ophthalmic surgery, and, most importantly, research into the mechanisms of vision constitute the main avenues through which medical neuroscience confronts the disability of blindness. In fact, vision research has been so successful that we can foresee a time when artificial sight restoration is as commonplace as limb prosthetics. Since the visual system is assailed by several diseases and subject to trauma beginning with the optical system of the eyeball and extending to the visual cortex, it is important that physicians become familiar with the processing pathways and neural mechanisms of the visual system.

The human brain gains more useful information about the environment through vision than any of the other senses. This importance is reflected by the number of axons in the optic nerves—each contains over a million—which exceeds the combined total of primary sensory fibers entering the CNS via the spinal nerves. Vision involves the transduction of light energy by the eye into neural impulses that are distributed to several parts of the brain to serve a number of functions, such as the guidance of movements, arousal, and the identification of objects in the environment. In this chapter, we will deal with the transduction process and survey the pathways that carry visual information from the retina to other parts of the brain. In the next chapter, we will focus on how visual information is coded by the brain.

The Optics of the Eyeball

Light that is reflected off surfaces in the environment is focused on the retina by the optical elements of the eyeball: the cornea and lens (Fig. 26.1A). Most light refraction occurs at the cornea; the adjustable lens enables the fine

314

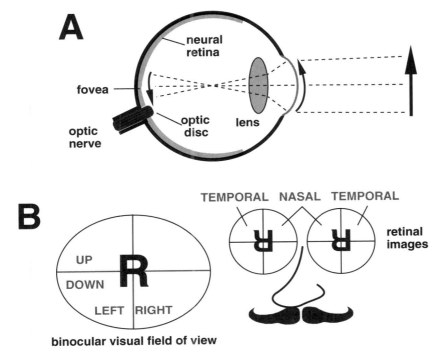

Figure 26.1 The optical system (**A**) of the eye, consisting of the cornea and lens, focuses light images on the retina (shaded red). As a result, the visual scene is inverted and reversed on each retina (**B**).

focusing of images. The other tissue elements that lie in the light path are essentially optically neutral; they have little effect on image quality. Although damage to these optical elements of the eyeball disrupts vision, the means to compensate are readily available to the ophthalmologist. The lens is subject to *cataracts* (opacities) that interrupt light transmission to the retina. Since its refraction is small, an opaque lens can be surgically removed and refraction assisted with eyeglasses. Deformations of the eyeball itself are very common and are also readily compensated with glasses.

The phototransducers lie in the **retina,** which lines the inner surface of the vitreous chamber. The retina is backed by a melanin-containing epithelium that absorbs any light that isn't first absorbed by the phototransducers. Because of the curved shape of the cornea and lens, images are inverted and reversed from right to left at the retina (Fig. 26.1B). As a consequence, the part of the visual field that is projected onto the nasal half of the retina (or nasal hemiretina) of one eye is projected onto the temporal hemiretina of the other eye. Since noses (especially big ones) block the light path from the extreme peripheral part of the visual field, there is a crescent-shaped monocular part of each nasal retina that sees what a similar crescent in the contralateral temporal retina does not.

The Structure of the Retina

There are two classes of phototransducer cell in the human retina, named according to their shape. **Rods** contain a highly absorbent photopigment and are responsive in dim light conditions; they mediate night vision. **Cones** are responsive in bright light and come with one of three photopigments that are differentially absorbant to photons in different parts of the spectrum; they mediate color vision. Cones are largely confined to the **fovea,** a dimple in the retina in line with the optical axis of the eyeball. Since the packing density of cones is far higher in the fovea than in more eccentric parts of the retina, the fovea is the site of highest acuity as well as of color detection. Eccentric to the fovea, the phototransducers are mostly rods, and their density decreases rapidly with radial distance from the fovea. A couple of millimeters nasal to the fovea is the **optic disc,** where the efferent axons of the retina accumulate to form the optic nerve. Since there are no phototransducers at the optic disc, it creates a blind spot in the monocular visual field of each eye.

The retina is a cortical structure with only five types of neuron, arranged in precisely ordered layers (Fig. 26.2). Outermost (with respect

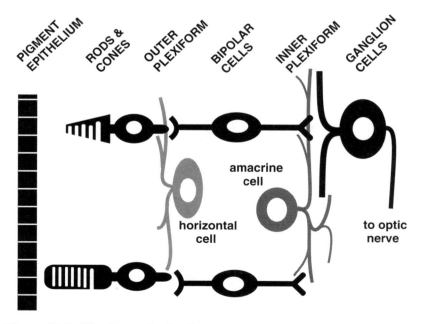

Figure 26.2 The five retinal cell types are arranged in a precise wiring pattern. In this schematic, the light path goes from right to left; the light must pass through all the layers of the retina to reach the rods and cones. The flow of neural signals proceeds in the opposite direction, but with a considerable amount of lateral spread mediated by the horizontal and amacrine cells.

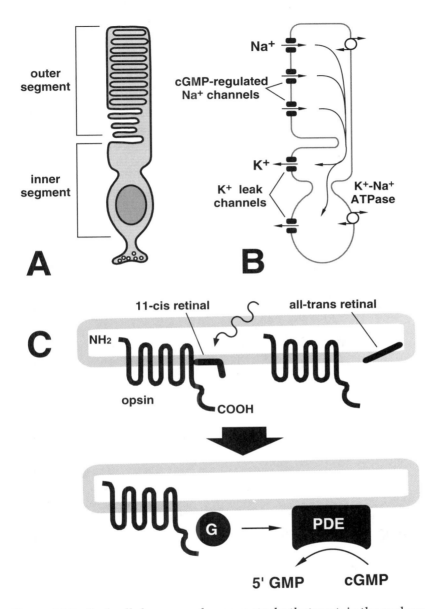

Figure 26.3 Rod cells have membranous stacks that contain the molecular machinery of phototransduction (**A**). They exhibit a resting potential that is determined by special channels in different parts of the cell (**B**). Photons absorbed by rhodopsin in the stacked membranes (red) trigger an initial biochemical reaction that ultimately alters the electrical activity of the cell membrane (**C**). Abbreviations: G, G protein; PDE, phosphodiesterase.

to the core of the eyeball) is the layer of rod and cone cells. Internal to this layer is a layer of **bipolar neurons** that mediate activity between the phototransducer cells and the innermost layer of **retinal ganglion cells.** The other two cell types are the **horizontal cells** and the **amacrine cells** that lie, respectively, at the outer and inner extremes of the bipolar cell layer. The synaptic network between the phototransducers and the bipolar cells is called the **outer plexiform layer,** and the one between the bipolar and retinal ganglion cell layers is called the **inner plexiform layer.** Whereas the simplest conduction route through the retina proceeds from the rods and cones to the bipolar cells to the ganglion cells, the amacrine and horizontal cells engage in a number of complex intraretinal connections that figure prominently in proper retinal processing.

Phototransduction

Rods and cones are highly specialized hair cells. Each gives rise to a single cilium, which differentiates as an elaborate structure—the **outer segment** of the mature phototransducer cell—characterized by stacked layers of membrane (Fig. 26.3A). In rods the membranes become fully internalized and appear as a stack of discs, like poker chips. In cones the membranous discs remain in continuity with the plasma membrane. The layers are continuously lost at the tips, where the sloughed membrane is scavenged by cells of the pigment epithelium, and renewed at the bases of the outer segments. *The membranes of the outer segments are enriched in photoresponsive pigments and a number of signal transduction proteins that mediate the response of the rods and cones to light.* The stacked arrangement of the photopigment-containing membranes increases the likelihood that photons will strike the photopigment molecules.

The outer segments are further characterized by a special form of Na^+ channel present in high density in the plasma membrane. *These Na^+ channels are directly regulated by cGMP.* Each channel is a single protein that exists in an open configuration while bound to three cGMP molecules. In contrast, the membrane of the cell body, the **inner segment,** has no such channels but instead is studded with leaky K^+ channels. In the dark, cGMP levels are high in the outer segment and sustain a steady inward Na^+ current, which is balanced by an outward K^+ current across the inner segment membrane (Fig. 26.3B). Na^+-K^+ ATPases constantly bail the ions back where they belong. This *dark current* keeps the membrane potential of the phototransducer cells resting near –40 millivolts. *When light is absorbed by the photopigments, the level of cGMP in the outer segment is rapidly and dramatically reduced, the cGMP-activated Na^+ channels close up, and the membrane of the phototransducer cell is hyperpolarized.*

How does light bring about a reduction in cGMP levels? Let's consider the transduction process in rod cells, where it is best understood. The photopigment of rod cells is **rhodopsin,** which consists of a GTP-binding protein (opsin; see Chapter 5) linked to a molecule that is an aldehyde form of vitamin A called **retinal** (Fig. 26.3C). The linkage occurs in the seventh transmembrane domain of the opsin protein. When light is absorbed by a retinal molecule, it changes from a bent configuration (*11-cis*) to a linear configuration (*all-trans*), which detaches from opsin. Free of its retinal burden, opsin undergoes a conformational change that allows it to interact with a G protein called **transducin.** The α subunit of transducin then activates a membrane-associated phosphodiesterase that catalyzes the conversion of cGMP to 5′-GMP. The decrease of cGMP—on the order of 10^5 molecules for each rhodopsin that absorbs light—decreases the probability that the cGMP-regulated Na^+ channels will open; the inward Na^+ current is diminished and the cell is hyperpolarized. We learned in Chapter 5 that G proteins such as transducin become inactive when they hydrolyze their bound GTP to GDP. Although this contributes to the termination of the light response, the major terminator is a protein kinase called **arrestin** that phosphorylates activated rhodopsin and deactivates it.

The photoactive mechanism in cones is the same as in rods, except that the opsin portion of the photopigment complex confers a spectral selectivity on the retinal molecule. There are three types of cone photopigment, which respond maximally to blue, green, or red-orange wavelengths (Fig. 26.4). Although each type exhibits a broad spectral sensitivity, the *number* of photons each will absorb varies with the wavelength of the photons. Thus, a red-orange cone will absorb twice the number of photons from a light of 560 nanometers wavelength as it will from a light of 490 nanometers. However, it will absorb the same number of photons from a light of 490 nanometers if the light is twice as bright. Light intensity can compensate for the photopigment's rate of

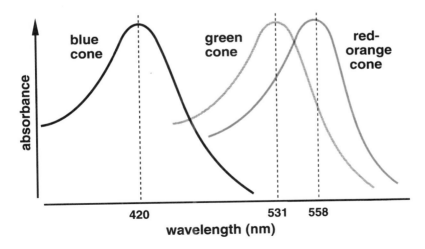

Figure 26.4 The spectral sensitivity curves for the absorbance of light by the three types of photopigments of cones.

absorbance. Thus, the response of the cones depends on intensity as well as wavelength, and the output of a single type of cone by itself cannot inform the brain about color. Instead, *color vision is achieved by comparing the relative outputs of at least two different cone types.* Such comparisons are performed by the retinal cells that receive an orderly input from the cone cells (see below).

The *all-trans* retinal that is generated by the phototransduction process is absorbed and degraded by the cells of the pigment epithelium rather than being recycled to *11-cis* retinal. Rod cells must, therefore, produce the *11-cis* form of retinal from a vitamin A precursor obtained in the diet. A dietary deficiency of vitamin A can cause a condition known as *night blindness,* in which the rods become unresponsive due to a lack of retinal. If prolonged, vitamin A deficiency can even lead to retinal degeneration.

Retinal Processing Circuits

There are two routes of information transfer from the cones to the ganglion cells. Relative to a given ganglion cell, there is a *direct* route, in which a cone cell contacts a single bipolar cell, which in turn contacts the ganglion cell. There is also a *lateral* route that involves *horizontal cells that transfer signals electrotonically via gap junctions* between distant phototransducer cells (Fig. 26.2). This horizontal cell conduction depolarizes phototransducer cells—the opposite effect to that of light. Like phototransducer cells, the bipolar cells and horizontal cells conduct membrane potentials passively, an adequate means for the short distances involved. The consequence of this wiring is that each bipolar cell has a characteristic type of receptive field that consists of a central circular region in which light will either depolarize (on-center) or hyperpolarize (off-center) the cell, and an annular surround, in which light will have the opposite effect to that of the center (see Fig. 26.6).

The cone cells release glutamate as a neurotransmitter onto bipolar cells. In the dark, when the cones are depolarized, they release a steady flow of glutamate, which inhibits on-center bipolar cells by opening K^+ channels and closing Na^+ channels, and excites off-center cells by opening Na^+ channels. Each effect is mediated by distinct glutamate receptor subtypes. If a cone that is connected to an on-center bipolar cell is hyperpolarized by light, it decreases its release of glutamate, and the reduction of inhibitory transmission allows the bipolar cell to depolarize. Off-center bipolar cells, which are continuously depolarized by cone cells in the dark, are hyperpolarized when the cone cells are struck by light and decrease their excitatory glutamate output. *The receptive field properties of a ganglion cell reflect those of the bipolar cell that contacts it because the synapse between bipolar cells and ganglion cells is excitatory.* The details of the ganglion cell response differ somewhat from those of the bipolar cell response, however, because of the additional input they receive from amacrine cells. Amacrine cells come in a wide variety of morphologies and express multiple neurotransmitters and cotransmitters. As a result of this complexity, their roles in retinal signal processing have been difficult to catalog.

Retinal Adaptation to Dim and Bright Light

Retinal conduction changes under different light conditions. In bright light, the direct route of signal conduction—cones to bipolars to ganglion cells—is dominant. Since cones are less sensitive in dim light, nearby rod responses are sent directly to cone cells via gap junctions between the two types of phototransducer cell. This use of the cone system preserves the receptive field properties of the ganglion cells. In very dim light, the receptive field properties and sensitivity of the ganglion cells are altered. In the dark-adapted state, they are no longer inhibited by light in the surround portion of their receptive fields, and their sensitivity increases dramatically. This phenomenon occurs because the gap junctions between the rods and cones close up. Instead, the rods convey their signals to bipolar cells that receive input *only* from rods. Moreover, these so-called *rod bipolar cells communicate with ganglion cells indirectly via synapses on amacrine cells.*

The mechanism of light adaptation, when the eyes go from darkness to bright light, depends mainly on a Ca^{2+}-mediated response of the phototransducer cells (Fig. 26.5). During the dark current, Ca^{2+} normally

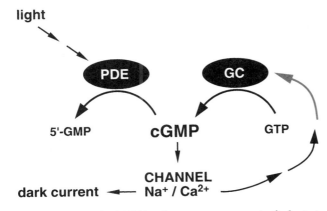

Figure 26.5 The level of cGMP in the outer segment of phototransducer cells is regulated by two mechanisms. The Ca^{2+} that enters through the cGMP-gated Na^+ channels during the dark current inhibits (red arrow) guanylyl cylclase (GC) via an intermediary protein pathway to dampen cGMP production. Light indirectly activates a phosphodiesterase (PDE) that rapidly and potently reduces cGMP. This dual mechanism enables rapid adaptation of the transducers to bright light after dark adaptation has occurred (see text).

enters the outer segments along with Na^+ and serves to indirectly inhibit guanylyl cyclase activity and lower the amount of cGMP produced from GTP. As usual, Ca^{2+} that enters the cytoplasm is rapidly pumped into organelles and back out of the cell. When the eye is suddenly brightly illuminated and the mobilized phosphodiesterases obliterate most of the extant cGMP, the cGMP-gated channels close and Ca^{2+} can no longer enter the outer segment. As a result, the Ca^{2+}-mediated suppression of guanylyl cyclase is lifted, cGMP synthesis goes up, and the Na^+ channels gradually begin to open again, restoring the dark current and hence light responsiveness.

Ganglion Cells Are Either Contrast or Change Detectors

Most ganglion cells have circular antagonistic center-surround receptive fields (Fig. 26.6). Those that receive input from the foveal region have *smaller* fields than those that receive input from regions of the retina eccentric to the fovea. An on-center ganglion cell increases its firing rate when a dot of light is shown in the center of its receptive field; a dot or, better still, an annulus of light shown in the surround decreases its basal firing rate. If its receptive field is evenly illuminated, it does not respond much at all, since the center and surround effects cancel one another.

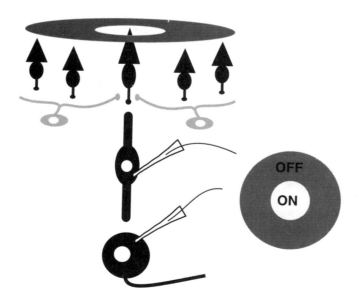

Figure 26.6 Recordings from a bipolar cell or a ganglion cell reveal their antagonistic center-surround receptive fields, as shown on the right. In this depiction, the cells happen to have on-center/off-surround fields.

Off-center ganglion cells decrease their activity when a light appears in the center of their receptive fields. When the light is turned off, the off-center cells respond with a brief burst of activity. The significance of this receptive field configuration is that *the firing rate of such a ganglion cell is a direct comparison of the amount of light in the center and surround of its receptive field*. Ganglion cells with small, antagonistic center-surround receptive fields have been designated *type X* to distinguish them from larger, faster-conducting *type Y* ganglion cells that respond briskly to illumination changes or to stimulus movement. *Type X ganglion cells are set up to detect subtle contrasts in the intensity of illumination and thus convey data about the fine grain of stationary images. Type Y ganglion cells are designed to detect the sudden appearance of, or rapid changes (movements) in, a visual stimulus.* In either case, the absolute level of ambient light intensity, though useful for pupillary reflexes and adaptive changes in retinal processing (as described above), is ignored. After all, it contains no detailed information about objects in the visible environment. The contrasts contained in an image, which are largely independent of the absolute light intensity, are laden with useful data about the dimensions, textures, locations, and movements of objects—the stuff of visual perception.

Different ganglion cells are attuned to different types of contrast in the visual scene. Serving each region of the retina are parallel sets of ganglion cells that receive input from the same subsets of phototransducer cells, but due to differential retinal processing circuits, analyze the color or the intensity of luminous surfaces. The ganglion cells we have considered so far are broad-band cells that are interested in *intensity contrast*. There are other X-type ganglion cells that have concentric *color*-opponent receptive fields. They receive antagonistic center-surround inputs from the green and red-orange sensitive cones. Such ganglion cells can be sorted into the four possible permutations: (1) green on-center/red-orange off-surround, (2) green off-center/red-orange on-surround, (3) red-orange on-center/ green off-surround, and (4) red-orange off-center/green on-surround. Ganglion cells that receive input from blue-sensitive cones have a uniform on or off field that is antagonized by inputs from both red-orange and green cones.

Thus, most ganglion cells are arrayed as a continuous matrix of intensity and color contrast detectors. In the next chapter, we'll see how higher visual processing centers exploit the rudimentary data acquired by this retinal matrix. But first let's examine the location and structure of the higher centers and the pathways over which visual signals are delivered to them.

The Central Visual Pathways

The three main synaptic targets of fibers in either optic tract are the **lateral geniculate nucleus** (LGN) of the thalamus, the superior colliculus

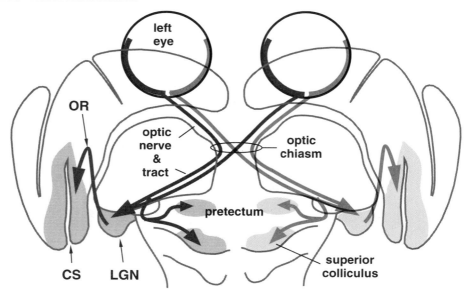

Figure 26.7 The pathways from the retina lead to the three main targets indicated on this drawing. The axons from the LGN form the optic radiations (OR) that extend to the V$_I$ area along both banks of the calcarine sulcus (CS).

of the midbrain, and some small nuclei in a region that lies between the colliculi and thalamus called the **pretectum** (Fig. 26.7). There are also a few smaller nuclei that receive some optic tract axons. The suprachiasmatic nucleus of the hypothalamus and a small nucleus in the base of the midbrain are examples. We have dealt with the functional organization of the superior colliculus and the pretectum in Chapter 17; the former processes signals regarding the position of stimuli in visual space to direct saccadic eye movements, and the latter is involved in pupillary light reflexes and the accommodation reflex. Not surprisingly, the superior colliculi and pretectum receive axons mainly from Y-type ganglion cells. Here, we'll focus on the pathway that leads ultimately to visual perception—the detailed identification of images. It is descriptively called the *retinogeniculostriate* pathway because of an obligatory synapse in the LGN from which the signals are relayed to the V$_I$ area (or "striate" cortex because of a uniquely dense axonal plexus visible to the unaided eye in layer IV) of each cerebral hemisphere.

Axons from the ganglion cells of each eye collect as the optic nerves, which extend to the optic chiasm, where the axons from the nasal hemiretinas cross the midline to continue with the uncrossed axons of the temporal retinas as the optic tracts (Figs. B8 and B9). As a result of this hemidecussation, the two halves of the binocular visual field are each represented on the opposite side of the brain, and the two repre-

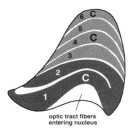

optic tract fibers
entering nucleus

Figure 26.8 The neurons of the LGN are arranged in layers that receive retinal input from either the contralateral (C) or the ipsilateral eye. The neurons of the first two (magnocellular) layers are larger than those in the remaining parvicellular layers (3 through 6).

sentations of the same visual hemifield, one from each eye, are brought into register before reaching the LGN (see also Fig. 26.9).

The cells in the LGN do not receive binocular input, even though the optic tracts carry axons from both eyes. Instead, the cells receiving input from the ipsi- or contralateral eye are segregated into three of the six discrete layers that make up the nucleus (Fig. 26.8). The layers are numbered beginning with the one closest to the incoming optic tract. Layers 1, 4, and 6 receive axons from the contralateral nasal retina; the other layers receive the axons of the ipsilateral temporal retina. Thus, within each layer there is a complete representation of the contralateral visual hemifield. And all six of these visuotopic maps are in register with one another. Because the highest density of transducers occurs in the fovea, its representation is disproportionately large within each layer of the geniculate. Incidentally, the temporal crescents do not project to the geniculate nuclei, explaining why we do not see our noses when we look straight ahead. The neurons in layers 1 and 2 are larger than those in the remaining layers. These magnocellular layers receive input from pre-dominantly Y-type ganglion cells and are therefore concerned with motion. The X-type ganglion cell axons terminate in the parvicellular layers, where the cells respond to steady illumination and color.

How visual signals are transformed by processing at the LGN remains largely speculative. There is an almost one-to-one correspondence of single ganglion cells with individual neurons in the LGN. Their receptive field properties are nearly identical. At most, there seems to be some sharpening of the contrast between the center and surround portions of the receptive fields. This correspondence holds despite the fact that most of the synaptic endings on neurons of the LGN come from sources other than the retina. Two of the more prominent are a feedback system from the V_I cortex and a functionally uncharacterized input from the brain stem reticular formation.

Most of the neurons in the LGN project by way of the optic radiations to the cortex lining each bank of the calcarine sulcus. The topography of the geniculostriate projection imposes a visuotopic map of the con-tralateral visual field on this primary visual area (V_I) of the cortex so that the upper visual field is represented below and the lower visual field above the calcarine sulcus (Fig. 26.9). The foveal representation, of course, occupies the lion's share of the V_I cortex nearest the occipital pole.

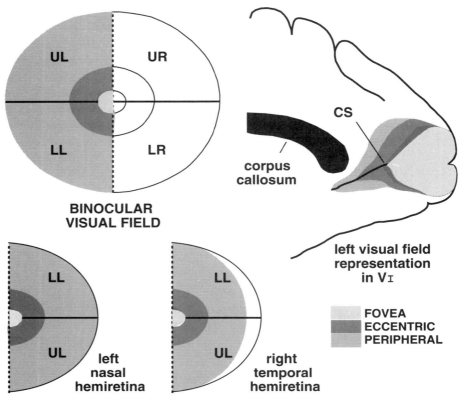

Figure 26.9 The visual pathways are organized so that an orderly map of the contralateral visual hemifield is imposed on the V$_I$ cortex. The horizontal meridian (solid lines in the visual field and on the retinas) corresponds to the calcarine sulcus (CS). The fovea and its disproportionate cortical representation are shaded red. The nose blocks light from the extreme lateral peripheries of the visual field, creating a monocular crescent-shaped area of the temporal retinas (shown as the nonshaded part of the right temporal hemiretina), which is not represented in V$_I$. The abbreviations refer to quadrants of the visual field, not the retina: LL, lower left; LR, lower right; UL, upper left; UR, upper right.

The organization of the central visual pathways is exposed by lesions (thankfully uncommon) that affect different segments of the pathways. Some of the more instructive lesions are summarized in Box 26.1.

The geniculostriate axons synapse on and excite stellate cells in cortical layer IV, the lower part of layer III, and layer VI. The Y-type cells in the magnocellular layers of the lateral geniculate project to one sublayer within layer IV, whereas the X-type, parvicellular neurons project to another. This excitatory input from the lateral geniculate engages a complex local circuitry that enables further extraction of information

Box 26.1   Lesions of the Visual Pathways

The visual system can be damaged by trauma or disease from the optical apparatus of the eyeball to the visual cortex and beyond. Despite the efficiency of the eyeblink reflex, the cornea can be scratched. The lens is subject to opacity from cataracts or *glaucoma* (increased intraocular pressure). *Retinitis pigmentosa* is a group of common congenital retinal dystrophies that are characterized by pigment deposition in the retina due to breakdown of the pigment epithelium. There is a progressive rod and cone degeneration that begins with night blindness in childhood, increasing tunnel vision through the fourth, fifth, and sixth decades of life, and complete blindness by the age of sixty. Since the disease mechanism remains unknown, little treatment is available. The retina can also be damaged by increased intraocular pressure, as in glaucoma, or by impact, such as that sustained by prize fighters.

In the diagram below, the left drawing shows some sites of the more instructive lesions (numbered black bars) to affect the visual pathways. The pairs of circles at the right depict the visual field defects (blackened areas) that result. Each circle in the pair represents the visual field of one eye. The common cause of the first lesion is papilledema (choked disc), which is caused by increased intracranial pressure or glaucoma. The compression of the optic nerve causes monocular blindness. Lesion 2

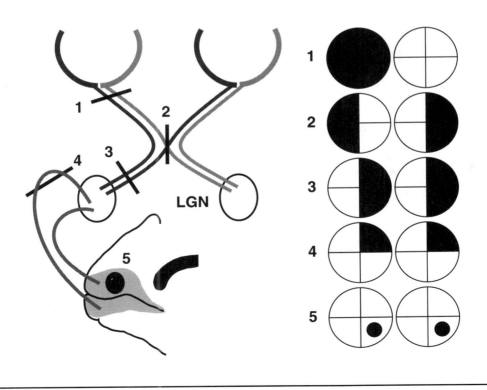

often results from a pituitary tumor that expands upward from the sella turcica to compress the optic chiasm. The visual field defect is called *bitemporal hemianopsia.* Damage to the optic tract on one side is rarely complete. When it is, as indicated by lesion 3, the defect is a *right homonymous hemianopsia.* Trauma or infarct of the temporal lobe can interrupt the fibers of Meyer's loop (a part of the optic radiation that sweeps forward in the temporal lobe before recurving toward the occipital lobe), as in lesion 4. The one depicted causes a *right bilateral upper quadrantanopsia.* Other lesions of the optic radiations or lesions of the visual cortex lead to blind spots in the contralateral visual field called *scotomas.*

Testing of the visual field for areas of blindness can be done quickly and without fancy equipment. The patient covers one eye and foveates the doctor's nose while the doctor moves a pencil inward from the periphery along the four main axes of the visual field. Abbreviation: LGN, lateral geniculate nucleus.

from the visual image. The V_I cortex is also the first site of significant binocularity. We'll take up the cortical processing of visual signals in the next chapter.

Detailed Reviews

Dowling JE. 1987. *The Retina: An Approachable Part of the Brain.* Belknap Press, Cambridge, MA.
Stryer L. 1987. The molecules of visual excitation. *Sci Am* 257: 42.

Visual Processing 27

How does the brain convert a tiny, two-dimensional, upside-down reflection of the outside world into a coherent image of the objects that lie before it, complete with meaningful textures, colors, shapes, sizes, spatial relationships, and movements? Although we are far from the final answer to this question, some promising inroads have been made, beginning with the mid-century discoveries of David Hubel and Torsten Wiesel on the processing of visual stimuli in the V_I cortex—discoveries that earned them a Nobel Prize in Physiology or Medicine. As a result of this work and the prodigious experimental effort it fostered, we now know that the rudimentary data sent from the contrast detectors of the retina are channeled over *parallel* pathways in which they are processed *serially* to extract increasingly complex details about stimulus attributes—the kind of processing that is requisite to the formation of perceptions. Early stages of this processing occur in the visual cortical areas of the occipital lobe. You have already been introduced to the primary visual cortex (V_I or Brodmann's area 17). More or less concentrically surrounding V_I are a second and a third visual area called V_{II} and V_{III}, which correspond, respectively, to Brodmann's areas 18 and 19. In this chapter, we'll consider the functional organization of these areas and the computations performed by their resident neurons. In the following chapter, we'll see how these early computations are further reshuffled and recombined by circuits of the association cortex to form perceptions.

The Basic Circuit of V_I

The V_I cortex, like the rest of the neocortex, contains two predominant morphological types of neuron—pyramidal and stellate—that are segregated in different layers (Fig. 27.1). Layer IV contains only stellate cells, the dendrites of which are either spiny or smooth. The spiny stellate cells

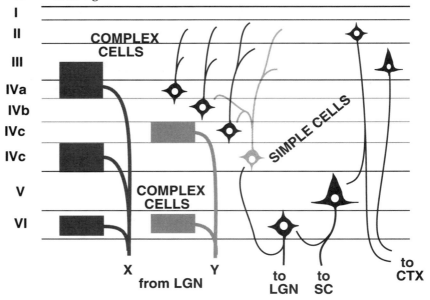

Figure 27.1   A simplified version of the geniculate inputs, local process-ing circuit, and outputs of the V$_I$ cortex is shown in this diagram. The terminal fields of the two categories of geniculate axons are indicated by shaded rectangles. The jet black cortical neurons are excitatory and the red one is an inhibitory, GABAergic, smooth stellate cell. Abbreviations: CTX, cortex; LGN, lateral geniculate nucleus; SC, superior colliculus.

and the pyramidal cells of the other layers use the excitatory neurotrans-mitter glutamate, whereas the smooth stellate cells use the neurotrans-mitter GABA and mediate local inhibitory connections. Geniculostriate axons distribute densely in layers IV and VI, where they have an excita-tory effect on the spiny stellate cells. The spiny stellate cells, in turn, activate pyramidal cells in the layers above them (II and III). The cells in layers II and III synapse on the pyramidal cells of layer V, which then activate smooth stellate cells in layer IV by way of recurrent collaterals of the axons they send to the superior colliculus. Finally, the smooth stellates inhibit spiny stellate cells in layer IV and cells in the supragranu-lar layers as well. As best as we can ascertain to date, this circuit, though surely an oversimplification, constitutes the basic processing route within each small volume of the V$_I$ cortex.

Geniculate Channels Remain Segregated in V$_I$

At least three separate functional axon channels can be identified in the geniculostriate projection. Axons of geniculate neurons with X-type receptive field properties (those in the parvicellular layers) synapse on spiny stellate cells in the deeper part of layer IV and in layer VI (Fig. 27.1). The Y-type axons from the magnocellular layers of the geniculate synapse on spiny stellate cells in a more superficial substratum of layer IV and in layer VI. Geniculate axons that carry color information are distributed to cylindrical clusters of cells called **blobs** that appear in the layers above and below layer IV. Thus it would appear that the integrity of the channels for fine contrast, motion, and color discrimination established in the retinas and maintained in the lateral geniculate nuclei is preserved in the V_I cortex.

Blob cells mimic the circular, color-opponent receptive fields of lateral geniculate cells. A relatively small number of neurons outside the blobs also share this geniculatelike receptive field pattern, but without much wavelength specificity. Such cells respond best to bright or dark dots in the centers of their receptive fields. More commonly, however, recordings from V_I cells disclose receptive fields that are considerably more elaborate than those of geniculate cells. The majority of such cells can be divided into two general types, called **simple cells** and **complex cells** because of their receptive field properties. None of them respond well to diffuse illumination of the retina; quite the contrary, most demand even more specific stimulus features than the cells we have so far encountered at lower levels of the visual system.

Simple cells are found exclusively in layers IV and VI and are characterized by receptive fields that are *elongated in one axis*. In other words, simple cells require not dots, but narrow *bars* of light or darkness to activate them. Moreover, *a bar will activate a simple cell only if it is presented in a particular orientation and in a particular part of the visual field.*

For example, the receptive field of the simple cell depicted in Figure 27.2 has a rectangular off-center that is oriented at 45° off the vertical and flanked on either side by rectangular on-surrounds. This particular simple cell is maximally activated only when all these criteria are met by the stimulus. Although bars near this orientation evoke some modulation of the cell's activity, the surrounding inhibitory zone ensures that the optimal change occurs only at the preferred orientation of 45°. As the orientation is shifted progressively away from 45°, the response of the simple cell rapidly declines. Likewise, if the dark bar is kept at 45° but moved from the central portion of the receptive field to the off-surround, the simple cell will decrease rather than increase its basal firing rate. A bar that is wide enough to cover both the center and the surround will have no effect on the cell's activity at all; neither will a bar in some other part of the visual field. Other simple cells prefer different orientations so that *all possible 180° of orientation are represented in different simple cells that "see" the same small part of the visual field.*

The receptive fields of simple cells and their dynamic properties can be explained by a hypothetical convergence of X-type geniculate axon input (Fig. 27.3). If the axons of a series of geniculate cells whose circular

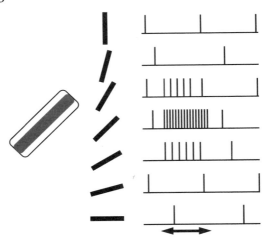

Figure 27.2 Recordings from a cortical simple cell upon a brief presentation (two-headed arrow along time axis) of dark bars in different orientations indicate that this particular cell has a preferred orientation of 45° in its receptive field (left).

receptive fields are aligned along a particular axis converge on a single cortical neuron, the areal and dynamic receptive field properties of the cortical cell will reflect the composite of the geniculate receptive fields. The example in Figure 27.3 shows a series of on-center off-surround geniculate receptive fields that are vertically arrayed. The collective input of these geniculate cells on the V$_I$ simple cell imposes a vertically oriented, linear receptive field with an on-center and an off-surround. A bar of light that exactly matches the on-center will maximally activate the simple cell. If the bar is lengthened beyond the receptive field ends, the discharge frequency of the cell will increase no further.

Not all simple cells have a symmetrical receptive field (Fig. 27.3). For some simple cells, one of the flanking surrounds is narrower than the other, whereas for others, antagonistic portions of equal width lie side by side. Presumably, different systematic arrays of X-type geniculate receptive fields can build the panoply of simple cell receptive fields that have so far been described in V$_I$. That all simple cells are aroused by short linear segments suggests that they function together as a huge battery of *edge detectors*—hardly a surprise, since edge detection is the essence of form and pattern perception.

Complex cells are found in the upper part of layer IV as well as layers above and below layer IV. Like simple cells, complex cells require linear stimuli of a particular orientation. However, *they are not very position-selective and do not exhibit flanking antagonistic regions.* Complex cells respond best to an *edge* of light (or darkness) in a particular orientation located *anywhere* within their receptive fields. Their receptive fields

LGN X   V$_I$ SIMPLE

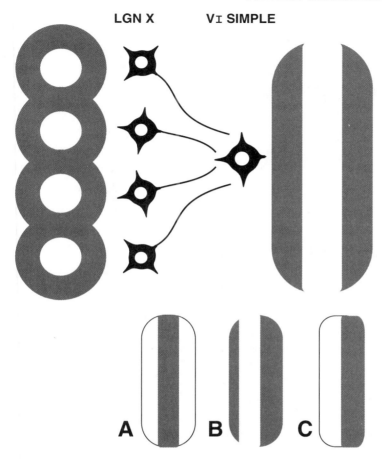

Figure 27.3 Simple cells respond to linear stimuli. The size and shape of their receptive fields and its antagonistic properties can be accounted for by convergent projections of center-surround geniculate cells. Some of the variations on simple cell receptive fields are shown at the bottom. The receptive field in **A** is a symmetrical, off-center/on-surround field, whereas **B** and **C** are examples of asymmetric fields. Abbreviation: LGN X, lateral geniculate nucleus type X neurons.

typically cover larger parts of the visual field than do those of simple cells. Unlike simple cells, cells with complex receptive fields are found in area V$_{II}$ as well as area V$_I$. Many complex cells exhibit stimulus requirements in addition to these defining ones. For instance, some are activated maximally when the light edge is stopped at one end to form a corner (Fig. 27.4). Others seem to be especially interested in edges that *move* across their receptive fields *in a certain direction*.

The receptive fields of many complex cells appear to be composites built up from the fields of a number of simple cells that provide them

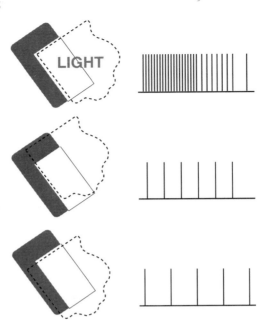

Figure 27.4 A special complex cell requires that a light (or dark) region be truncated at one or both ends. The light stimulus—in this case a lighted corner—is passed over the receptive field (left) of the cell whose response is shown in the recordings of action potentials to the viewer's right.

with synaptic input. The properties of the special complex cells may depend also on inputs from other complex cells of lesser complexity. Thus, complex cells, the special ones in particular, represent an end point in a stepwise processing scheme, beginning with the bipolar cells in the retina, in which each successive level imposes additional stringency on the next. Many complex cells, however, *receive a direct synaptic input from Y-type geniculate cells,* which dictate their receptive fields, much as the simple cell receptive fields are composites of X-type geniculate cell receptive fields. Remember also that the blob cells receive direct input from a population of geniculate cells concerned with color-coding. Thus, there are at least three *parallel channels that serially process different stimulus parameters in the V_I cortex.* The principles of serial and parallel processing are robust but certainly not unique to the visual system; we have seen them before in the somatosensory and auditory systems. Indeed, they form the basis of all perceptions.

Slabs and Blobs in the V_I Cortex

We know from the topography of the geniculostriate projection that cortical cells that "look at" the same part of the visual scene must be located near one another. We now also know that within that small area of cortex, there must be a complement of simple and complex cells responsive to all possible edge orientations. Likewise, there must be an ample set of blob cells for color detection. A question immediately leaps to mind: *How are all these cells arranged in the V$_I$ cortex?* Certain as we are by now that Mother Nature is an orderly housekeeper, you can bet their distribution is not helter-skelter. If a recording electrode were lowered into V$_I$ perpendicular to the pial surface and the stimulus-evoked responses of cells were sampled every few micrometers along the electrode path through the depth of the cortex, all of them would exhibit the same orientation specificity regardless of their laminar affiliation. Move the electrode a few dozen micrometers to one side and repeat the experiment, and the orientation preference changes by a few degrees, but again, all the cells respond maximally to the same stimulus orientation as long as the electrode is advanced radially through the cortex. If the penetration is made tangentially, however, the orientation preferences of the cells would change by a few degrees every 35 micrometers or so. It was through the tedium of just such recording experiments that a revolutionary principle of cortical organization was discovered by Hubel and Wiesel. *The cells of V$_I$ are organized in radial units* (originally thought to be columns) *each of which contains cells that have the same preferred stimulus orientation.* And *progressing systematically across adjacent units, the orientation preference changes gradually and systematically through all of the 180° of possible orientation* (see Fig. 27.6). Orientation slabs can also be demonstrated by radioactive 2-deoxyglucose labeling of metabolically active cells. If a monkey is exposed to a visual stimulus consisting of black and white stripes in a single orientation (say, 45°) for several minutes and a bolus of radiolabeled 2-deoxyglucose is injected systemically, the V$_I$ cells selectively activated by lines at 45° drink up the 2-deoxyglucose and can be seen in autoradiograms of tissue sections. As predicted by the physiological studies, the labeled cells appear in sinuous *slabs* about 20 to 50 micrometers in width that course for several millimeters along the cortex.

Radial electrode penetrations also reveal "columns" of cells that respond more strongly to input from one eye than the other. These *ocular dominance columns,* as they were originally (and are still commonly) called, are wider than the orientation slabs, on the order of 250 to 500 micrometers across, and occur in an alternating pattern of ipsi- and contralateral domination (Fig. 27.5). An anatomical basis for the ocular dominance columns was demonstrated soon after their initial discovery. If a high concentration of radioactive proline is injected into the vitreous body of one eye, the ganglion cells will take it up and transport it along their optic tract axons (see Box 1.1). Upon arrival at the lateral geniculate nuclei, some of the radiolabel escapes across the synapses and gets into the geniculate cells, which in turn send it along their own axons to the V$_I$ cortex. Autoradiograms of tissue sections through the V$_I$ area reveal the labeling in layer IV in a striking pattern of high-density terminal

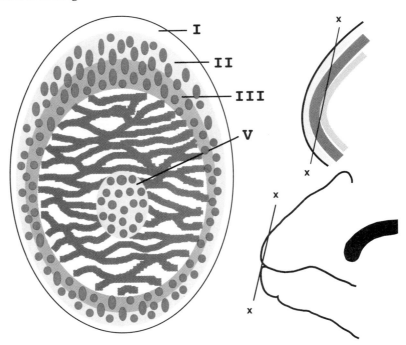

Figure 27.5 The patterns of the ocular dominance slabs and color-analyzing blobs can be seen in a tangential slice taken through the V$_I$ cortex (lower right image) when the slice (left) is stained appropriately. In this case, the slice passes through all but the deepest (VIth) layer of the cortex (indicated in the upper right image).

zones in alternation with equal-size stretches devoid of labeled terminals—a pattern resembling a fingerprint. This anatomical experiment demonstrated that the ocular dominance "columns" were actually sinuous slabs like the orientation slabs. Interestingly, *the general patterns of ocular dominance slabs and orientation slabs are arranged so as to intersect one another at more or less right angles* (Fig. 27.6).

Continued analysis of this radial organization eventually led to a hypothetical modular plan of the V$_I$ cortex: *For each small area of the visual field, there is a unitary volume of V$_I$ cortex that is essentially cuboidal, in which both eyes* (two ocular dominance columns) *and a complete 360° of orientation axes are represented.* The cube is called a **hypercolumn** (Fig. 27.6). Also within each hypercolumn are cylindrical arrays of neurons dedicated to the analysis of color information. Because the cylinders stain heavily for cytochrome oxidase and appear as dark polka dots in tangential tissue slices, they were given the name *blobs* (Fig. 27.5). The blobs are discontinuous through the depth of the cortex; they are evident in the supra- and infragranular layers but are interrupted in layer IV. Thus, *each hypercolumn appears to have the minimum necessary allocation of cells neces-*

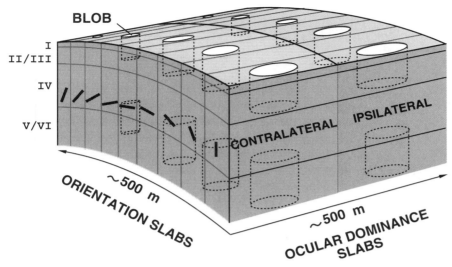

Figure 27.6 A prism of V$_I$ cortex shows the hypothetical orthogonal relationship of the orientation slabs to the ocular dominance slabs. The blobs (cylinders in the supra- and infragranular layers) associated with color processing are scattered throughout. The prism corresponds to a so-called *hypercolumn*.

sary to detect edges of every orientation and color in either eye. In reality, things are not so neat. As mentioned, the slabs tend to squirm about the V$_I$ cortex, and the intersection of orientation and ocular dominance slabs is not perfectly orthogonal most of the time. But the model is conceptually useful and may be functionally, if not structurally, accurate.

Binocularity

We learned in the preceding chapter that information from the two retinas is kept separate in the layers of the LGN. Neither is there any convergence of data from the two eyes in layer IV of V$_I$; ocular domination is total. Cells that are activated by stimulation of either eye are first encountered in V$_I$ in the layers above and below layer IV. Although one eye always has a greater influence than the other on such binocular cells (ocular dominance), the inputs from the two eyes are usually additive. The visual field locations of the receptive fields are identical for many such binocular cells, but not for all. Some have slightly disparate receptive fields, so that stimuli slightly in front of or behind the plane of focus will maximally excite them. Such cells receive mainly Y-type input and play an important role in both depth perception (stereopsis) and triggering vergence movements of the eyes by way of connections with gaze control regions of the upper brain stem (see Chapter 17).

Color-Coding

Remember that color-concerned cells in the LGN are coded according to *single* color opponencies, as are retinal ganglion cells. That is, if a cell has a red-orange on-center and a green off-surround, it will not respond to green in its center or red-orange in its surround. Although the receptive fields of cortical blob cells share the concentric circular configuration of the geniculate cells, the dynamic properties of their color opponency are very different. Cortical cell receptive fields exhibit a *double-opponent* relationship of their centers and surrounds. That is, they experience two-color opponency in *both* the centers *and* the surrounds of their receptive fields (Fig. 27.7). For instance, if a cell is activated by red-orange light shone on its center, it will be inhibited by green light shone there; in its surround, green light will excite and red-orange light will inhibit the cell. In all, there are four types of color double-opponent cells found in the blobs, which respond best to a red-orange dot on a green background, a green dot on a red-orange background, a blue dot on a yellow (red-orange plus green) background, and a yellow dot on a blue background. The connectivity that generates double-opponent cells from single-opponent cells remains unknown. Blob cells are not concerned with stimulus orientation, but are selective for a precise location in the visual field. Many blob cells send corticocortical axons to cells in V$_{II}$.

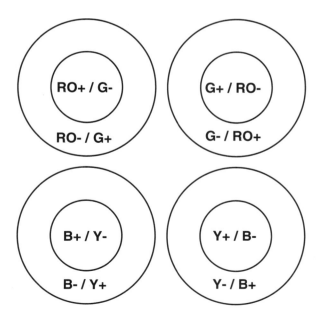

Figure 27.7 Blob cell receptive fields are double-opponent. Abbreviations: B, blue; G, green; RO, red-orange; Y, yellow; +, excitation; −, inhibition.

Many of these V_{II} color cells (which happen to be arranged in slabs) are selective for orientation and show far less selectivity for visual field location. They seem to be the color-processing counterpart of the achromatic complex cells found in V_I.

Higher-level cortical cells compute the *hue, saturation*, and *brightness* of colors. Hue refers to the many variations of the basic colors that we can detect. Although only paint manufacturers and interior decorators seem able to name them, most of us can discriminate about two hundred hues. Presumably, cells somewhere in the cortex are able to compute the proportion that each of the three chromatic phototransducers contributes to the detection of light reflected from an object. The cortex can also compute how much a hue is diluted by grayness—its level of saturation. Such a computation requires cortical cells to monitor the degree to which the three receptors are stimulated equivalently by the object and its background. Finally, there are cells that monitor the total effect of light reflected from an object on all three chromatic phototransducers—its brightness.

Several Cortical Areas Process Visual Signals

The three channels we have followed to and through the V_I cortex—one for form discrimination, one for motion detection and stereopsis, and one for color—continue to V_{II}. The cortex of V_{II} is highly structured into thick and thin slabs that stain positively for cytochrome oxidase and that are called *stripes* to distinguish them from the slabs (or columns) in V_I. The thick and thin stripes are separated by *interstripe zones*. The cells of the *thick* stripes receive motion information conveyed over the Y-type channel, and after processing it in some unknown way, they send it along to V_{III}. The brain centers that control eye movements depend heavily on data provided to them from V_{III}. The *thin* stripes of V_{II} receive color information from blob cells and rearrange it, at least in part, in a way that contributes to form discrimination. The cells of the interstripe zones receive input mainly from the interblob X-type cells of layers II and III in V_I. Beyond these striate and peristriate areas, *there are some two dozen independent cortical areas discovered so far in which the cells respond with some immediacy to visual stimulation.* From the peristriate regions, the channels lead outward across the vast, largely uncharted reaches of the association cortex to these distant visual processing stations of the temporal and parietal lobes. As they go, we soon lose track of what exactly they are up to. For as the channels reach each new synaptic way-station, their messages are once again transformed. Although we are ignorant of those transformations, we know that ultimately the signals carried over these multipartite channels are integrated to form images of the external world that are coherent in time and space, and that at still later stages of cortical processing, the images become associated with perceptions based on signals arriving over parallel processing channels of the other sensory modalities—sound and touch and, perhaps, smell—leading to perceptions rich with meaning.

Detailed Reviews

Cohen B, Bodis-Wollner I (eds). 1990. *Vision and the Brain: The Organization of the Central Visual System.* Research Publications, Association of Nervous and Mental Disease, vol 67, Raven Press, New York.

Hubel DH. 1988. *Eye, Brain, and Vision.* Scientific American Library, New York.

Rose D, Dobson VG. (eds). 1985. *Models of the Visual Cortex.* Wiley & Sons, New York.

Perception, Cognition, and Language

28

It is well established by now that certain areas of the cerebral cortex serve as the primary recipients of sensory information, and that another area, the primary motor cortex, is synaptically close to the lower motor circuits. In other words, these areas differ from other parts of the cortex in that they are specialized for particular and rather immediate sensory or motor functions. You will also recall that within the primary sensory and motor areas there are subregions that are dedicated to a particular part of the receptive surface or a particular muscle. This *areal specialization* in the cerebral cortex is an example of a principle of cortical organization that is commonly referred to as the *localization of function*. It means simply that not all of the cortex participates equally in every brain function; *certain areas appear to be specialized to contribute more than other areas to particular brain computations.* It is also the case that *the cortices of the two hemispheres are functionally asymmetrical;* each side has its limitations and particular capabilities, and a number of cerebral functions are *lateralized* to a greater or lesser degree.

In this chapter, we will explore the localization of mental functions in the cerebral cortex and review evidence that suggests that the cortices of the two hemispheres differ in their functional capacities. Since our knowledge of the computational circuits remains obscure, and cortical areas are but components in extensive brain circuits, the assignment of higher functions is necessarily coarse. Moreover, although the words we use to describe mental functions—*thinking, imagining, predicting,* and so on—have subjective meaning for us all, they are not rigorously defined scientific terms. We cannot be sure that an instruction to *imagine* the launch of the space shuttle elicits the same mental activity from one subject to the next or even from one experiment to the next in the same subject. Moreover, despite the recent advances in brain imaging with PET in vivo (see Box 18.1), we still cannot directly study the neural correlates of mental activity in humans with any great resolution. Instead, we rely

341

on animal models, indirect measures, and inference from clinical manifestations in cases of stroke (see Appendix A). As in all the previous chapters of this survey, the discussion of cortical functions will cling closely to brain structure. For material realism—the philosophy of science—compels the assumption that the "mind" is nothing more than an emergent property of brain circuit activity. Although psychological constructs often provide a convenient means to talk about subjective mental phenomena, the ultimate goal of neuroscience is to elucidate and understand the neural mechanisms that support the human cognitive process.

What Makes Cortical Areas Distinct from One Another?

It is important to appreciate that localization and lateralization of function do not stem from *intrinsic* differences between cortical regions. In fact, we have at several points promoted the notion that the basic cortical processing circuit is everywhere the same and performs the same routine computations on the data it receives. If the neurons are counted in core samples taken from disparate regions of cortex, the number of neurons per cylinder will be essentially identical. Moreover, the numbers, locations, and connections of neurons of a particular phenotype are also remarkably similar from region to region; the content and laminar distribution of local circuit GABA neurons, for instance, are the same in every cortical sample. Even the differences that enable histological identification of different cortical areas—mainly the relative thickness of the layers—depend on the afferent axons that innervate the cortex more than any other determinant (see Chapter 30). Thus, *the functional specialization within the cortex is a product of the manner in which the different cortical areas are specifically connected with one another and with subcortical structures such as the specialized thalamic nuclei.*

Information Flow across the Cortex

You learned in the preceding chapter on central visual processing that sensory information is channeled outward from area 17 over divergent and convergent pathways to sets of cortical neurons in the adjacent cortex of areas 18 and 19, and that this selective distribution results in a hierarchical processing of visual information that allows for feature extraction and accounts for the complex response properties of the target neurons. A comparable sensory processing occurs as a result of information transfer between the other primary sensory cortices (auditory and somatosensory) and the surrounding perisensory areas. What happens to information flow beyond these perisensory areas? We know that *the signals must travel across the cerebral cortex in parallel and serial channels*

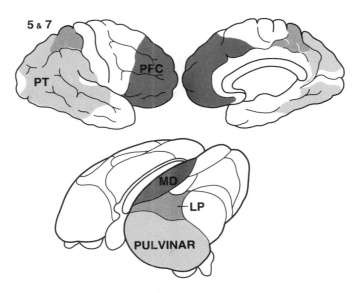

Figure 28.1 The association areas of the cerebral cortex and their thalamic partners are diagrammed. Abbreviations: LP, lateral posterior nucleus; MD, mediodorsal nucleus; PFC, prefrontal cortex; PT, parietotemporal cortex.

leading into those cortical sheets still further removed from the primary sensory and motor areas—the vast and largely uncharted association areas of the cerebral cortex, which are the regions considered to mediate higher mental functions such as perception, imagery, language, and analytical thought (Fig. 28.1). Since we already have examined the primary and secondary sensory and motor regions, our focus in the present chapter will be these mysterious and intriguing association areas of the cortex.

Perception Depends on Serial Cortical Processing

We now know the posterior reaches of the cortex to be, in a sense, the end point for ascending sensory data. What does the cortex do with this sensory information that it receives? Outward from each primary sensory area there is a stepwise march of short axonal projections leading to the ever-widening and more distant cortical sheet comprising the parietotemporal association areas (Fig. 28.2). Over the course of this series of connections, the raw sensory data are refined and compiled into meaningful and multiparametric images of the world called *perceptions*. Thus, the parietotemporal association cortex receives a great deal of conver-

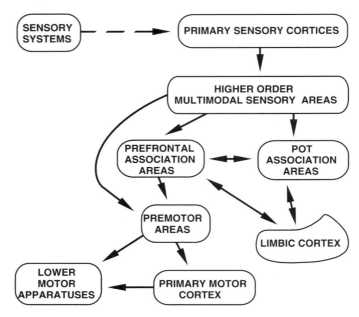

Figure 28.2 The flow of information through the cerebral cortex is summarized. Abbreviation: POT, parieto-occipito-temporal.

gent sensory information from the somatosensory, auditory, and visual cortices. As this information marches across the cortex, cells at each successive level of reprocessing are able to extract increasingly complicated features of the sensory data. You have seen how the hierarchical processing of convergent information from cells with simple response properties leads to relatively complex feature extraction, even within the primary and secondary visual cortices. After reprocessing through several synaptic stations and recombination with sensory information in other modalities, it would not be too surprising to find cells that respond only to very complicated and unique constellations of stimulus features.

The Perception of Space and One's Place in It

Humans depend heavily on visual data to assemble perceptions of the world around them. Recent experimental work has shown that visual data are channeled over two main streams from the peristriate areas: a dorsal stream that leads stepwise through area 19 to area 7 on the caudal part of the superior parietal lobule, and a ventral stream that leads to the inferior parts of the temporal lobe (Fig. 28.3). Whereas the ventral stream deals with visual recognition (see below), the dorsal stream appears to

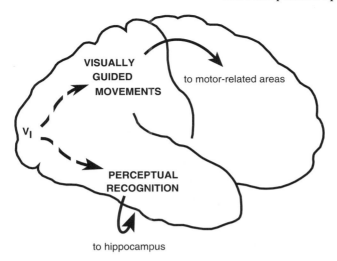

VISUALLY
GUIDED
MOVEMENTS

to motor-related areas

V₁

PERCEPTUAL
RECOGNITION

to hippocampus

Figure 28.3 Two important avenues of stepwise information processing from the primary and secondary visual cortices have so far been described. The same sort of progressive processing occurs for auditory, somatic, and other sensory information.

be involved primarily with the reconstruction of the three-dimensional space surrounding a subject. Studies of regional cerebral blood flow using PET have provided compelling evidence for this dual stream hypothesis. If a normal human subject is asked to examine a complex scene, both the inferior temporal cortex and the superior parietal lobule show an increased blood flow, along with the peristriate visual areas (Fig. 28.5A).

David Lee Robinson recorded the activity of area 7 neurons in awake, behaving monkeys. He found many neurons that responded to visual stimuli. Most had large but well-circumscribed receptive fields. The neurons showed no preferences for the orientation of the stimulus, and although they modulated their activity according to stimulus movement, they were unconcerned with the direction of movement. Importantly, about half of the visually responsive neurons in area 7 responded more vigorously when a visual stimulus was involved in a behavioral response, that is, when it guided the monkey's movements. Moreover, to activate the area 7 neurons, a monkey did not have to direct its movement to the stimulus, but only had to *attend* to the stimulus as a trigger for its conditioned response. Thus, the response of the neurons suggested that they were dually concerned with *where* the stimulus was located with respect to the monkey and whether or not it merited *attention* by virtue of its relevance to the monkey's immediate interests.

Clinical observations in patients with parietal lobe stroke support the notion that the superior parietal lobule is important for space perception and attention. Such patients show very complicated deficits in perceptual processing, especially of their body image and its relationship to objects. The deficits often take the form of contralateral *neglect* syndromes after unilateral parietal damage. The patient will fail to pay attention to objects in the contralateral visual field, even though the stimuli are reliably delivered to the visual cortex. If asked to draw the face of a clock, the patient might put only half the hours in a circle, or cram all twelve into one-half of the clock face. The neglect often includes the patient's own body. In severe cases, a patient might fail to dress the side of his body contralateral to the lesion or leave the contralateral jaw unshaven. This is especially likely if the lesion includes area 5 (the more rostral part of the superior parietal lobule), which receives a dense input from S_I and S_{II} and is important for object localization by feel. The neglect is sometimes so extreme that the patient denies ownership of his own contralateral limbs!

When area 7 is intact but deprived of its appropriate visual data, it will commit perceptual errors. But in this instance, the patient is fully aware of the errors. Cerebrovascular accidents of the occipital lobe can result in localized necroses of larger or smaller size in the visual cortex. The blind spots or *scotomas* that can be revealed through careful testing usually go unnoticed by the patient, even when they are quite large. The patient is not aware of a dark area in her visual field, just as normal people are not aware of a "darkness" beyond the periphery of their vision. But the patient with a scotoma frequently complains of a failure to "notice things" in the environment and a tendency to "bump into things" more often. The patient may not be aware of the blind spot but is fully cognizant of missing stimuli that ought to be salient.

The segregation of visual space perception within the parietal cortex is emphasized by a phenomenon called *blind sight* in patients with complete occipital lobe destruction. Such patients are clinically blind. However, if they are forced to say whether a visual stimulus is displayed on the left or the right side of a screen, after some complaints about the futility of the task they will make the correct response at a greater than chance level. Moreover, if their eye movements are monitored during the experiment, they are sometimes observed to make saccadic movements to the correct side when the stimulus is presented. Both the tell-tale saccade and the conscious response are probably mediated by an intact retinotectal system. You'll recall that the superficial layers of the superior colliculi that receive axons from the optic nerves project to part of the pulvinar nuclei of the thalamus (Fig. 28.1). The tectal recipient zone of the pulvinar provides the major thalamic input to much of the peristriate parietal cortex (area 19), which in turn influences area 7 of the superior parietal lobule. Are the positional data garnered by the superior colliculi relayed over this transthalamic pathway to the superior parietal lobule? Although they are not the full package of sensory data typically supplied, perhaps they are enough raw data for a modest computation to be performed by the parietal microprocessor.

Recognition of an Object's Abstract Meaning

If the inferior temporal cortex of a monkey is destroyed bilaterally, the monkey displays a syndrome called "psychic blindness" in which it approaches, handles, and even mouths objects such as lighted cigarettes and live snakes that it would ordinarily avoid. Could it be that the monkey no longer recognizes what it sees? Charles Gross and his collaborators have explored the ventral visual stream in monkeys by recording the activity of neurons in this inferior temporal cortex upon presentation of various visual stimuli (Fig. 28.4). Unlike the cells of the primary and peristriate visual areas, the cells are not arranged in any kind of visuotopic order. In fact, virtually all the cells have large receptive fields that include the center of gaze and are usually bilateral. In other words, visual space is unified in the inferior temporal cortex. The cells show stimulus equivalence across retinal translations—they don't care *where* the stimulus is located. In contrast to this relaxed requirement for place specificity, the neurons are very choosy about the shape, color, texture, and orientation of the stimulus, and many are concerned with all four stimulus parameters at once. Some neurons have even been found with stimulus requirements so strict as to suggest that they exist to detect singular objects. Figure 28.4 shows two examples of such punctilious cells: one neuron that responds well to an actual hand but not to its rectilinear approximation, and another that discharges vigorously to a frontal view of a monkey's face but only lamely to the same face lacking its eyes.

Clinicopathological data indicate that the inferior temporal cortex in humans, as in our fellow ape, relates the integrated visual qualities of a stimulus object to its more abstract characteristics (Fig. 28.3). For example, a patient who sustains bilateral damage to the inferior temporal gyrus during surgical treatment for temporal lobe epilepsy might fail to recognize her husband of 30 years when he enters the recovery room. It is not unusual, however, for her to smile in immediate recognition when he utters a greeting. The clinical term *agnosia*—literally, to "not know"—applies to such an inability to recognize a stimulus that ought to be familiar. In the special case of faces, it's called *prosopagnosia*. Other times the visual agnosia is not for faces but for objects. The attending physician might query the patient about a common object he shows in his hand—say, a house key—only to find that the patient cannot name it. When he hands it to her, she may suddenly remember not only its name but its use as well. There is no auditory agnosia in the first case—voice recognition is intact—and no tactile agnosia in the case of the key.

In fact, agnosias are more often than not limited to a single sensory modality. Different parts of the temporal lobe cortex may carry out associations of object meaning with auditory or tactile sensory qualities

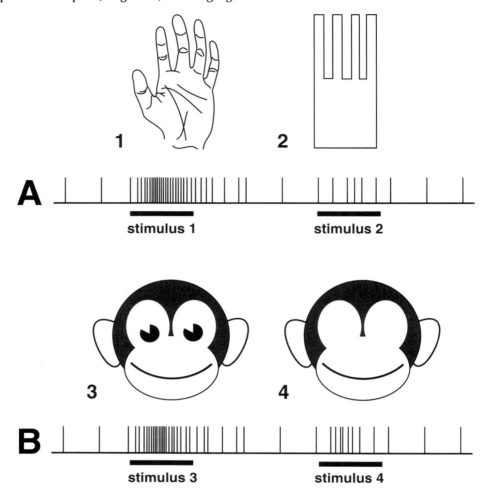

Figure 28.4 Charles Gross has identified the inferior temporal convexity as a cortical area important for visual recognition in the brains of rhesus monkeys. The neurons display response properties that demand highly specific visual inputs. In **A,** a cell that responds vigorously to the appearance of a hand (its activity is shown as spikes on the horizontal time line) is relatively impassive to the appearance of a rectilinear approximation of the stimulus. In **B,** a different neuron that is excited by the face of another monkey is far less responsive when the same face is presented without eyes.

analogous to those performed on visual data by the inferior temporal cortex. That agnosias can be separated into visual, auditory, and tactile indicates a considerable modality independence during advanced stages of perceptual processing. Where, or even if, stimulus qualities are reassembled in the human cortex is not known for certain. What is known,

however, is that the temporal cortex delivers the products of its computations to the prefrontal cortex, where they may be fully reconstructed,
and to the threshold of the hippocampus, where they may be assigned
an affective quality (see Chapter 24) and (or) consolidated for storage in
memory (see Chapter 29).

Planning Behavioral Strategies Based on Predictions

The prefrontal cortex, the broad association area that occupies the rostral
part of the frontal lobe, is synaptically the furthest removed from the
sensory cortices (Fig. 28.1). That the prefrontal cortex increases in volume
from lower mammals to primates and attains its largest size in humans
suggests that it shelters many of the metal capabilities that are maximally
refined in humans. Bilateral damage to the prefrontal cortex results in a
mysterious syndrome that cannot be characterized as perceptual, linguistic, or motor. Yet one or more parts of the prefrontal cortex come on
line in many, if not all, mental activities performed by humans, whether
it be analyzing a road map, listening to a song, dealing from a deck of
cards, or lecturing to knowledge-hungry students (Fig. 28.5). Only
through careful testing can subtle deficits be revealed in people who
have suffered prefrontal damage.

Rhesus monkeys are a good experimental model for prefrontal disorders. If a normal monkey is required to retrieve a grape placed beneath
one of two overturned cups, it unfailingly reaches for the loaded cup.
Monkeys with bilateral prefrontal lesions do just as well as intact monkeys. However, if a delay of more than a few seconds is introduced
between the loading of a cup and the opportunity to respond, the
prefrontally damaged monkeys are as likely to choose the grapeless cup
as the loaded one. Does the prefrontal cortex, through its extensive
corticocortical connections and reverberations through the basal ganglia
system, bridge the gap of time between the perception (of grape place-
ment) and action (cup selection and grape acquisition)? Are the neurons
of the prefrontal cortex able to retain the blueprints of *prospective* actions?

A good test of prefrontal function in humans involves a card-sorting
task in which the subject is challenged to sort playing cards one at a time
according to shape, color, or number. The tester decides on the sorting
criterion, but does not directly inform the subject. Instead, she declares
the subject's choices to be "right" or "wrong." The subject soon discovers
the criterion through trial and error and sorts correctly thereafter. If the
tester now secretly changes the criterion, normal subjects, after a brief
snit, will glean the new criterion and again sort correctly. Subjects who
have sustained bilateral prefrontal damage are less able to change their
original strategy; they persevere in sorting the cards according to the old,
incorrect criterion, despite clear indications of their errors from the tester!

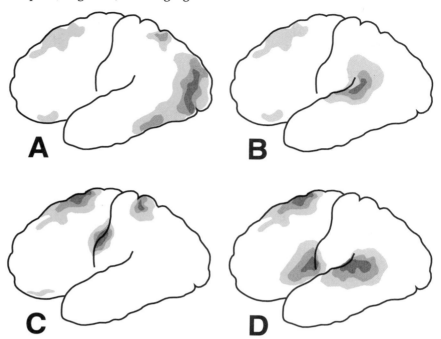

Figure 28.5   With PET techniques, cortical activity patterns can be monitored in vivo in human subjects while they engage in particular activities. In these examples of activity changes that might be expected to occur, the areas of increasing activity are indicated by progressively darker red shading. In **A**, the subject is examining complex pictures; in **B**, he is listening to a speaker who reads passages from a novel. The subject in **C** is palpating a skeleton key, and the subject in **D** is reading aloud.

That they are aware of their mistakes is clear; some subjects will declare their own errors to be "wrong" as they slam down card after card. Interestingly, behavioral *perseveration* is a common sign in schizophrenic patients who appear to suffer from frontal lobe hypoactivity (see below).

Magnetic resonance imaging and postmortem examination suggest that the loss of the ability to anticipate the consequences of one's actions and an inability to adjust one's behavioral strategies in the face of certain failure require that the damage include the dorsolateral prefrontal area on the convexity of the frontal lobe; remember that the prefrontal cortex extends onto the orbital and medial surface of the hemisphere as well. Thus, *the dorsolateral prefrontal area is involved with the selection of appropriate plans of action that are likely to succeed in achieving desired goals.* To accomplish this computational task, the prefrontal circuits must compare the plans of action with their projected outcomes. We have seen analogous computations before in the hierarchy of behavioral control systems arrayed along the frontal lobe. The primary motor cortex (area 4) compares commands that encode the force of muscular contractions with relatively raw *somatosensory* data that are fed back to it (from S$_I$ and VPLo) about the actual movements that are made. The rostrally adjacent premotor cortex (area 6) compares the sequence of movements with *perceptual* feedback (arriving from areas 5 and 7) about their accuracy in three-dimensional space. Since area 6 projects densely to area 4, one might surmise that the results of this computation are forwarded and translated into detailed movement commands. Finally, the still more rostral prefrontal cortex compares alternative behavioral scenarios with an *imaginary* feedback about their consequences. Presumably, it then selects a plan of action that, based on past experience, is likely to produce the desired outcome and sends the computed data along to the premotor cortex. To use a military metaphor, M$_I$ serves as a complement of field marshals who direct troop (motoneurons and muscles) movements. The premotor cortex is a tactical staff that coordinates the maneuvers of the field marshals in a broader arena. Finally, the prefrontal cortex is the strategist whose priority is ultimate victory and who orchestrates the long-range activities of the tactical staff.

The dorsolateral prefrontal cortex also sends its output to the parietal cortex—the very area from which it derives much of its behaviorally relevant perceptual data. What is the message sent back to the parietal cortex? According to a theory proposed by Hans-Lukas Teuber, the message informs the perceptual areas about the movements to be executed. This "corollary discharge," as Teuber called it, enables the perceptual areas to "realize" that the changing sensory data are due to the commanded behavior rather than agents in the environment. The oft-cited example of this phenomenon is the observation that the scene does not appear to "jump" about when we make voluntary eye movements, but does so when the eyeball is pushed with a finger.

Personality and Socialization

The other major part of the prefrontal cortex, which covers the orbital and medial surface of the frontal lobe, is more closely linked to the limbic system than to the skeletomotor control system. The thalamic input to this orbitomedial area comes from a part of the mediodorsal nucleus that relays data from the amygdala, the piriform cortex, and the brain stem reticular formation. It also receives a direct amygdalar projection and corticocortical axons from the cingulate gyrus. If the orbitomedial cortex is destroyed in a monkey, the normal aggressiveness and emotional responsiveness of the animal will decrease. If this cortex is stimulated in an intact monkey, a set of visceral responses ensues that is associated with a state of generalized arousal and vigilance. The observation that lesions of the orbital prefrontal cortex had a calming effect on monkeys led to attempts to treat violent emotional disorders and antisocial behavior in humans by the neurosurgical procedure of *prefrontal lobotomy*. The side effects of this procedure—epilepsy, personality changes, and loss of initiative, to name a few—were so severe that it was quickly discontinued. Although the lobotomy patient becomes less aggressive and insouciant, he also becomes impetuous and short on social decorum. He may, for example, urinate in public without a moment's hesitation or tell risqué jokes at the church choir meeting. As in the case of convexity damage, the patient cannot seem to fit his performance to the social context, nor can he anticipate the rebuke that must certainly follow his mischief.

Schizophrenia May Involve Prefrontal Malfunction

Schizophrenia remains among the most devastating and elusive psychotic disorders. As we have seen above, the prefrontal cortex is involved in constructing motor strategies and behavioral initiatives (by way of connections to the premotor and motor cortex), personality (perhaps in part by virtue of its connections with the limbic system), and social behavior. Deficiencies in these very functions resemble what have been described as the *negative* (loss of usually present functions) symptoms of schizophrenia (an example of a positive symptom is the hallucinations common to schizophrenics). Schizophrenic patients are poorly motivated, plan poorly, and show little emotional responsiveness. It is no small coincidence that among all neocortical areas, the prefrontal cortex and the limbic-related cortices receive the densest dopamine innervation (from the paranigral cell groups in the ventral tegmental area of the midbrain). That altered dopamine transmission plays a key role in schizophrenia comes from convergent clinical and experimental observations. In the first place, the most potent antischizophrenic drugs are those that block the dopamine receptors. Secondly, an overdose of methamphetamine, a dopamine agonist, induces a mental state that is indistinguishable from clinical psychosis. Even the L-DOPA taken for Parkinson's disease can induce mild psychotic symptoms. In monkeys, prefrontal function is dependent on normal dopaminergic transmission; when it is experimentally reduced through the administration of receptor antagonists, a monkey's performance declines in a manner similar to that seen after lesions. Interestingly, although hallucinations are reduced by treatment with dopamine D_2 (and D_4) receptor antagonists, the negative symptoms of schizophrenia do not respond well to such antischizophrenic drugs and may be exacerbated. You'll recall that the dopamine receptor subtypes mediate opposing effects in postsynaptic neurons.

Language

Perhaps the best illustration of *both* localization *and* lateralization of function is the cortical control of language (verbal and written symbols

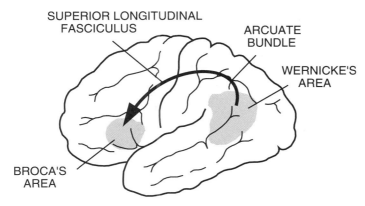

Figure 28.6   The areas of the cortex that control speech are located in the left hemisphere of most humans. Wernicke's "sensory" speech area is localized in the parietotemporal region, whereas Broca's "motor" speech area occupies the cortex of the frontal opercula. The curved arrow represents the association projections from Wernicke's area to Broca's area, which travel with the subcortical bundles labeled in the drawing.

arranged according to a syntax). Because language is so uniquely human, the finding that certain areas of association cortex in the *left* hemisphere appear to be dedicated to linguistic functions was met with great interest when first reported by two pioneering nineteenth-century neurologists. In 1861, Paul Broca published an account of a speech impediment that resulted from damage to the cortex on the basolateral part of the left frontal lobe (Fig. 28.6). Broca's *aphasia* is characterized by slow, telegraphic, effortful speech without any paralysis or even paresis of the vocal apparatus. Such patients are fully capable of understanding spoken and written language. Standing in contrast to Broca's aphasia is a syndrome discovered by Carl Wernicke a few years later. He noted that a lesion of the cortex on the convexity of the left parietotemporal region produced speech that sounded quite fluent and rhythmic, but lacked meaning. A patient with Wernicke's aphasia is properly grammatical but uses many nonwords and phrases of little specificity. Her discourse is *"like a tale told by an idiot, full of sound and fury and signifying nothing."* Moreover, such a patient appears to have difficulty in language comprehension as well. The distinction between these two syndromes led neurologists to consider Broca's area as a focus for "motor" speech and Wernicke's area to mediate "sensory" speech. Significantly, the two areas are interconnected with one another by way of axons that travel in the arcuate bundle and superior longitudinal fasciculus (Fig. 28.6). Recent PET analyses show that both of these areas are highly active during normal speech (Fig. 28.5D). They also show that the supplementary motor area of the frontal lobe is concurrently active, perhaps providing the motor sequencing necessary for articulation. Although Broca's and

Wernicke's areas are commonly found in the left hemisphere, a few cases have been reported in which these functions may be located in the cortex of the right hemisphere.

Only recently has evidence been obtained to suggest that the emotional aspects of language are localized in homotopic regions of the right cerebral cortex. There is a region of the right frontal lobe, corresponding in location to Broca's area, that is crucial to the generation of intonation and the emotional qualities in speech. A lesion in this area results in flat monotonic speech. A region in the right hemisphere corresponding to Wernicke's area enables a listener to interpret the emotional qualities of another speaker's words.

Other Lateralized Functions

There is evidence that other higher-order processes, such as perceptual abilities (e.g., music production and appreciation), performance skills (e.g., handedness), and many personality traits, are lateralized as well. Several psychologists today consider such abilities and traits as analytical reasoning and social skills to be expressions of the left cerebral hemisphere, whereas creativity and spatial skills are dominated by the right hemisphere. Unfortunately, the story is less complete on these processes than it is for language.

Much of what we know about the lateralization of function comes from the work of Roger Sperry who shared (with Hubel and Wiesel) the Nobel Prize in Physiology or Medicine for his studies of *split-brain* patients. Several patients with intractable epilepsy were treated neurosurgically by transection of the corpus callosum and sometimes the anterior commissure as well. Such patients could not be distinguished from normal individuals in the performance of their daily chores. However, with special testing procedures that take advantage of the fact that the sensory and motor systems are largely crossed, Sperry was able to demonstrate several abnormalities in the patients and a profound division of labor between the two hemispheres. He used a *tachistoscope,* a device that allows brief visual stimuli to be presented in one or the other visual field. When split-brain patients were shown a common object such as an apple in the left visual field (which, you'll recollect, is conveyed to the visual cortex of the right hemisphere) and asked to name the object, they often were unable to do so, even though they could readily select it from among a group of dissimilar objects. If the object was displayed in the right visual field, however, they had no problem naming it. This experiment shows that perceptual processing by the right hemisphere must ultimately be transmitted to the left in order to access the language control areas.

Michael Gazzaniga has carried on this line of investigation and made additional inferences about the role of the two hemispheres in cognitive functions. He was able to show that the two hemispheres can even be brought into conflict. In one particularly intriguing case, the tachistoscope was used to present a photograph of a seagull in flight to the left hemisphere and a picture of an orange to the right. The patient, who happened to be left-handed, had been previously instructed to draw what he saw. He chose an orange crayola and sketched a circle, adding a few dots for effect. When the experimenter engaged him in casual conversation and asked for a verbal account of what he had seen, the patient replied that he had seen a bird. Forced to confront his contradiction, the patient began to rationalize and finally protested that he had not yet finished the sketch. He hastily added wings and a beak to the orange circle and insisted that what he had actually drawn was a likeness of the puppet character Big Bird, whom he had seen in the tachistoscope! While inconclusive, this observation suggests that the left hemisphere takes precedence over the right when it comes to the imposition of logic and order on our experiences and behaviors.

For many decades, students of the brain have recognized that the association cortex is a challenging puzzle that holds the key to understanding the human cognitive process. Vast and remote, it resists exploration. Already in the latter half of the nineteenth century, the great Santiago Ramón y Cajal, despite his enviable success at unraveling the complex tangle of neurons in many parts of the brain, was stymied by the cell and fiber forest that is the cerebral cortex. In a prescient moment, the great cartographer of the brain remarked on the cortex in his autobiography, "the supreme cunning of the structure of the gray matter is so intricate that it defies and will continue to defy for many centuries the obstinate curiosity of investigators." That defiance continues to this day.

Detailed Reviews

Damasio AR, Geschwind N. 1984. The neural basis of language. *Annu Rev Neurosci* 7: 127.

Fuster JM. 1989. *The Prefrontal Cortex: Anatomy, Physiology, and Neuropsychology of the Frontal Lobe.* Raven Press, New York.

Gross CG, Desimone R, Albright TD, Schwartz EL. 1983. Inferior temporal cortex as a visual integration area. In Reinoso-Suarez F, Ajmone-Marsan C (eds), *Cortical Integration.* Raven Press, New York.

Pandya DN, Seltzer B. 1982. Association areas of the cerebral cortex. *Trends Neurosci* 5: 386.

Robinson DL, Petersen SE. 1983. Posterior parietal cortex of awake monkeys: visual responses and their modulation by behavior. In Reinoso-Suarez F, Ajmone-Marsan C (eds), *Cortical Integration.* Raven Press, New York.

29

Memory and Learning

We can all agree by this advanced stage in our survey of neuroscience that the primary biological function of a nervous system, whether it is the simple nerve net of a jellyfish or the exquisitely complex human brain, is to monitor the conditions in and around its owner and to orchestrate its owner's behavior to optimize its chances for survival and reproductive success. Much of this information processing and response selection capability of nervous systems appears to be built in to the basic plan through genetic evolution (e.g., respiratory reflexes). But nervous systems are endowed also with the ability to undergo some degree of *physical alteration* to provide flexibility in coping with the variability that might arise in their environment. The degree to which such alteration is possible is determined by the complexity of the nervous system, which in turn is determined by the ecological niche that its owner has come to occupy through the evolutionary process. The nerve net of the jellyfish will support only a handful of sensorimotor responses, which are stubbornly resistant to change. By contrast, a broad ecological domain such as our own guarantees a variable environment, which favors a more complex and adaptive nervous system.

In other words, the narrow ecological niche of the jellyfish requires the ability to detect but a few environmental energies, say temperature and touch, and to produce only one response—undulation. These meager abilities are relatively easy for Mother Nature to hard wire into the jellyfish nervous system during evolution; the jellyfish sticks to its ecological niche, and the need for adaptive mechanisms is minimal. In contrast to the jellyfish, an organism such as a baboon must negotiate an irregular terrain, experience seasonal variations in food supply and climatic conditions, evade a host of cunning predators, and engage its conspecifics with proper social decorum if it is to gain reproductive opportunity. In this case, Mother Nature can still hard wire those fundamental sensorimotor mechanisms that sustain the basic rhythms

356

of life, but can she hope to anticipate the great variety of situations a curious baboon might encounter during its lifetime in order to preprogram the necessary responses? Although it would appear that she favors this approach for animals as high in the evolutionary tree as the birds and never fully abandons it, even in the human brain, she wisely increases the ability for fine tuning of the basic plan according to experience.

A Neurobiologist's View of Learning and Memory

As is often pointed out, we can learn to identify letters on a printed page, but we cannot learn to see infrared light. Thus, *adaptive changes in a nervous system support the enhancement of an already existing capability* (the Kantian a priori?). Such changes potentially can occur through any number of physical alterations at any level of the signaling process in the brain. Although evidence is beginning to accumulate, the actual cellular mechanisms that support such adaptation are still unknown. They could be structural (synaptic remodeling), chemical (changes in the amount of interneuronal signaling substances), any of a number of other signal-related cellular events (regulation of receptors, ion channels, and intracellular mediators), or any combination of such events. The behavioral scientists have shown us that such an adaptive change is not an all-or-none phenomenon, but rather an alteration in the *probability* that certain patterns of neural activity will lead to certain other patterns (i.e., shifts in the associations of neural events). The enhanced capability can be manifest as performance (hitting a baseball), knowledge (the names of the seven dwarfs), or feelings (fear of spiders).

In the vernacular, we often refer to these adaptive changes as *learning*, particularly if we are aware of them. It is important, however, to keep in mind that we undergo many such changes that do not enter into consciousness. Even the simple nerve network of a snail is capable of learning, but it is difficult to argue that such a nervous system supports consciousness as we experience it. Thus, *learning refers to a more or less gradual progression of physical change resulting in an increase in the probability that certain neural events will lead to certain other neural events. Forgetting refers to a process of physical change that results in a decrease in this probability.* Forgetting may be, but is not necessarily, the reversal of the same physical change that occurs during learning. *Learning is a function of most, if not all, parts of the nervous system.* For example, classical conditioning of certain simple reflexes in mammals can be achieved by the spinal cord even after it has been surgically detached from the brain. The extent of the nervous system that is involved in a given learning process can vary greatly from as little as one neuron and an effector to a large but finite number of widespread brain circuits.

Are the mechanisms that support learning similar to those that guide development of the basic patterns of brain circuitry? As asserted above, the fundamental capabilities of a nervous system are determined by genetic programming and developmental expression of specific inter-connections of nerve cells with one another. However, the final adult pattern is greatly dependent on the experiences of the nervous system as it develops. Even where the basic patterns are established by genetic direction, there are *critical periods* of development during which the nervous system requires exposure to particular experiences in order to achieve normal adult function (see Chapter 31). A dramatic example comes from infants who are born with cataracts that are not treated for several months or years. When the condition is later corrected, such children suffer persistent deficits in pattern recognition.

Thus, both the evolutionary molding of nervous system capabilities and their developmental dependence on experience are analogous to learning in that they are progressive phenomena (although vastly differ-ent in time scale), dependent on experience (selective pressures and epigenetic factors, respectively), and manifest as physical alterations. For this reason, it is useful to search for common cellular and molecular mechanisms that guide, on the one hand, the formation of synaptic connections in the brain during development and, on the other hand, the reorganizations that occur during learning. Since mushy brains decay rapidly and leave no fossil records, we cannot directly study them from an evolutionary perspective except perhaps through the metaphor of phylogeny.

To summarize, we can reasonably consider that *the total of the capabili-ties possessed by a nervous system at any point in time is a function of three factors: genetically programmed expression, developmental variability, and learning.* This total can be considered the nervous system's *memory.* Nervous systems do not store information in a literal sense, but they do maintain circuitry conditions, such as *specific interconnections and differ-ential potencies of synaptic transmission,* that affect the probabilities linking neural events to one another. Today, there is much evidence to suggest that the state of these specific synaptic conditions within a brain is plastic and in constant, though limited, flux. In learning (the formation of memories) or forgetting (the deterioration of memories), the synaptic conditions are modified and the probabilities change. For this reason, *learning and forgetting can be considered simply as phases or degrees of memory.*

Some Simple Forms of Learning

One approach to the cell biological events that underlie learning and memory has been to employ invertebrate model systems. Invertebrates offer the advantages of simplicity and accessibility, and they exhibit a limited number of rather stereotyped and therefore quantifiable and reproducible sensorimotor responses. The marine mollusk *Aplysia* has

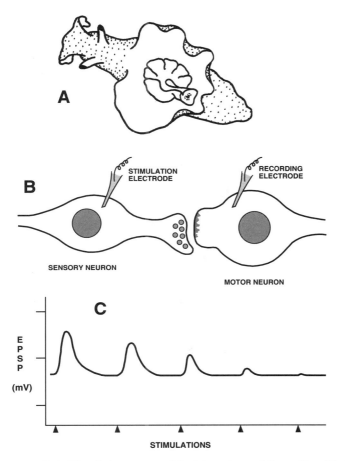

Figure 29.1 Eric Kandel promoted habituation of the gill withdrawal reflex in the sea slug *Aplysia* (**A**) as a simple and useful model of learning. Part **B** shows one of several experimental paradigms used to study habituation in Aplysia. Part **C** shows the EPSPs recorded from the motoneuron upon repeated stimulations of the sensory neuron. See the text for an explanation of habituation.

served well in this capacity (Fig. 29.1A). One of the reflexes *Aplysia* exhibits is the so-called gill withdrawal reflex: the gill is infolded upon weak mechanical stimulation of the animal's siphon. Repetition of the stimulation leads to a progressive reduction in the vigor of gill infolding, a process termed **habituation,** which lasts only a minute or two after the stimulus train is halted. Some investigators consider habituation to be an elementary form of learning. Since the neurons of *Aplysia* (found in abdominal ganglia) are large, their activity patterns can be recorded easily and they can be readily manipulated and observed in other ways. Now, it so happens that the nervous system of *Aplysia* has been nearly fully described; almost every one of its neurons is identifiable from animal to animal and the interconnections that the individual nerve cells make with one another are known. Twenty-four sensory neurons innervate the siphon and make direct synaptic connections with six motoneurons that control the gill muscles. There are a number of interneurons in the ganglia, but let's disregard them for the moment.

If the activity of a gill motoneuron is recorded while a siphon sensory neuron is electrically stimulated at a frequency that mimics the mechanical stimulation of the siphon (Fig. 29.1B), the response of the motoneuron progressively decreases in temporal correlation with the decrease in gill folding behavior (Fig. 29.1C). Recording the activity of the siphon sensory neuron shows that it remains responsive throughout the repeated stimulations. Moreover, the motoneuron's threshold to electrical stimulation is not changed by the stimulations. Nevertheless, the EPSP elicited in the motoneurons is progressively reduced in amplitude upon repeated stimulation. This entire pattern of habituation can be reproduced in a sensory and motor neuron that have been dissected from the animal with their synapse intact. By controlling the Ca^{2+} concentration in the medium bathing such a preparation, it is possible to measure the amount of neurotransmitter released upon stimulation of the sensory neuron. The results indicate that the decreased motoneuronal response seen in behavioral habituation is a result of decreased release of neurotransmitter by the sensory nerve ending. Further work has shown that this decrease can be accounted for in large part by a reduction in the Ca^{2+} current at the synaptic ending. Since vesicle exocytosis is dependent on, and proportional to, local Ca^{2+} concentration, neurotransmitter release is decreased.

Aplysia also shows behavioral **sensitization,** whereby a noxious stimulus to, say, the head a few seconds before siphon stimulation will produce a more vigorous gill withdrawal than would the siphon stimulation alone (Fig. 29.2). As with habituation, recording the motoneuronal activity in this paradigm reveals an augmented response to siphon stimulation that is a consequence of increased neurotransmitter release from the sensory nerve ending. This change, however, is dependent on an interneuron. The head stimulation activates a serotonergic interneuron, which makes synaptic contact with the presynaptic ending of the sensory neuron. The serotonin released interacts with G protein-coupled receptors in the membrane of the sensory nerve ending and triggers the production of cAMP, which in turn activates a protein kinase (PKA). One substrate for this enzyme is a subtype of K^+ channel that is shut off when phosphorylated by PKA. Since K^+ outflow is responsible for restoring the resting membrane potential when an action potential has invaded the axonal ending, this partial reduction (partial because there are other, serotonin-insensitive K^+ channels around as well) in the K^+ outflow prolongs the action potential. This leads to a longer open time of the voltage-sensitive Ca^{2+} channels, increased Ca^{2+} influx, and augmented exocytosis of neurotransmitter.

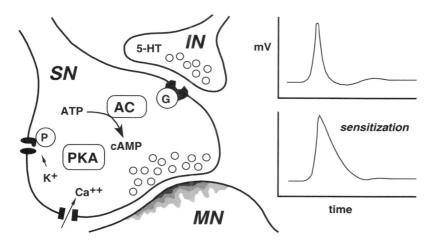

Figure 29.2 Sensitization of the gill withdrawal reflex is mediated by a serotonergic interneuron. See the text for an explanation of these synaptic events. Abbreviations: AC, adenylyl cyclase; G, GTP-binding protein; IN, interneuron (serotonergic); MN, motoneuron; PKA, cAMP-dependent protein kinase; SN, sensory neuron; 5-HT, 5-hydroxytryptamine (serotonin).

Sensitization, like habituation, decays in minutes. If, however, it is repeated some four to six times in a row, the augmented response will persist for more than 24 hours. The intracellular events described above for short-term sensitization do not persist long enough to account for this long-term form of sensitization. Since this persistent form of sensitization can be blocked by inhibitors of transcription and translation, it clearly depends on protein synthetic events in the cell body of the sensory neuron. Thus, it would appear that local synaptic events may mediate short-term changes in behavioral response, whereas long-term changes require a more elaborate mechanism. You may recall that second messenger activation sets in motion a cascade of parallel regulatory events in a neuron. It seems that whereas one avenue in this cascade may lead to changes that are of fast onset, transient, and local, other avenues can produce more slowly developing, persistent effects at longer range. Interestingly, our own everyday learning experiences indicate two analogous phases of memory that the psychologists cleverly refer to as **short-term** and **long-term memory** (Fig. 29.3). The former lasts for seconds to minutes, whereas the latter can last a lifetime. In mammals, including humans, there also appears to be an intermediate form of memory, called **middle-term,** that, unlike short- and long-term memory, depends on mental rehearsal of the newly acquired learning. It lasts from minutes up to a few hours and is easily disrupted.

Cellular and molecular correlates of learning such as those described above for *Aplysia* are being revealed also in the more complex nervous

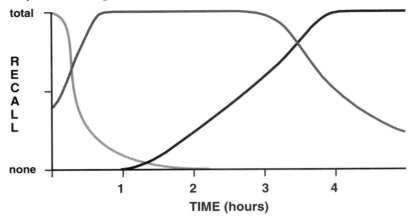

Figure 29.3 Memory consolidation in mammals, including humans, appears to proceed in at least three phases. The area under the light gray line indicates the time span of decay of short-term memory. The medium gray curve indicates the decay of learned material that is processed into middle-term memory, and the black curve indicates the consolidation of new knowledge into indefinite storage or long-term memory.

systems of vertebrates. Although these events are wildly exciting and essential to the understanding of learning and memory, one cannot hope to fully explain learning and memory without analysis at the level of circuitry. Even in *Aplysia*, the events in the sensory and motor neurons during the execution of even the simplest reflexes are accompanied by activity modulations in scores of other neurons—those many interneurons we conveniently left aside. Moreover, most learning of importance to us differs from habituation and sensitization in that it involves the *association* of two (or more) stimulus events. This so-called **associative learning** is better approached using animals with more highly evolved nervous systems.

Fun With Chicks

Hatchling chickens are fast learners. If chicks are presented just once with a tiny chrome bead (they instinctively peck beads) that has been flavored either with water or with a highly noxious chemical, most of the chicks that peck the bitter bead will not do so upon retesting 24 hours later. This paradigm of *one-trial* learning can be employed to discover many things about brain cells and regions involved in learning and remembering. It has recently been exploited to show that molecular and cellular mechanisms of the kind described for *Aplysia* are probably also at work in the vertebrate brain (Fig. 29.4). In addition, when combined with selective lesions of brain nuclei, this model has revealed some of the rudiments governing the formation, transferral, and storage of memories.

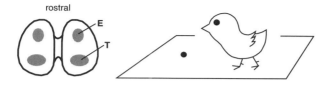

TIMESCALE	INCREASE	NUCLEUS
sec - min	2DG uptake	T & E
min - hrs	PKC translocation	T
	IEG induction	T
few hrs	tubulin synthesis	T
	glycoprotein synthesis	T & E
	synthesis	T
	bursting activity	T
hrs - days	spine density	T
	spine head diameter	T
	synapse number	T
	vesicle number	T & E
	subsynaptic density	T & E

Figure 29.4 Marked metabolic and morphologic events take place in certain nuclei of the chick brain after it experiences one-trial learning. Abbreviations: 2DG, 2-deoxyglucose; IEG, immediate early gene; PKC, protein kinase C.

Steven P. R. Rose and his collaborators have identified two cell groups in each hemisphere of the chick brain as critical to one-trial learning (Fig. 29.5). To avoid the finer points of chicken neuroanatomy, let's call one the *T* nucleus (for **t**ransient) and the other the *E* nucleus (for **e**nduring). If the *T* nucleus of the left hemisphere is destroyed prior to the first exposure to the learning paradigm (training), the chicks do not learn to avoid the chrome beads. Destruction of the right *T* nucleus or either *E* nucleus is without effect when done before training. Conversely, destruction of both *E* nuclei will obliterate the memory (not to peck chrome beads) if the ablation is done after the learning trial; ablation of the left *T* nucleus has no effect on memory if the ablation is done after the initial learning trial. These data suggest that the left *T* nucleus is essential to memory formation and can store the memory, if necessary. The *E* nuclei are sufficient to store the memory and ordinarily do. If the *E* nuclei are ablated before the learning trial and the right *T* nucleus is ablated after the learning trial, the chicks fail to demonstrate recall upon retesting. Apparently, the right *T* nucleus serves the *transfer* of the learning from the left *T* nucleus to the two *E* nuclei. The right *T* nucleus also has the capacity to store the memory if the *E* nuclei are destroyed.

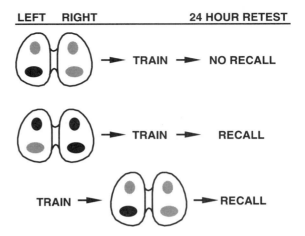

Figure 29.5 Steven Rose used variations on the one-trial learning paradigm combined with selective lesions of the relevant nuclei in the chick forebrain to study memory transposition and consolidation. See the text for an explanation of the experiments.

The results of these and many similar studies indicate that some brain regions are specialized for the *formation* of memories; without them, learning cannot occur. Moreover, *recently formed memories are transferred from one brain region to another, often over multiple pathways.* Finally, further experimental work has shown that *different versions of the same memory can be stored in different parts of the brain.*

The Hippocampus Is Special for Memory Formation in Mammals

In the rat brain (arguably more like our own than the chick's), the structure that appears to be most critical to memory consolidation is the hippocampus. To illustrate this, rats can be challenged with a task that requires them to learn the location of a slightly submerged platform in a pool of opaque water (Fig. 29.6). Their performance can be assessed by measuring the time it takes them to reach the platform in subsequent trials. Rats don't like to swim, and their performance time will rapidly asymptote to the brief time it takes to steer a straight course to the platform. If previously trained rats receive a bilateral lesion of the hippocampus, their performance remains as good as it was before the lesion. However, if the hippocampuses are destroyed prior to training, the rats show no improvement in their swim times: they don't learn the location of the platform. Thus, *the hippocampus appears necessary for the consolidation of learning into a new memory.* But once learned, the memory does not

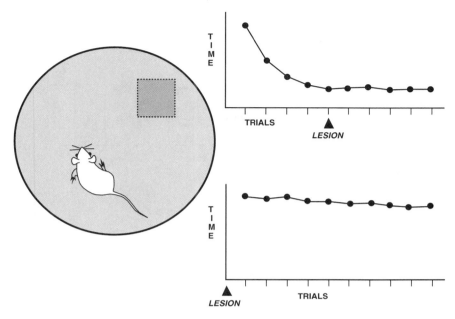

Figure 29.6 The hippocampus is necessary for the acquisition of new knowledge by mammals, but not for its retention. See the text for details.

depend on the integrity of the hippocampus. As in the chick, it is probably transferred to other parts of the brain. It seems highly plausible that *the storage sites include those neocortical areas where the perceptions of the environment are originally formed.*

Long-Term Potentiation in the Rat Hippocampus

When the main axonal pathway from the entorhinal area to the hippocampus is rapidly stimulated (tetany), the response of the postsynaptic cells in the dentate gyrus to a subsequent stimulation is augmented with respect to its pretetanic state and remains so for many minutes, hours, or even days (Fig. 29.7). The potential significance of this phenomenon, called **long-term potentiation,** for memory formation was immediately recognized and has been studied extensively. The critical neurotransmitter in long-term potentiation is an excitatory amino acid; most likely it's glutamate. Further, the glutamate must interact with a particular subtype of excitatory amino acid receptor (there are several) named the N-methyl-D-aspartate (NMDA) receptor after the compound first used to reveal it. As receptors go, the NMDA receptor is fantastically complex (Fig. 29.8). It has multiple regulatory sites where different kinds of signal

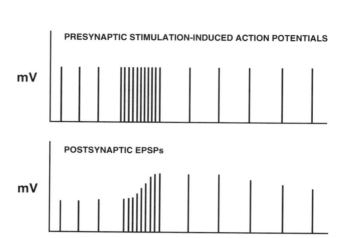

Figure 29.7 Tetanic stimulation of axons (top recording) that end on hippocampal pyramidal neurons causes an augmentation of the post-synaptic response (bottom record).

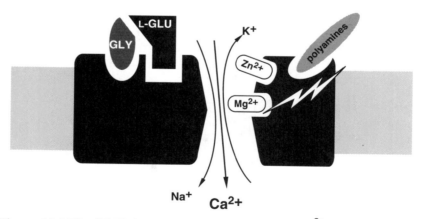

Figure 29.8 The NMDA receptor is a ligand-gated Ca^{2+} channel that is also voltage-gated (symbolized by the lightning bolt). Depolarization eliminates the Mg^{2+} block from the ionopore and allows the channel to conduct Ca^{2+} (and monovalent cations) upon activation by glutamate. Additionally, it has a number of binding sites for substances that can modulate its responses to glutamate. Abbreviations: L-GLU, L-glutamate; GLY, glycine.

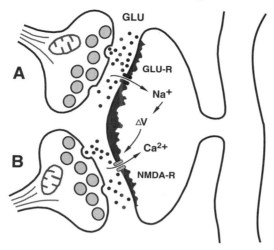

Figure 29.9   The phenomenon of long-term potentiation is widely accepted as a synaptic model of associative learning. See the text for an explanation of the events in long-term potentiation.

molecules can bind to modulate the responsiveness of the receptor to glutamate. Moreover, the NMDA receptor is a Ca^{2+} channel that is both voltage- and ligand-sensitive. The voltage sensitivity is conferred by a Mg^{2+}-binding site within the lumen of the channel that blocks Ca^{2+} flow.

When the membrane is depolarized, the Mg^{2+} is kicked out of the channel. Because of its dual sensitivity, the NMDA receptor will open to allow inward Ca^{2+} flow only if it already happens to be depolarized to a certain threshold level when glutamate binds to it. Such depolarization is most likely to occur when the cell membrane near the site of the NMDA receptor is activated by an excitatory synapse. Glutamate as a neurotransmitter has a powerful excitatory effect on postsynaptic cells, which is mediated by any of the four (known so far) non-NMDA receptor subtypes.

Now, imagine that two excitatory synapses contact a region of postsynaptic membrane close to one another, say on the same dendritic spine (Fig. 29.9). At synapse *A*, the receptors are of the non-NMDA type, and at synapse *B*, they are NMDA receptors. Inflow of Ca^{2+} will not occur through the NMDA receptor when either synapse is active alone. But if synapse *A* is active just before synapse *B*, the NMDA channel will have been depolarized, removing the Mg^{2+} block, *and* glutamate can produce an allosteric change that opens the channel, allowing Ca^{2+} to enter the postsynaptic cell. If synapse *B* is active just before synapse *A*, the Ca^{2+} channel of the NMDA receptor will remain blocked by Mg^{2+}. Thus, the

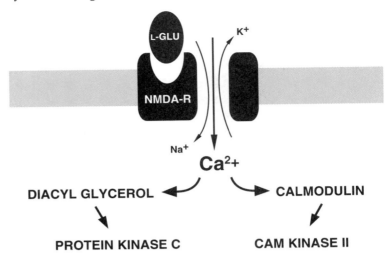

Figure 29.10 The Ca^{2+} that enters the postsynaptic cell through the NMDA channel can bind to calmodulin, which activates CAM kinase II, as well as to diacyl glycerol, which activates protein kinase C.

NMDA receptor–channel complex appears to fulfill two criteria of *associative* learning: *two separate signals* (events) must occur with a *precise temporal relationship.*

Recall that in Pavlovian conditioning, the bell (conditioned stimulus) must precede the food (unconditioned stimulus) by no more than a few seconds if the dog is to subsequently salivate upon hearing the bell alone. Present them in reverse order and no learning occurs. Does a neural impulse representation of the bell arrive at the hippocampus to depolarize cell membranes just prior to the arrival of an impulse representation of the food, thus allowing Ca^{2+} to flow into the cell? Although direct evidence for this is lacking, it seems to be a compelling hypothesis.

After Ca^{2+} entry, we are again lost in the morass of events triggered by the second messenger systems (see Chapter 5). Is the enhanced postsynaptic response due to channel phosphorylation? Changes in the sensitivity of the neurotransmitter receptors? Changes in the morphology of the dendritic spine? There is evidence for each. We know that the Ca^{2+} that enters through the NMDA channel can bind to calmodulin to activate a Ca^{2+}-calmodulin-dependent kinase called **CAM kinase II** (Fig. 29.10). Mice that have been rendered deficient in CAM kinase II by a genetic engineering technique called a gene *knockout* perform poorly in the water maze paradigm and exhibit a diminished capacity for long-term potentiation induction. But CAM kinase II has several substrates that it regulates. Which are relevant to long-term potentiation and learning remains a mystery.

It has been suggested that the enduring change that manifests as long-term potentiation occurs in the presynaptic element. In this scheme, a signal molecule (not a neurotransmitter) that is capable of diffusing through neuronal membranes (e.g., arachidonic acid or nitric oxide) is released by the postsynaptic cell and induces changes in the presynaptic ending that enhance the efficiency of impulse–release coupling, leading to increased neurotransmitter release. One plausible model for such a phenomenon again begins with the postsynaptic influx of Ca^{2+} through the NMDA receptor channel. The Ca^{2+} can bind to diacyl glycerol that has been generated, for instance, by the recent activation of a metabotropic receptor linked to the hydrolysis of inositol phospholipids (Fig. 29.10). Diacyl glycerol, in the presence of Ca^{2+} (and phosphatidyl serine), activates protein kinase C, which in turn activates the enzyme *nitric oxide synthase*. The nitric oxide that is produced can diffuse across the synaptic cleft and into the presynaptic element, where it stimulates the production of the second messenger cGMP (Fig. 29.11). Thus, nitric oxide is quite capable of influencing local protein function, perhaps even that involved in neurotransmitter release (see Chapter 4). It has been shown experimentally that blocking the activity of nitric oxide synthase interferes with long-term potentiation induction. Whether this mechanism or one of the many others is ultimately responsible for long-term potentiation (and learning) must await further research. Given the present interest in this phenomenon, the definitive answers to these important questions may be found shortly.

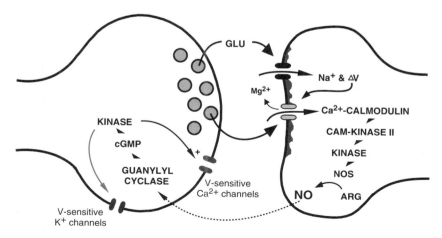

Figure 29.11 Nitric oxide (NO) gas may be a retrograde messenger that is generated during long-term potentiation induction. NO is hypothesized to influence the release of neurotransmitter from the presynaptic ending. See the text for details.

Clinicopathological Insights into Human Memory

We cannot study the human hippocampus with the same resolution that can be achieved in the rat. However, we can be confident that Mother Nature would conserve the cellular mechanisms of memory formation. Not surprisingly, therefore, the hippocampus seems as essential for learning in the human as it is in the rat. In many famous cases of hippocampal damage (as ultimately determined post mortem), careful clinical observation and testing indicate its critical role in memory formation. The most extensively studied case was that of a patient known by his initials, H. M., who was surgically treated for intractable epilepsy with bilateral removal of the medial temporal lobes, including much of the hippocampuses. H. M. experienced *anterograde amnesia:* he could not form any lasting memories from the time of his surgery onward. His premorbid memories remained intact. Many other descriptions of anterograde amnesias appear in the clinical literature. In nearly all of them, the postmortem examination of the brain indicated damage in either the hippocampus or a closely associated part of the limbic system, such as the mammillary bodies or anterior nuclei of the thalamus (see Chapter 24).

These limbic system-associated deficits stand in marked contrast to those in cases of generalized head trauma. In the latter, the patient can form new memories but is unable to conjure up images of past events. Call it *retrograde amnesia.* It tends to erase recent memories more completely than older, longer-sustained memories. Retrograde amnesias do not show a strong correlation with the hippocampus or other limbic structures, perhaps because older memories are already consolidated and represented in nonlimbic regions of the brain, especially the cerebral cortex. Not surprisingly, one of the primary cognitive deficits in Alzheimer's dementia—a disease characterized above all by a generalized atrophy of the cerebral cortex (including the hippocampal cortex)—is a progressive and profound obliteration of all memories. In such cases, self identity is lost with the fading of personal history. Mental life is stripped of its depth and texture. The victim is doomed to dwell in the vacant house of the existential present.

Detailed Reviews

Andrew RJ (ed). 1991. *Neural and Behavioral Plasticity.* Oxford University Press, New York.

Dudai Y. 1989. *The Neurobiology of Memory: Concepts, Findings, Trends.* Oxford University Press, New York.

Kandel ER, Schwartz JH. 1982. Molecular biology of learning: modulation of transmitter release. *Science* 218: 433.

Nicoll RA, Kauer JA, Malenka RC. 1988. The current excitement in long-term potentiation. *Neuron* 1: 97.

Squire LR. 1987. *Memory and Brain.* Oxford University Press, New York.

BUILDING AND REBUILDING THE CNS VI

Using the power of molecular technology, neuroscientists have begun to identify the relevant molecules and illuminate the cellular mechanisms that control the formation of cell groups, the emergence of cellular phenotypes, and the growth and specificity of axonal connections in the CNS. A full understanding of such mechanisms may one day enable medical neuroscientists to repair brains and spinal cords that have been damaged by trauma or disease. It is also likely that at least some of the same molecular events that occur during generation and repair of the CNS are at work when brain circuitry is changed by noninjurious experience. From what has already been achieved on the research frontier, it seems an inescapable conclusion that molecular neurobiology will eventually deliver the keys that will unlock such mysteries as our proclivities to learn, to think creatively, and to sink into the abyss of psychopathology. In the chapters that remain, let's examine what is known about the formation of the CNS from a molecular perspective and explore the possible therapies that will certainly dominate the medical neurosciences in the near future.

The Histogenesis of Nuclei and Cortices 30

In Chapters 7 through 9, we examined the differential expansions and foldings that produce the gross shape of the adult CNS. While this morphogenesis proceeds, the interior of the CNS is busy establishing its various cell groups and fiber tracts. Since all neurons are postsynaptic targets, they must arrive in the right place at the right time in order to receive the axons growing to meet them. Once in their proper positions, neurons elaborate their characteristic shapes and express the proper complement of neurotransmitters and receptors. In this chapter, we will consider some examples of how neurons migrate and assemble themselves into cortices and nuclei. We'll examine the molecular determinants of neuronal and glial shape and other phenotypic traits. Finally, we'll examine the factors that regulate the ultimate number of neurons that survive to CNS maturity. We'll learn that the histogenesis of the cortices and nuclei of the CNS depends on interactions between its cells, interactions that are mediated by specific cell surface molecules. Any failure to express such molecules in the proper time and place will lead to malformed and dysfunctional central structures. The question of how axons bundle into tracts, find their way to their appropriate targets, and form synapses on the newly generated cortices and nuclei will be discussed in Chapter 31.

How Do Neuroblasts and Glioblasts Move?

After leaving the mitotic cycle in the ventricular zone of the developing neural tube, the neuroblasts and glioblasts migrate to an eccentric position by crawling over the substrate (Fig. 30.1). They extend a process called a **pseudopodium,** which adheres firmly to the substrate near its leading end. The bulk of the cell is then pulled along the pseudopodium

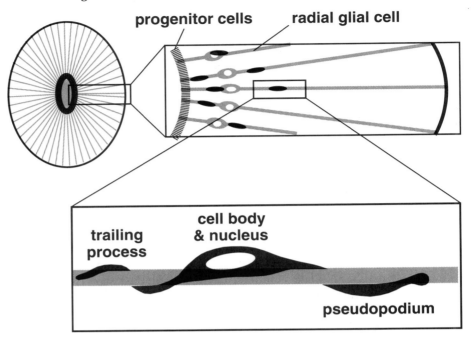

Figure 30.1   Radial glia (shaded red) extend processes that·span the depth of the neural tube (upper left) from the lumenal to the pial surface. Pashko Rakic has shown that many neuroblasts migrate by crawling along the radial glia (lower panel).

to the point of adherence. A new pseudopodium extends and the movement cycle is repeated until the cell reaches its destination. *This crawling process is based on the dynamic interaction of contractile cytoskeletal elements such as actin and myosin.* The first several waves of cells to become postmitotic need migrate only a short distance to the intermediate zone of the neural tube. Some of these "early birds" differentiate into specialized glia that extend processes from the ventricular to the outer surface of the neural tube (Fig. 30.1). These **radial glia** serve as a scaffold for later cell migrations (see below). To end up in the right places, *neurons and glia require a signal to activate their contractile machinery, an adhesive surface, directional signals by which to navigate, and perhaps a signal to stop moving when they arrive at their final position.* In the CNS, most of the molecules responsible for these phases of migration occur as cell surface proteins.

Cell Adhesion Molecules

Because migration depends on interactions of cell surfaces with one another and with the extracellular matrix, much effort has gone into the identification of relevant cell surface and extracellular matrix molecules.

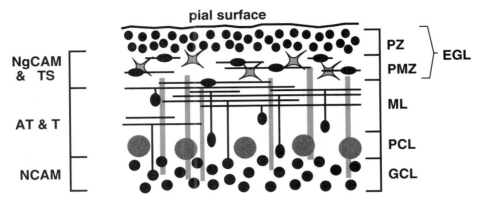

Figure 30.2 The migration of granule cells (jet black) from the transient external granular layer (EGL) to their adult position in the granule cell layer (GCL) deep to the Purkinje cell (large red circles) layer (PCL) depends on a number of differentially distributed cell adhesion molecules (listed at left). The gray cells are astrocytes. Abbreviations: AT, astrotactin; ML, molecular layer; NCAM, neural cell adhesion molecule; NgCAM, neuron–glial cell adhesion molecule; PMZ, premigratory zone; PZ, proliferation zone; T, tenascin; TS, thrombospondin.

Several **cell adhesion molecules** that appear to influence migration have now been identified in nervous tissue. Cell adhesion molecules have been implicated as activators and deactivators of the intracellular motile machinery, as providers of surface traction, and as signals that guide migration in particular directions.

The interactions of various cell adhesion molecules in neuronal migration have been extensively studied in the formation of the cerebellar cortex, where the spatiotemporal parameters of migration have been described in detail (Fig. 30.2). The large Purkinje neurons are the first neurons generated and migrate radially through the intermediate zone to form a monolayer. A second wave of neuroblasts then migrates beyond the Purkinje cell layer to form a *transient external granule cell layer* at the pial surface. The neuroblasts in the external granule cell layer retain their mitotic competence and give rise to the stellate and basket cells, which will migrate but a short distance to reside in the molecular layer, and the numerous granule cells, which must migrate to their final home deep to the Purkinje cell layer. Once postmitotic, the granule cells move a short distance to a *premigratory zone* and extend two processes (which become the parallel fibers of the adult molecular layer) in opposite directions tangential to the pial surface. When these processes reach their final length, the cell bodies of the granule neurons migrate downward along previously elaborated radial glia called Bergmann glia, whose processes plumb the depth of the cerebellar cortex. The migration results in the depletion of the external granule cell layer and the formation of the inner granule cell layer of the mature cerebellum.

Two glycoproteins—**neuron–glial cell adhesion molecule,** expressed only by postmitotic neurons, and **thrombospondin,** expressed by astrocytes—appear to function in the migration of newly postmitotic granule cells from the proliferative zone to the premigratory zone of the external granule cell layer. These two cell adhesion molecules may also mediate the later migration of granule cells through the ever-expanding molecular layer. The continued migration of granule cell bodies past the layer of Purkinje cells depends on **astrotactin,** another cell surface glycoprotein expressed by granule neurons. Another cell adhesion molecule associated with the membrane of the Bergmann glial processes, **tenascin,** is also essential for granule cell migration. Thus, *granule cell migration in the developing cerebellum requires the interaction of several distinct cell adhesion molecules acting at different sites and stages.*

Neuronavigation

The initial direction of migration may be determined by purely mechanical factors. It has often been noted that the daughter cells tend to move apart in opposite directions after cell division. Thus, in the ventricular zone of the neural tube, the radial movement of a postmitotic daughter cell may be a consequence of the tangential plane of the final cleavage (see Fig. 7.3). Given an adherent substrate, no further directional cues are needed; the newborn cell simply crawls radially until signaled to stop. Although this simple mechanism may serve the early stages of spinal cord and brain stem formation, later migration to distant sites appears to rely on environmental signposts. For instance, cell adhesion molecules and other cell surface and extracellular matrix molecules may guide migration. Different populations of migrating cells may have preferences for particular types of cell adhesion molecules that lead cells along certain paths depending on their distribution. In our example of the migration of cerebellar granule neurons, the same cell adhesion molecules that enable adhesion may also to some extent guide migration by their differential distribution (Fig. 30.2). Barrier molecules that may keep cells from straying from their proper paths have also been identified recently. Finally, concentration gradients of specific tropic substances that are released from cells in destination zones may also be Sirens' songs that entice migratory cells to their end points by chemotaxis. Although chemotaxis is apparently essential for properly directed axon extension (see Chapter 31), there is little experimental evidence for chemotaxis during neuron migration.

What Causes Neurons to Stop Migrating?

Very little is known about the signals that instruct neurons to stop where they do during histogenesis. Since neurons seem to require signals such as cell adhesion molecules to activate their motile machinery, it may be that migration stops when the neurons enter a zone devoid of such signals. Alternatively, the interruption of migration may be an active process mediated by the appropriate disposition of particular cell adhesion molecules. Circumstantial evidence suggests that a cell adhesion molecule called **neural cell adhesion molecule** (NCAM) may serve such a purpose. NCAM appears to be a widely distributed molecule that enables cells to aggregate by "sticking" together. For instance, antisera to NCAM inhibit the reaggregation of dissociated retinal neurons that typically occurs shortly after such cells are dispersed in a culture dish. NCAM is an integral membrane glycoprotein that exists in embryonic cells in forms of various molecular weight (180–250 kilodaltons) that result from alternative splicing of a single NCAM gene. The embryonic forms of NCAM have unique polysialic acid chains that are lacking in mature neurons and glia. Since the homophilic binding of NCAM molecules is decreased by the polysialic acid chains, NCAM function appears to be modulated during development by the amount of polysialic acid residue attached to the core protein. Since the NCAM expressed in mature cells is "stickier" than the embryonic forms, its expression at the proper time during histogenesis may be important for the termination of migration and the aggregation of neurons into cortical layers and nuclei.

Layering the Cerebral Cortex

Progenitor cells that elaborate the cerebral cortex reside in the ventricular zone of the telencephalic vesicles (Fig. 30.3). The postmitotic cells migrate along radial glia to form the layers of the cerebral cortex. The first wave of postmitotic neurons form the deepest layer of the cortex (layer VI), and the subsequent waves form the remaining layers by migrating past the previously formed layers. Thus, the cortex is often described as forming from the "inside out."

In the last several years, it has become possible to examine neuron migration by infecting progenitors with a recombinant, replication-defective retrovirus that carries the histochemically detectable β-galactosidase gene from *Escherichia coli* (Fig. 30.4). Since only the progeny of infected cells will express the viral genes, this method allows one to locate the neural cells that originate in a particular proliferative zone of the developing nervous system. This method has been used to show that cortical neurons derived from a common progenitor remain close together in a column and do not intermix extensively with the clonal cells of adjacent columns. Later in cortical development, a tangential migration occurs, but mainly of glioblasts that differentiate into astrocytes.

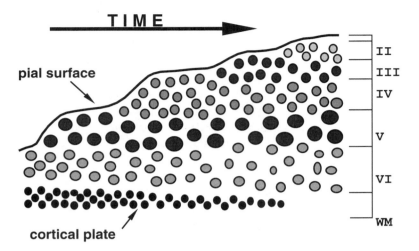

Figure 30.3 Richard Sidman showed that the cells that occupy the deeper layers of the cerebral cortex are born and migrate first. Later-born cells must negotiate the cells already laid down in existing layers. Abbreviation: WM, white matter.

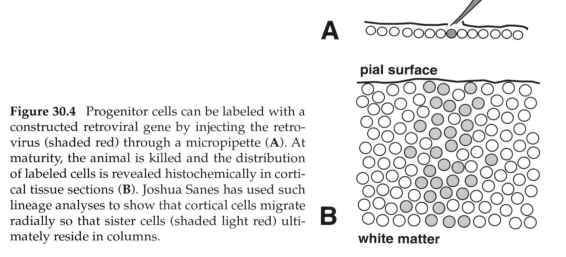

Figure 30.4 Progenitor cells can be labeled with a constructed retroviral gene by injecting the retrovirus (shaded red) through a micropipette (**A**). At maturity, the animal is killed and the distribution of labeled cells is revealed histochemically in cortical tissue sections (**B**). Joshua Sanes has used such lineage analyses to show that cortical cells migrate radially so that sister cells (shaded light red) ultimately reside in columns.

The Emergence of Areal Specialization in the Cortex

Lineage analyses show that many of the unique properties of different cortical areas are not genetically predetermined, but are instead the product of epigenetic influences. The newly formed cortex—exclusive of archicortical areas—is uniform in structure. Everywhere there are six layers with the same cell types arrayed in the same densities in each layer. Even in the mature brain, much of this consistency persists. For instance, the total number of neurons in a cylinder of cortex is remarkably consistent regardless of the cortical area from which the cylinder is obtained. Moreover, the number and deployment of particular neuronal phenotypes (e.g., GABAergic stellate cells) are similar. But the architectural distinctions from one area to another are sufficient to enable neuroanatomists (like Brodmann) to distinguish them. And these structurally distinct areas correlate to a high degree with functional domains. How do these structural–functional differences emerge?

You'll recall that the primary visual area develops in the occipital cortex. As the visual cortex of a rat matures, it initially extends axons into the spinal cord. But such axons are only temporary; they eventually retract (you recall that the occipital cortex does not contribute to the corticospinal projection of the adult). The frontal cortex also sends axons to the spinal cord that will remain in place throughout the rat's lifetime. In rat embryos, it is possible to transplant a prism of occipital cortex to an excavated site in the frontal lobe shortly after the cortex has completely formed but before its axons have grown out. Do the cells in this displaced cortex retract their spinal axons, as they certainly would had they remained in the occipital lobe, or do they behave like frontal cells and sustain their spinal axons into adulthood? Retrograde tracing of corticospinal neurons in transplanted rats that have grown up shows that the layer V neurons in the transplant retain their corticospinal projections. In other words, the occipitally derived cortex takes on the areal specializations of true frontal cortex, suggesting that *the location, not the lineage, of cortical cells determines their ultimate identity.* What is it about the local milieu that instructs the differentiation of neurons?

The local factors that instruct cortical cells to organize in particular ways are poorly understood at present. However, experimental evidence indicates that thalamic inputs are of paramount importance for the differentiation of areal properties of the cerebral cortex. The so-called *barrel field* of the rodent S_I cortex presents a vivid example (Fig. 30.5). Barrels were initially described as anatomically and functionally identifiable entities in the part of the rodent S_I cortex that receives input from the snout. Each barrel consists of three concentric cylinders composed of, from outside in, astrocytes, neuronal cell bodies, and neuropil. Each barrel receives input from one and only one of the major snout whiskers

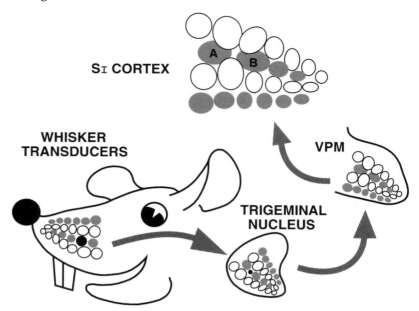

Figure 30.5   Tom Woolsey showed that when one of the major whiskers is plucked from the snout of a neonatal rat (jet black disc), the whisker representation (barrel) in the contralateral S_I cortex disappears and is taken over by the expansion of adjacent barrels (labeled A and B). Abbreviation: VPM, ventral posteromedial nucleus.

via the trigeminothalamocortical pathway. Developmentally, the barrels do not appear until after the thalamic axons have entered the cortex. Moreover, if a whisker is removed at birth, its barrel never forms in the cortex. Instead, the site at which the barrel would have formed is taken over by an expansion of adjacent barrels. Thus, the barrel formation is not programmed by the cortical cells themselves, but is induced in some way by thalamic axons.

Programmed Cell Death

Neuronal cell death is a normal process in development and is observed in most areas of the nervous system (Fig. 30.6). Although cell death may at first seem inefficient, it is likely that the superfluity of neurons produced in early development actually reduces the genetic load on the nervous system by making it unnecessary to program the specific number of pre- and postsynaptic cells. Theoretically, neurons that are unsuccessful in competing for the limited number of target cells (synaptic sites) do not survive. It is important to note, however, that *the cells that die are neither intrinsically defective nor fated to die.* This has been shown

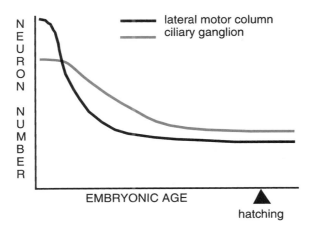

Figure 30.6 Viktor Hamburger showed that neuronal cell death occurs normally in the developing chick nervous system. A massive loss of neurons occurs in the early embryo.

experimentally by grafting an extra limb onto chick embryos. During subsequent development of the embryo into a hatchling chick, the additional limb is innervated by motoneurons that otherwise would have died.

The synaptic activity of the neurons at their target cells also plays a role in their own survival. If neuromuscular transmission is blocked in chick embryos by treatment with curare, there is an increase in the number of motoneurons that survive to hatching. However, when the curare is withdrawn and neuromuscular transmission restored, the surplus neurons soon expire. These data suggest that *functional innervation regulates the production by the target cell of a retrogradely acting agent necessary for survival of the neuron: a trophic factor.* Deductively, one might reason that as muscle is innervated, its production of trophic factor is diminished, reducing the number of neurons it can sustain. Consequently, only a finite percentage of the neurons that originally innervate a muscle are able to survive. The most convincing demonstration of the dependence of neurons on trophic factors for their survival during development was obtained by injecting antibodies to a trophic substance called **nerve growth factor** (NGF) into neonatal mice. This treatment interferes with the ability of target-released NGF molecules to bind to their neuronal receptors and results in the death of neurons in the sympathetic ganglia. Such neurons are thus dependent on a steady supply of NGF for continued survival.

NGF was the first neurotrophic factor discovered and is the best characterized to date. The major function of NGF is the regulation of neuron survival in the sympathetic and sensory ganglia. Some phenotypes of central neuron are also NGF-dependent. Cloning and sequencing of the NGF gene indicates that it is part of a large family of related trophic factors called **neurotrophins.** Other members of this family are **brain-derived neurotrophic factor** (BDNF), **neurotrophin-3** (NT3), and **neurotrophin-4** (NT4). These factors exist as homo- and perhaps heterodimers in the brain and PNS. Which neurons are influenced by the different trophins depends on their expression of specific receptors that selectively bind the trophins. The intracellular domains of the trophin receptors are *kinases that selectively phosphorylate tyrosine residues* on protein substrates; such receptor tyrosine kinases are called "tracks" (spelled **Trk**). NGF binds to TrkA. Retinal ganglion cells and certain sensory neurons bind BDNF and NT4 via TrkB. The survival of dorsal root ganglion cells and the neurons of the mesencephalic trigeminal nucleus is supported by NT3 binding to TrkC.

Although the mechanism of action of NGF is not completely understood, it is known that when NGF binds to TrkA present in the membrane of an axon terminal, the NGF–receptor complex is internalized and transported retrogradely to the cell body. The internalized complex does not interact directly with the nucleus, as do steroid hormone–receptor complexes, but instead activates enzymes through the tyrosine kinase action of TrkA. It is likely that some of the enzymes are nuclear regulatory proteins (see Chapter 5). If neurons of the sympathetic ganglia are deprived of NGF, the neurons exhibit a variety of ultrastructural changes (disintegration of neurites, formation of lipid droplets, changes in the nuclear membrane, dilation of the rough endoplasmic reticulum) that end in death and decomposition. However, if the NGF-deprived cells are exposed to agents that inhibit protein synthesis (e.g., cycloheximide or puromycin) or RNA synthesis (actinomycin D), cell death is prevented. Thus, it appears that NGF-dependent neurons must synthesize one or more "killer" proteins in order to die when deprived of NGF. Presumably, *NGF-activated nuclear regulatory proteins suppress an otherwise active self-destruct program in certain neurons.*

Determinants of Neuronal Phenotype

Throughout this book, we have seen examples of how the shape of a neuron determines in part its signal integration properties. Many neurons, such as hippocampal pyramidal neurons and cerebellar Purkinje neurons, elaborate their characteristic morphology even when grown in culture dishes, indicating that their general shape is genetically programmed. The finer details of neuronal morphology, however, depend also on epigenetic factors, such as contact by afferent axons and successful establishment of synapses with target tissues. Remember also that nuclear groups are characterized as much by their specific complement of neurotransmitters and receptors as by their morphology. As you certainly appreciate, the appropriate expression of neurotransmitters and receptors is essential to proper brain function. What factors instruct neurons to express a particular neurotransmitter phenotype is a central question in developmental neurobiology and may hold the key to understanding the dynamic aspects of synaptic signaling that support higher brain functions such as learning and memory.

Most efforts to identify the factors that determine neurotransmitter phenotype have focused on the trophic factors. NGF, for instance, which sustains sympathetic neurons, also induces them to synthesize the neurotransmitter norepinephrin and suppress the synthesis of ACh. Other signals that may instruct neurotransmitter phenotype are the neurotransmitters that are synaptically released onto a postsynaptic neuron. For most neurotransmitters, one or more receptor subtypes exist that mediate transcription changes in genes encoding neurotransmitter-related enzymes, receptor proteins, and peptide cotransmitters (see Chapter 5).

Differentiation of Glial Phenotypes

The differentiation of the three glial cell types of the CNS [protoplasmic (type 1) and fibrous (type 2) astrocytes and oligodendrocytes] is an exquisitely regulated process. A series of experiments has provided strong evidence that *type 2 astrocytes and oligodendrocytes arise from a common ancestor, the O2A progenitor* (Fig. 30.7). In contrast, *type 1 astrocytes arise from a different progenitor.* The type 1 astrocyte is the first glial cell to

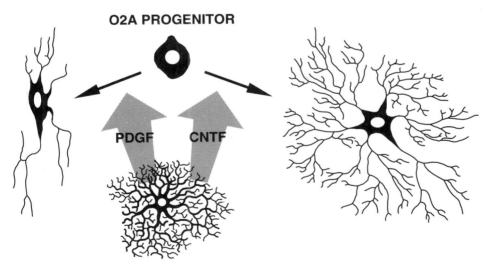

Figure 30.7 The daughters of O2A progenitors differentiate into either oligodendrocytes (left) or fibrous astrocytes (right) depending on the diffusible signal released by protoplasmic astrocytes (bottom). Abbreviations: CNTF, ciliary neurotrophic factor; PDGF, platelet-derived growth factor.

appear prenatally in the optic nerve of the rat. The O2A progenitors then appear around the time of birth and are stimulated to divide by a growth factor first isolated from platelets called **platelet-derived growth factor** (PDGF), which is also synthesized and released by type 1 astrocytes. All of the glia derived from the O2A progenitors will differentiate into oligodendrocytes during the time that the type 1 astrocytes express PDGF. For unknown reasons, the type 1 astrocytes eventually switch off their expression of PDGF and instead begin to synthesize and release a second growth factor called **ciliary neurotrophic factor** (CNTF). CNTF induces all subsequent daughters of the O2A progenitors to differentiate into type 2 astrocytes. Thus, glial cell differentiation is regulated during early development by a complex cascade of closely timed interactions.

What instructs the glioblasts to stop their proliferation? A clue comes from observations on human astrocytoma cells grown in vitro. When plated alone in the culture dish, such cells divide rapidly. But proliferation is all but halted within 12 hours of the addition of cerebellar granule neurons to the culture dish. If granule neurons that are deficient in their expression of astrotactin are used instead of normal granule cells, astrocytoma cell proliferation is not inhibited. These data suggest that neuron–glia interactions are necessary to prevent aberrant glial proliferation in the adult and, further, that uncontrolled glial cell division (tumor formation) may be the result of perturbation of such interactions.

Detailed Reviews

Altman J. 1992. The early stages of nervous system development: neurogenesis and neuronal migration. In Björklund A, Hökfelt T, Tohyama M (eds), *Handbook of Chemical Neuoranatomy. Vol 10: Ontogeny of Neurotransmitters and Peptides in the CNS.* Elsevier, Amsterdam.

Lindsay RM, Wiegand SJ, Altar CA, Di Stefano PS. 1994. Neurotrophic factors: from molecule to man. *Trends Neurosci* 17: 182.

O'Leary DDM. 1989. Do cortical areas emerge from a protocortex? *Trends Neurosci* 12: 400.

The Formation and Modification of Connections

31

The CNS must be wired correctly if it is to work properly. The wiring up of synaptic connections can be divided into three distinct phases: axon *extension,* synapse *formation,* and synapse *selection* (the maintenance or elimination of formed synapses). As in histogenesis, the formation and maintenance of connections in the nervous system is thought to depend upon a variety of adhesion, tropic, trophic, and recognition processes. Many of these molecular guide-ons for axonal growth are the products of astrocytes, suggesting that to a large extent, axons steer by the "stars." Let's examine axon growth and how these molecular factors regulate the three phases of synapse formation.

The Growth Cone

The tips of growing axons (and dendrites) are characterized by a specialized motile element called a **growth cone** that determines the direction of neurite elongation and the branch points of collaterals (Fig. 31.1). Growth cones actively explore the local environment by repeatedly extending and retracting spikelike **filopodia** and ruffled **lamellipodia.** These movements are produced by the interaction of actin filaments with myosin, both of which are abundant in the filo- and lamellipodia. The actin filaments are continuously and rapidly polymerized at the leading end of the podia, moved in a retrograde direction by myosin, and depolymerized at the trailing end. Under certain conditions, this tread-mill-like process can be halted by a series of cytoplasmic proteins that bind to the actin filaments. Such anchoring proteins are presumed to be activated by membrane surface proteins called cell adhesion molecules (see Chapter 30) when they bind to similar adhesion proteins on the surface of adjacent neurons or glia. Continued polymerization of the stabilized filaments thus extends the podium. Podial membranes that do

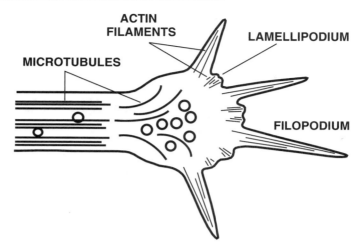

Figure 31.1   The essential elements of the growth cone. Axon extension is driven by actin filaments and microtubules. Vesicles that concentrate mainly in the central region of the growth cone supply the additional plasma membrane to the growing axon.

not encounter adhesion molecules soon collapse back into the growth cone body. If more than one podium discovers a favorable surface for growth, the growth cone may split, creating a branch in the axon.

As the growth cone moves forward, it places linear tension on the trailing axon. Through an unknown mechanism, the curved and randomly scattered microtubules of the growth cone assemble themselves into orderly arrays along the axis of tension. Since the microtubules polymerize at the distal end, they too serve to lengthen the axon. This microtubule-based mechanism contributes to axon lengthening. If actin filaments are poisoned with the drug *cytochalasin*, the growth cone collapses, but the axon continues to lengthen due to polymerization of its microtubules. Likewise, if the growth cone is severed from its axon, it will continue its vigorous exploratory actin-based movements.

Needless to say, a growing axon requires a steady supply of plasma membrane to maintain its integrity. This demand is met by vesicles that are transported anterogradely along microtubules to accumulate in the central region of the growth cone (Fig. 31.1). The vesicles can add to the plasma membrane by fusion.

Adhesion Molecules Are Required for Axon Extension

A series of complex and often overlapping processes is required for successful target location and synapse formation by growing axons.

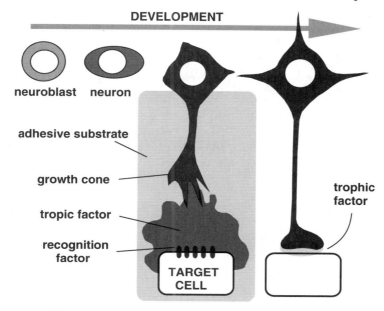

Figure 31.2 Developing neurons require both an adhesive substrate and a tropic signal for axonal guidance. A target-released trophic factor sustains many neurons after synapse formation.

Because a growing axon requires both traction and guidance for its extension across unfamiliar terrain, it depends on both cell adhesion molecules and tropic molecules (Fig. 31.2). Cell adhesion molecules are expressed by all cells and appear on the external surface of cell membranes or as secretion products in the extracellular matrix. As mentioned in Chapter 30, there are several different kinds of cell adhesion molecules—NCAM, NgCAM, fibronectin, laminin, thrombospondin, and a host of others—expressed in the developing CNS that play a critical role in neuronal migration. Such CAMs also modulate axonal growth. Given a choice between substrates containing different cell adhesion molecules, a growing axon will exhibit a preference for one or another depending on the constellation of cell adhesion molecules it expresses itself (Fig. 31.3). Thus, it follows that a differential distribution of cell adhesion molecules can serve to guide growing axons in particular directions. The growth of retinal ganglion cell axons through the optic nerve serves as a good model for studies of cell adhesion molecules because the system is accessible to embryonic manipulation. Such a model system has been used successfully to show that retinal ganglion cell axons grow along a pathway of NCAM that is present on the processes of astrocytes and the end-feet of epithelial cells in the optic tract. If antibodies to NCAM are infused along this pathway, blocking the homophilic binding of the NCAM molecules of the axon membrane with those in the epithelial cells, axon growth fails.

PORN   LAMININ

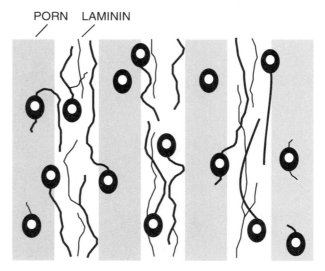

Figure 31.3 Axons exhibit a preference for particular types of adhesion molecules. In this case, axons grow along a laminin-coated surface rather than one coated with polyornithine (PORN).

Cell adhesion molecules *may also be involved in the molecular recognition that enables growing axons to establish synapses on appropriate neurons and not others.* Evidence for a possible role of NCAM in target cell recognition has been provided in experiments on the neuromuscular junction. When muscle cells and spinal cord neurons are cocultured, the interaction of axons with the muscle cells can be blocked by antibodies to NCAM but not by antibodies to other cell adhesion molecules. NCAM is expressed on the surface of embryonic, but not mature, muscle cells. In fact, NCAM colocalizes with α-bungarotoxin (a marker for cholinergic receptors), indicating that it might assist the formation of synapses at specific sites on myocytes. If a previously innervated muscle is denervated by treatment with tetrodotoxin, NCAM is re-expressed in the muscle. Following reinnervation, NCAM expression is once again down-regulated (Fig. 31.4), implying that it serves primarily to assist synapse formation.

Barrier Molecules Prohibit Axon Extension

Recent studies indicate that although the presence of growth-promoting cell adhesion molecules is important in guiding axons, the presence of growth *prohibitory* molecules is also crucial to the process. Such molecules exhibit a highly restricted pattern of expression in the developing CNS; the regions that express them are referred to as *barriers*. The

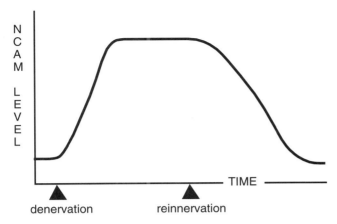

Figure 31.4 Neural cell adhesion molecule (NCAM) expression by skeletal muscle is augmented following denervation and is down-regulated subsequent to functional reinnervation of the muscle.

best-characterized barrier structure is the roof plate of the neural tube. The roof plate is a clump of modified astrocytes that extends the length of the brain stem and spinal cord (Fig. 31.5). The prohibitory molecules elaborated by the roof plate astrocytes appear to be neural forms of keratan sulfate and chondroitin sulfate *proteoglycans*. Because of this

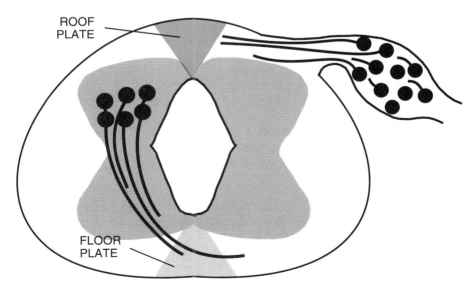

Figure 31.5 During early spinal cord development, the roof plate (shaded red) prevents axons from crossing at the dorsal side of the cord. Rather, axons grow ventrally to cross the floor plate. At later stages of development, the roof plate prevents sensory axons of the dorsal root ganglion from crossing the midline.

barrier, the axons of dorsal horn neurons that cross the midline during early spinal cord development must do so in the ventral white commissure rather than dorsally. Interestingly, it appears that the floor plate secretes a diffusible tropic molecule that lures these axons toward the floor plate. Once the axons have crossed the floor plate, they form the longitudinal tracts that ascend to the brain. Later in development, the roof plate precludes the crossing of axons that ascend in the dorsal columns.

It has been postulated that the olfactory and optic projections are segregated by a barrier structure. During development, when the optic axons reach the optic chiasm at the base of the diencephalon, they have many potential pathways from which to choose. The axons can follow the other optic nerve to the contralateral eye, turn dorsally into the hypothalamus, or continue on their way to the lateral geniculate nuclei. The olfactory axons also grow in close proximity to the diencephalon but never enter this brain region. Rather, the olfactory axons are confined to the telencephalon. It has now been suggested that a glial barrier containing the same molecules as the roof plate prevents the optic axons from entering the telencephalon and the olfactory axons from growing into the diencephalon.

A final example of a molecular barrier is one that regulates the development of the PNS. In embryonic chicks, the spinal nerves grow out through the rostral halves of the laterally adjacent somites and avoid the caudal halves. When a somite is rotated 180°, the axons seek out and enter the rostral somite even though it is now in the caudal position. This indicates the presence of either growth-promoting molecules in the anterior somite or growth-prohibitory molecules in the posterior half. Recent studies support the latter hypothesis. Two proteins have been isolated from posterior somites that prohibit axonal growth in vitro. In addition, many of the biochemical properties of these somite proteins are shared with the identified CNS barrier molecules.

Thus, a combination of growth-promoting and growth-prohibiting influences will contribute to the ultimate pathway selected by a growing axon (Fig. 31.6). In addition, the *rate* of axonal growth may be regulated by the relative abundance of promoting and prohibiting influences.

The Role of Neurotrophic Factors in Synapse Formation

In the last chapter, evidence was presented to support the notion that trophic factors are required for survival of at least some neuronal populations. Since dead neurons don't retain synapses, neurotrophins de facto sustain synapses. But *trophic* molecules like NGF may also have a *tropic* action during synaptogenesis. For instance, an axon will alter its direction of growth in pursuit of a source of NGF (Fig. 31.7). It seems compelling that when axons grow near to a prospective terminal zone,

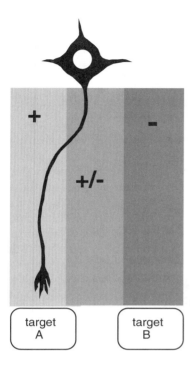

Figure 31.6 Both growth-promoting (+) and growth-prohibiting (−) molecules contribute to pathway selection by growing axons. This schematic diagram shows a growth cone of an axon that is to innervate target **A.** As the environment becomes more positive, this results in steering of the growth cone to the proper target cell. Likewise, extremely inhibitory environments will cause growth cone collapse (inhibition of further axon extension), preventing the incorrect innervation of target **B.**

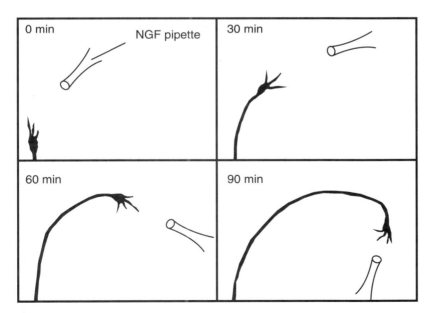

Figure 31.7 R.W. Gundersen and J.N. Barrett showed that the neurotrophin NGF can also serve a tropic function. An axonal growth cone follows the source of NGF as it is released from a moving pipette.

NGF or another trophic factor released by the target tissue may serve to attract the axon.

The Selection of Synapses during Development

Rearrangement of synaptic connections is a normal process during development. Synaptic rearrangement has been well studied in mammalian skeletal muscle, in which each neonatal muscle fiber is innervated by several axons but each adult fiber is innervated by a single axon terminal. This demonstrates that synapses have been eliminated during development. Synapse elimination in muscle begins before birth and continues into the first few postnatal weeks (Fig. 31.8). Because there is little evidence of nerve terminal degeneration, the superfluous synapses must be eliminated by withdrawal of collaterals into the parent axons.

Synaptic rearrangement has also been studied in the CNS, with the developing visual cortex serving as an excellent model. In adult cats and in some species of monkey, the primary visual cortex is organized into ocular dominance columns that form during development. The afferent terminal fields of geniculate axons initially overlap extensively. The zones of overlap are gradually eliminated to yield the adult columns.

This progression can be artificially duplicated using the retinotectal system of the frog. In the frog visual system, there is a complete crossover of optic axons at the chiasm; the retinal axons from the two eyes never encounter one another at the optic tectum. If an extra eye is implanted into a tadpole, however, the optic axons of the third eye must compete with

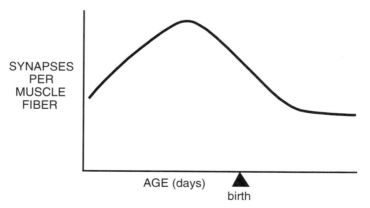

Figure 31.8 Dale Purves showed that synapses are eliminated during the late embryonic development of the rat nervous system. Early on, the number of synapses increases as more growing axons reach the muscle. Beginning before birth and continuing into early postnatal life, however, synapses are lost from the muscle.

one of the host eyes to innervate one of the optic tectums. The outcome of this forced competition is the formation of ocular dominance columns!

There is a *critical period* for the formation of ocular dominance columns, which indicates that synaptic rearrangement only occurs during a finite period of development. This was demonstrated in experiments by Hubel and Wiesel in which one eye of a 2-week-old monkey was sutured shut. Eighteen months later, the geniculostriate axons were labeled transsynaptically by an intravitreal injection of radioactive proline into either the open eye or the closed eye. The columns receiving input from the open eye were expanded at the expense of those receiving input from the deprived eye, providing direct evidence that sensory deprivation early in life alters the synaptic arrangement of the cerebral cortex.

One possible mechanism for the retraction of synapses is the competition for trophic molecules secreted by cortical neurons in response to stimulation. Accordingly, afferent fibers with more extensive contacts will take up more trophic factor and will be able to maintain synaptic contacts. Alternatively, the *activity* of the afferent axons and receptor stimulation of the target cells may regulate column formation. It has been shown recently that the formation of ocular dominance columns is inhibited by drugs that selectively antagonize NMDA receptors. Likewise, stimulation of NMDA receptors leads to a sharpening of the column borders. Since the axons of the LGN use glutamate as a neurotransmitter, their activity will activate NMDA receptors expressed by cortical neurons. These two possible mechanisms of column formation are not mutually exclusive and can be combined to generate a model whereby NMDA-mediated target cell stimulation by temporally correlated axon activity would augment neurotrophic factor production and release.

Because the somatotopic map is an excellent example of the precision of target cell innervation during cortical development, recent studies have tried to identify molecular mechanisms that control the formation of whisker representation barrels in the rodent S$_I$ cortex. Molecular boundaries seem to play a critical role in assembling neurons into functional units such as cortical barrels. These boundaries formed by glial cells are analogous to the barrier structures described above for the spinal cord. Interestingly, molecules that prohibit axon growth have been isolated from the barrel field boundaries. Such molecules are expressed prior to entry of thalamic afferent axons into the cortex and disappear once the barrels are fully established. Thus, the boundary molecules may promote barrel formation by preventing the extension of axons from one barrel to another.

Maintenance and Modification of Synaptic Connections

It is now apparent that synapses are not static. The modification of synapses may even be the physical basis of learning (see Chapter 29).

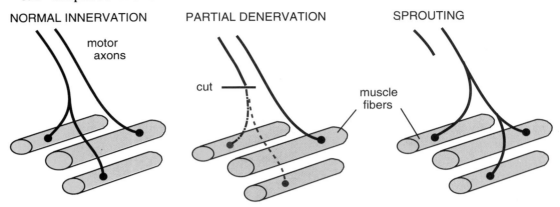

NORMAL INNERVATION   PARTIAL DENERVATION   SPROUTING

motor axons

cut

muscle fibers

Figure 31.9 Michael C. Brown showed that partial denervation of mammalian skeletal muscle leads to sprouting of intact motor axon terminals. The sprouting axons appear to interact with the denervated endplates. Sprouts that do not form the proper connections at endplates will soon disappear.

The status of synaptic connections can be considered as a balance between axonal *sprouting* and *retraction*. Figure 31.9 illustrates axonal sprouting at the neuromuscular junction, where it is readily demonstrated. It occurs after partial denervation (i.e., injury to a subset of motor axons in a nerve). The uninjured axon sprouts new terminal branches to maintain an appropriate level of innervation to the target. Sprouting can also occur without denervation; paralysis of muscle with α-bungarotoxin induces sprouting. Sprouting has been shown to occur in the CNS as well. When one source of afferent axons to a nucleus is eliminated, the axons from other sources sprout locally to occupy the nearby vacated synaptic sites.

The opposite of axonal sprouting is retraction. Axons will withdraw their synaptic endings from a neuron that is injured. We will revisit these phenomena of retraction and sprouting in the next chapter on recovery of the CNS from injury.

Representational Maps Can Be Altered by Experience

We learned in the chapter on histogenesis (Chapter 30) that the adult forms of the representational maps in the sensory cortices are determined in large part by epigenetic factors during the embryonic phase of development and by the early neonatal experience of the animal. But are they static in the adult? Recent experiments suggest that features of the cortical maps can change, to some extent, throughout life.

The representation of the plantar surface of the hand in the S_I cortex of an adult monkey can be mapped in great detail by recording with microelectrodes while selectively stimulating each digit (Fig. 31.10). Once the map is obtained, the monkey can be trained in a task that

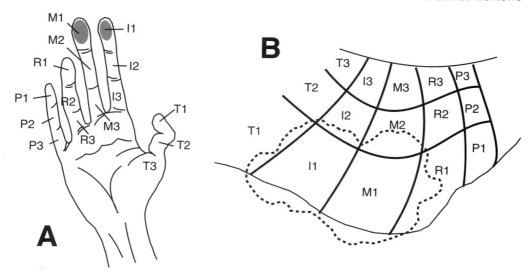

Figure 31.10 Michael Merzenich has studied experience-induced alterations of the somatosensory representation in the primate S_I cortex. Plotting the representation of the fingers (**A**) leads to a map like the one schematized in sketch **B**. The solid lines are the normal boundaries found by recording from the cortex while stimulating the plantar surface of the fingers. After the monkey has been forced to preferentially depend on fine tactile sensations from the tips of its index and middle fingers (shaded) for an extended period, replotting the S_I cortex shows an expanded area (dashed line in **B**) that has become responsive to these finger regions.

requires it to use the tips of its index and middle fingers to precisely control the speed of a rotating turntable. When the monkey exerts exactly the right amount of pressure and the turntable rotates at the correct speed, the monkey is rewarded with a sip of fruit juice. Re-examination of the S_I cortex after several weeks of daily performance of this task reveals that the cortical zone responsive to stimulation of the index and middle finger tips has increased in area at the expense of adjacent digit representations. There are many other examples now in the experimental literature that support the notion that adult cortical maps are continuously subject to change. The mechanisms for such changes have not been entirely worked out as yet. They may range from the unmasking of existing inhibitory circuits to anatomical reorganization in the patterns of convergence of cortical afferent axons. Also, the changes are limited in scope, occurring over only a few millimeters in distance.

Detailed Reviews

Bray D, Hollebeck PJ. 1988. Growth cone motility and guidance. *Annu Rev Cell Biol* 4: 43.

Haydon PG, Drapeau P. 1995. From contact to connection: early events during synaptogenesis. *Trends Neurosci* 18: 196.

Reichardt LF, Tomaselli KJ. 1991. Extracellular matrix molecules and their receptors: functions in neural development. *Annu Rev Neurosci* 14: 531.

Smith SJ. 1988. Neuronal cytomechanics: the actin-based motility of growth cones. *Science* 242: 708.

Stirling RV, Dunlop SA. 1995. The dance of the growth cones—where to next? *Trends Neurosci* 18: 111.

Wall JT. 1988. Variable organization in cortical maps of the skin as an indication of the lifelong adaptive capacities of circuits in the mammalian brain. *Trends Neurosci* 11: 549.

Damage Control and Recovery 32

Following injury to its axon, a neuron undergoes a stereotyped series of changes that play an important role in its ability to regrow the axon. Peripheral neurons and motoneurons (whose axons travel in the periphery) are often quite successful at axon regeneration. Likewise, interneurons in the CNS of lower vertebrates such as fishes and frogs are frequently able to re-establish synaptic connections after axon transections. In contrast to these success stories, and much to the chagrin of physicians who treat patients with CNS damage, axon regeneration is insubstantial at best within the bounds of the mature human CNS. Let's examine the changes that occur when neurons are injured, the putative reasons why central neurons are so sadly incompetent at regeneration, and what might be done to help them along.

The Neuronal Response to Injury

Within a few hours after an axon is injured, visible changes begin to appear in the parent cell body, which are collectively termed *chromatolysis* (Fig. 32.1). There is dispersion of the Nissl substance with an increase in free polyribosomes. The soma swells up and the nucleus migrates to the periphery of the cell. RNA and protein synthesis increase overall, but synthesis of proteins related to neurotransmission declines precipitously. In the axon, the rate of slow transport and the amount of material conveyed by rapid transport increase. Presumably, this bulk flow consists of material necessary to regrow the axon. Axosomatic and axodendritic synapses from afferent axons are withdrawn from the injured cell.

Since the latency between axon damage and the onset of chromatolysis is directly related to the distance the injury occurs from the cell body, *the signal that triggers the chromatolytic reaction appears to be carried to the cell*

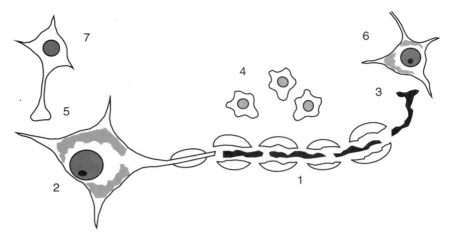

Figure 32.1 Neurons undergo degenerative changes following injury. After axotomy, the injured neuron undergoes anterograde axon degeneration (**1**) and chromatolysis (**2**); its terminals also retract and degenerate (**3**). Phagocytic cells invade the damaged area (**4**). Afferent synapses to the damaged neuron are often retracted (**5**). Transneuronal degeneration of target neurons (**6**) and, less frequently, afferent neurons (**7**) can also result.

body via retrograde axonal transport. According to one attractive hypothesis, the interruption of a steady retrograde flow of trophic substance sets off the chromatolytic reaction.

After an axon is cut or crushed, the segment distal to the site of damage, now separated from its source of raw materials and metabolic machinery, undergoes a relatively slow degenerative process called *Wallerian degeneration* (Fig. 32.1). The myelin sheath, if present, detaches from the axon membrane, and the axon membrane itself begins to disintegrate. Synaptic contacts of the axon with target neurons are withdrawn. Axoplasmic transport declines as microtubules depolymerize. Wallerian degeneration occurs any time an axon is injured, even in PNS neurons that ultimately regrow their axons.

Sometimes neurons that are not immediately damaged themselves, but that are in synaptic contact with killed neurons, will exhibit *transneuronal* degenerative changes. This happens most commonly in the anterograde direction to neurons that received their predominant input from the killed neurons. A prominent example is the degeneration of layer IV cortical neurons in primary sensory areas after loss of their corresponding thalamic input. Transneuronal degeneration has been observed less frequently in neurons that project to the killed population; thalamic neurons are likely to die when their cortical targets have been eliminated.

Glia May Serve to Isolate an Injury Site in the CNS

Large numbers of macrophages and astrocytes arrive rapidly at an injury site in the CNS (Fig. 32.1). The macrophages invade the damaged tissue from nearby capillaries and seem preoccupied solely with the removal of cellular debris. The local increase in astrocyte density is due to a sudden mitotic episode—a process called **reactive gliosis**—for which the trigger remains unknown. Although reactive astrocytes may contribute to the removal of debris to a limited extent, it seems more probable that they serve to seal off the damage site by forming a glial *scar.* Obviously, damaged neurons will liberate a number of neuroactive and toxic substances that can diffuse to nearby tissue elements (such as synapses) and disrupt their normal functioning or even damage tissue that had escaped the initial injury. Because astrocytes are ready and willing consumers of many substances liberated from injured neurons, the glial scar may protect adjacent tissues from exposure to potentially harmful substances. This protection, however, comes at a price. Unlike immature astrocytes in the developing nervous system, *reactive astrocytes offer a poor substrate for growing axons.* The astrocytic scar presents a mechanical and biochemical barrier to axon regeneration. Since there are no astrocytes in the PNS, no such glial scar forms at a lesion site in a peripheral nerve.

Why Don't CNS Axons Regenerate?

The failure of central axons to regenerate is of cardinal interest to medical neuroscientists. Blame for this failure has been laid both on the central neurons themselves and on the local environment. Whereas it seems that some populations of neuron are unable to muster the molecular resources to regenerate an axon, it is certain that the CNS is everywhere inhospitable to significant axon regrowth. If CNS neurons are presented with an opportunity to grow axons into *peripheral* nerve tissue—say, a stretch of sciatic nerve grafted so as to bridge the gap along a severed tract—their axons can accomplish distances of some thirty millimeters (Fig. 32.2). Upon re-entry into the CNS from the graft, however, the axons grow only a few micrometers further. In this case, the axons are intrinsically capable of growth, but the extracellular environment of the CNS apparently lacks the growth-promoting properties of the peripheral tissue. Accordingly, a segment of optic nerve inserted between the stumps of a transected sciatic nerve does not support the axonal growth of PNS axons.

The observations above derive largely from experiments by Albert Aguayo and his coworkers. Encouraged by the results, they propose the therapeutic use of peripheral nerve grafts to circumvent the problem of CNS recalcitrance. To explore this possibility, Aguayo's group has used

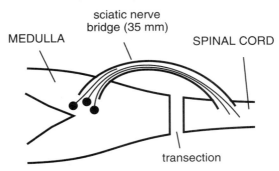

Figure 32.2 Albert Aguayo used peripheral nerve grafts to entice regeneration of CNS axons. Here, the sciatic nerve bridges the surgical gap placed between the medulla and spinal cord. Axon growth from the medulla occurs only in the graft; upon re-entry to the CNS, the fibers cease to grow.

the adult rat retinotectal projection as a model system to determine whether or not functional recovery could be achieved. In these experiments, one optic nerve is removed and replaced with a peripheral nerve graft that extends from the optic nerve stump at the eyeball to the contralateral superior colliculus (Fig. 32.3). Anatomical tracings in such preparations indicate that retinal ganglion cell axons indeed transit the graft to enter the tectum. From electron micrographs of tectal sections, it was confirmed that some retinal axons had established synaptic contacts on tectal neurons. Perhaps most importantly, microelectrode recordings revealed EPSPs in some tectal neurons upon illumination of the contralateral retina. Although these results are promising, the renewed synaptic contacts were dismally few in number and restricted to a small

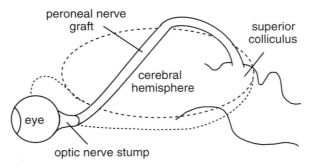

Figure 32.3 In a series of experiments, a peroneal nerve graft was attached to the orbital stump of the transected optic nerve close to its emergence from the eyeball. The other end of the graft was inserted into the superior colliculus. The axons of retinal ganglion cells were able to traverse the graft and partially reinnervate the superior colliculus.

volume of tectal tissue near the adhesion site of the grafted nerve. It appears, therefore, that even though a graft of peripheral nerve can be used as a conduit to deliver central axons to their appropriate target area, the unfavorable environment within the CNS does not allow sufficient reinnervation for adequate functional recovery.

Several laboratories, most notably that of Anders Björklund in Lund, have used a different approach to study the mechanisms regulating CNS regeneration. Their working hypothesis is that *embryonic* CNS neurons have the competence to grow axons in the adult CNS. This hypothesis is based on the premise that fetal CNS neurons express the growth-associated molecules required for axon extension that are forever lost to mature CNS neurons. Early studies by Björklund's group were based on transplants of fetal rat dopaminergic neurons into the striatum of adult monkeys in an attempt to reverse the effects of Parkinsonism that had been previously induced by destruction of the pars compacta of the substantia nigra with selective toxins. Suspensions of dissociated fetal neurons were infused through a micropipette into the deep-lying caudate and putamen. Subsequent fluorescence microscopy (remember that dopamine neurons are autofluorescent when exposed to formaldehyde vapors) showed that the fetal neurons survived in the adult striatum and extended local processes. More to the point, some of the characteristic motor dysfunctions were partially reversed in monkeys that received the transplant. This transplant therapy has now been applied on a trial basis in several Parkinsonian patients. The outcomes have varied, but in some cases there was significant amelioration of the motor signs. Thus, there appears to be considerable promise for using embryonic neural tissue transplants to repair damaged CNS circuitry. It should be kept in mind, however, that in Björklund's experiments the dopamine axons did not have to grow more than a few micrometers to reach their targets, and further, there may be no spatial constraints on striatal dopamine transmission. Can other types of fetal neuron be used to reconnect more distant parts of the CNS? If so, can the fetal axons re-establish the spatial pattern of connections necessary for full functional recovery?

In the rat, the pupillary reflex has been used as a test of functional recovery following CNS damage and fetal tissue transplantation (Fig. 32.4). Restoration of this reflex requires that relatively specific connections be re-established. In the normal reflex circuit, light activates retinal ganglion cells that send axons through the optic nerve to certain nuclei in the pretectal area on each side of the brain. Axons from these nuclei in turn project to the Edinger–Westphal nuclei, whose neurons send their axons with the 3rd cranial nerve to the ciliary ganglion, where the ganglionic parasympathetic neurons control the pupilloconstrictor muscles. Thus, activation of this reflex pathway results in an obvious end point, constriction of the pupil. Ray Lund and his colleagues have exploited this pupillary reflex to study the feasibility of fetal tissue transplantation. The experimenters removed retinas from donors on

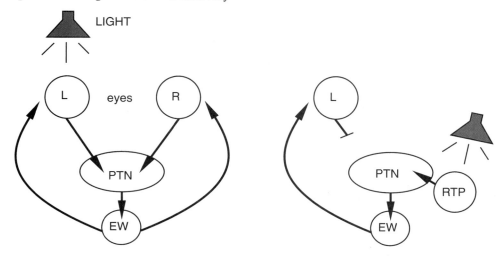

Figure 32.4   Ray Lund used the pathway that mediates reflexive pupil-loconstriction (diagrammed on the left) to assess whether or not transplanted neurons could form functional connections. Normally, light activates retinal ganglion cells that project convergently to pretectal nuclei (PTN). The PTN neurons project to the Edinger–Westphal nuclei (EW). Preganglionic efferent axons from the EW nuclei project via the oculomotor nerve to the parasympathetic neurons of the ciliary ganglion, which innervate the pupilloconstrictor muscles. In the experimental rat (diagrammed on the right-hand side), the right retina (RTP) is transplanted near the PTN and the left optic nerve is cut. A pupillary response cannot be elicited from illumination of the left eye, but illumination of the RTP leads to a constriction of the left pupil, because the retinal ganglion cell axons from the RTP have grown to and innervated the neurons of the PTN.

embryonic day 13 and laid them atop the left superior colliculi of neonatal rats from which the right eye had been removed. After 5 months, the optic nerve of the remaining left eye was severed to preclude any transmittal of photic data from the eyes to the rest of the brain. Parasympathetic input to the eye remained intact, since it is carried by the oculomotor nerve. The rats showed no pupillary reflex when light was shone in their eyes (Fig. 32.4). However, when the transplant was exposed by craniotomy and illuminated, a pupillary reflex was observed! All animals tested exhibited the pupillary reflex, indicating that the ganglion cell axons of the transplanted retina had formed functional synaptic contacts with the "correct" population of neurons in the pretectum. Thus, transplantation of fetal CNS tissue in perinatal mammals may be effectively used to restore some functions. Alas, it is more troublesome for repairs in the adult CNS; only six of twenty-four animals recovered a pupillary reflex when the same experiment was repeated with adult rats.

Despite the restoration of a simple behavioral response, it appears that the use of transplants to restore more complex functions, such as patterned vision in blinded individuals, is not as promising. This is because transplants in the visual cortical area in blinded animals have not been shown to form the spatially appropriate connections that are required for pattern detection. In summary, although transplantation and peripheral nerve grafting studies demonstrate a therapeutic potential for the restoration of CNS circuitry in mammals, the molecular mechanisms are still too poorly understood for therapeutic application. A number of different approaches have been employed in an attempt to understand these mechanisms.

What Are the Molecular Requirements for Regeneration?

By comparing the proteins expressed in neurons engaged in active axon growth with those in a nongrowing population, several growth-associated polypeptides (**GAP**s) have been identified. GAP-43 is the best characterized and is expressed primarily during the active regeneration of optic nerve axons in toads. This correlation suggests a role for GAP-43 in successful regeneration. Could the failure to induce GAP-43 after injury account for the lack of CNS regeneration in adult mammalian neurons? It has been shown that GAP-43 is present in *neonatal* rabbit optic nerve (during axonal growth) but is absent in adulthood. Induction of GAP-43 does not occur following injury to the rabbit optic nerve. In contrast, injury of the rabbit hypoglossal nerve, which does regenerate, induces the expression of GAP-43 by hypoglossal neurons. These data support the notion that the failure of mammalian CNS neurons to regenerate may be due to their inability to re-express GAP genes and are consistent with the observation that fetal, but not adult, CNS neurons will extend axons into adult CNS tissue. However, it has been demonstrated recently that GAP-43 is a cytoplasmic protein associated with the inner surface of the membrane and the cytoskeleton—an observation that raises questions about its interaction with the molecules of the growth substrate (see Chapter 31). It may serve as part of an intracellular link between membrane surface proteins and the cytoskeletal components that mediate axon elongation. Unfortunately, the issue is further confused by the fact that some neurons *lacking* GAP-43 can grow axons perfectly well, while others cannot.

Because neurotrophic factors are likely to play a role in the establishment of neuronal connections during development, it is possible that these factors are also important during nerve regeneration. Following transection of the adult rat optic nerve, the majority of retinal ganglion cells die. However, in the presence of a growth factor derived from fibroblasts, a fourfold increase in ganglion cell survival is observed; NGF has no effect, although another member of the neurotrophin family, brain-derived neurotrophic factor, can promote retinal ganglion cell survival. It is not yet known whether these growth factors promote axon regrowth.

Peripheral nerve grafts to transected optic nerves also promote retinal ganglion cell survival, suggesting that one or more trophic factors from the graft enhance the ability of the retinal ganglion cells to survive axotomy. One candidate is NGF, which is known to be released by Schwann cells present in the graft. The retinal ganglion cells may also survive because of surface interactions of their axons with the cellular and extracellular matrix elements at the graft junction; those that extended axons into the graft survived the longest. Laminin, for instance, will promote neurite extension in vitro. Many cell types, including Schwann cells, secrete a laminin–heparan sulfate proteoglycan complex that promotes neurite outgrowth from both PNS and CNS neurons. In vitro, regenerating axons from superior cervical ganglia explants will grow on sections of sciatic nerve but not on sections of laminin-deficient optic nerve. The outgrowth on sciatic nerve is blocked by pretreatment of the tissue with an antibody to the laminin–heparan sulfate proteoglycan complex.

Is laminin necessary for axonal growth in vivo? Laminin is expressed on the surface of radial glial fibers in the embryonic neural tube and may, therefore, participate in neuron–glia interactions. In addition, the ability of retinal ganglion cell axons to grow on laminin is developmentally regulated. Such axons only respond to laminin during periods when the ganglion cells are normally extending axons. Laminin is continually expressed in the olfactory nerve, which exhibits continual turnover of its neurons (thus resembling a developmental system). In normal frog optic nerve and tectum, which readily regenerate, laminin is continually expressed by astrocytes and up-regulated after optic nerve transection. Is it possible that the unfavorable environment for CNS regeneration is due to the failure of glia to express laminin following injury?

The precise role of cell adhesion molecules in nerve regeneration, particularly in the CNS, remains to be determined. A preliminary analysis of the response of NCAM and NgCAM to sciatic nerve injury has been performed. The level of expression of both CAMs increased after injury, particularly in the area of the lesion. NCAM and NgCAM were detected on neurites, on Schwann cells, and in the perineurium. Although correlative, these data suggest that the up-regulation of CAMs is requisite to axon regeneration.

Clearly, the absence of growth-promoting factors may account for the inability of CNS neurons to regrow their axons. But evidence has now emerged to suggest that the adult CNS also may produce molecules that actively prohibit growth. For instance, two polypeptide fractions have been isolated from mammalian CNS myelin that appear to impede axon growth. Application of antibodies to the two suspected prohibitors enables axons to grow on CNS myelin, a substrate that is ordinarily unfavorable to such growth. Conversely, addition of these CNS myelin proteins to substrates conducive to growth hampers the ability of axons to grow. The prohibitory polypeptides are not expressed in CNS myelin from lower vertebrates, which, you'll recall, do exhibit CNS regeneration. Recent studies have attempted to provide support for the hypothesis that the oligodendrocyte-derived prohibitors of axonal growth impair CNS regeneration in adult mammals. The experimental approach employed to test this hypothesis involved the injection of hybridoma cells that secrete monoclonal antibodies to these proteins into the frontoparietal cerebral cortex of adult rats. The corticospinal tract was then severed at the mid-thoracic level, and regeneration of its axons was monitored with tracing techniques. In antibody-treated rats, massive axonal sprouting occurred at the lesion site and axons grew some 7 to 11 millimeters caudal to the lesion. In control rats, growth did not exceed 1 millimeter. How the immunoglobulins traveled from the cortex to the throacic spinal cord is unclear, but travel they apparently did! Thus, some regeneration could be effected by blocking interaction of the axon membranes with the prohibitory polypeptides. At present it is not known if these oligodendrocyte-derived proteins are the only inhibitors of CNS regeneration; there may be many others. You'll recall from Chapter 31 that prohibitory molecules present in barrier structures regulate axonal growth during the development of the nervous system. Thus, barrier molecules, if present, could also impair nerve regeneration in the adult CNS.

Taken together, these many studies raise the intriguing possibility that higher vertebrates have evolved two mechanisms to actively prevent exuberant regeneration within the confines of the CNS: a constitutive expression of prohibitory factors and a suspension of the expression of growth-promoting molecules. You know from the CNS odyssey you have just completed that with increased computational complexity comes a greater demand for large numbers of neurons with precise spatial ordering to their synaptic relationships. You know further that the complex architecture of the brain is necessarily assembled according to an elaborate sequence. As in the construction of a house, the plumbing must be done before the walls are in place; later renovations or repairs require at least some costly destruction. Mother Nature cannot temporarily remove the red nucleus to rewire the cerebellothalamic tract, as a plumber would tear out a floor to access a leaky drain pipe. Although the precise wiring of the CNS is readily accomplished during the developmental phase, it is fiendishly tricky to recreate distant connections once the brain is built. Could it be that the price we pay for the fantastic abilities of our complex brain is the forfeiture of regenerative capacity? Did Mother Nature intend that our brains not repair themselves?

Detailed Reviews

Bähr M, Bonhoeffer F. 1994. Perspectives on axonal regeneration in the mammalian CNS. *Trends Neurosci* 17: 473.

Gorio A (ed). 1993. *Neuroregeneration.* Raven Press, New York.

Seil FJ (ed). 1989. *Neural Regeneration and Transplantation.* Alan R Liss, New York.

Sivron T, Schwartz M. 1994. The enigma of myelin-associated growth inhibitors in spontaneously regenerating nervous systems. *Trends Neurosci* 17: 277.

Epilogue

A student who embarks on a career in neuroscience necessarily spends a portion of her early training in the classroom, where she is like a snorkeler on the surface of the sea surveying the contours, objects, and activities on the reef below. In some places the reef is clear; in others it recedes from view or is obscured by a pocket of turbid water. Occasionally, the snorkeler submerges for a closer look at something interesting, but she is not equipped to linger for very long. If you have read this far without skipping too many chapters, then you may feel a bit as though you've been snorkeling over the reef of neuroscience. Like the snorkeler who, eager to remain longer at depth and examine a part of the reef more closely, swaps her snorkel for an air bottle and regulator, you may soon trade your textbooks for the healing implements of the neurology clinic or the analytic instruments of the research lab. But you would be well advised to remember that although it is rewarding, working at depth comes at the expense of the panoramic view. It is important to surface from time to time and snorkel, so that the broader context of your focused explorations is not forgotten and so that you might identify new sites for future exploration.

APPENDIXES

Blood Supply and Drainage A

Cerebrovascular accidents or *strokes* are the most common cause of brain dysfunction. Since different parts of the brain are served by different major arteries, the particular clinical signs and the degree of impairment in stroke cases depend on *where* the vascular system is compromised.

The major vessels that supply arterial blood to different parts of the brain and spinal cord are reviewed in Figure A1. Two main arterial trunks approach the ventral surface of the brain on either side—the **internal carotid** and **vertebral** arteries—where they unite to form the **circle of Willis.** At the optic chiasm, the internal cerebral artery branches to form the **anterior** and **middle cerebral** arteries, which together serve the rostral two-thirds of the brain, including most of the basal ganglia and internal capsule. Several branches come off the middle cerebral arteries as they approach the lateral fissures. These **striate** arteries penetrate the base of the forebrain at the **anterior perforated substance** to supply blood to the lentiform nucleus and internal capsule. The middle cerebral arteries continue through the lateral fissure, where they branch profusely to supply blood to most of the cortical convexity (Fig. A2). The anterior cerebral arteries pass between the frontal lobes and recurve along the dorsum of the corpus callosum. They serve the medial surface of the hemisphere and the corpus callosum (Fig. A2).

The vertebral arteries unite near the pontomedullary junction to form the midline **basilar** artery. Proximal to this juncture, the vertebral arteries give off two branches. The first two branches form the **posterior inferior cerebellar** arteries (PICA), which serve the inferior surface of the cerebellum. The next two branches immediately fuse to form the **anterior spinal** artery, which feeds into a network of arteries along the length of the spinal cord. The network is also served lower down by the several **radicular** arteries. The **anterior inferior cerebellar** arteries (AICA), several long and short **pontine** arteries, and the **superior cerebellar** arteries branch from the basilar artery before it splits rostrally to form the

411

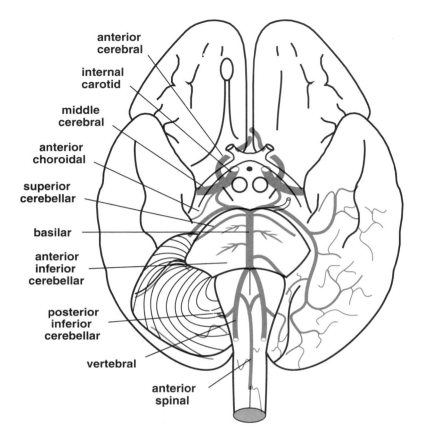

Figure A1   The major arterial supply to the brain and upper spinal cord and the circle of Willis is shown against a drawing of the ventral surface of the brain. The cerebellar hemisphere has been removed on the viewer's right to expose the distribution of the posterior cerebral artery (unlabeled) on the inferior surface of the temporal lobe.

posterior cerebral artery of either side. The **posterior communicating** arteries branch off the posterior cerebral and extend rostralward to fuse with the middle cerebral arteries. The circle of Willis is completed by the short **anterior communicating** artery, which connects the anterior cerebral arteries. The vertebral-basilar system thus supplies all of the brain stem and cerebellum, the caudal thalamus, and the posteromedial part of the cerebral cortex (Fig. A2).

The three cerebral arteries serve different functional areas of the cortex (Fig. A2). With respect to the somatic sensorimotor areas, all but the leg representation depends on the middle cerebral artery for oxygenated blood. The primary auditory, olfactory, and taste areas are also served by the middle cerebral artery. The leg representation, which lies on the

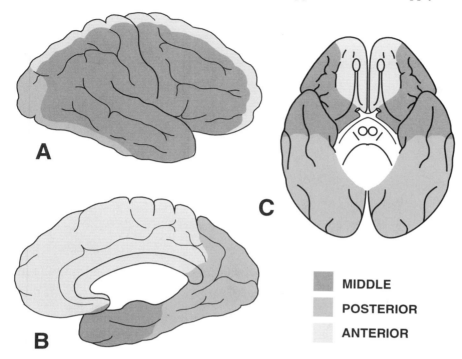

Figure A2 Three views of the cerebral hemispheres—lateral (**A**), medial (**B**), and ventral (**C**)—indicate the areas of cortex and subcortical white matter supplied with blood by each of the three cerebral arteries.

medial wall of the hemisphere, is served by the anterior cerebral artery. The primary visual cortex receives arterial blood from the posterior cerebral artery.

Several larger or smaller veins drain the cortex and internal structures of the brain. The cortical venous system empties blood into the **superior sagittal** and **transverse sinuses.** Much of the interior of the forebrain is drained by the **septal** veins and the **thalamostriate** vein of either side. The latter lie in the floor of the lateral ventricles in the groove between the thalamus and the caudate nucleus. The septal and thalamostriate veins converge on the two **internal cerebral** veins, which lie side by side in the transverse fissure. The internal cerebral veins converge near the caudal end of the thalamus to form the **great cerebral vein** (of Galen) dorsal to the midbrain. The great vein feeds into the **straight** sinus along the dorsum of the cerebellum. The straight sinus meets the superior sagittal sinus at the midline to empty blood into the **transverse** sinus of either side. Each transverse sinus leads into the **sigmoid** sinus, which in turn empties into the **internal jugular** vein of either side.

B Ten Representative Sections

Presented below are ten representative sections through the human brain taken in the frontal or coronal plane. The tissues were treated with a stain that is selective for the myelin coatings around axons. Accordingly, the white matter is stained jet black. The gray matter has a paler appearance that is more or less gray depending on the relative density of stained myelinated fibers present within the individual nuclei. Only landmark structures (major nuclei and fiber tracts) have been labeled. Although ten sections cannot do justice to the intricate anatomical organization of the brain, they do give a flavor of the general organization that occurs at different levels and allow the student to appreciate the levels at which certain nuclei are found and to follow some of the functionally important ascending and descending paths. There are several excellent texts and atlases of neuroanatomy available for those who hunger for greater detail.

Abbreviations

AC	anterior commissure
ah	anterior hypothalamic area
ALT	anterolateral tract
an	abducens nucleus
ant	anterior thalamic nuclei
CC	corpus callosum
cd	caudate nucleus
CF	cuneate fasciculus
cl	claustrum
cn	cuneate nucleus
CP	cerebral peduncle
F	fornix

fmn	facial motor nucleus
GF	gracile fasciculus
gn	gracile nucleus
gp	globus pallidus
hn	hypoglossal nucleus
IML	internal medullary lamina
IC	internal capsule
ICP	inferior cerebellar peduncle
ins	insula
ion	inferior olivary nucleus
ivn	inferior vestibular nucleus
LL	lateral lemniscus
lt	lateral thalamic nuclei
lvn	lateral vestibular nucleus
MCP	middle cerebellar peduncle
mgn	medial geniculate nucleus
ML	medial lemniscus
MLF	medial longitudinal fasciculus
mn	mammillary nucleus
MT	mammillothalamic tract
mtn	motor trigeminal nucleus
mvn	medial vestibular nucleus
na	nucleus ambiguus
nst	nucleus of the solitary tract
OC	optic chiasm
omn	oculomotor nucleus
OT	optic tract
pg	periaqueductal (or central) gray substance
PT	pyramidal tract
ptn	principal trigeminal nucleus
pu	putamen
RST	rubrospinal tract
s	septum
SCP	superior cerebellar peduncle
SCT	spinocerebellar tracts
SM	stria medullaris
sn	substantia nigra
son	superior olivary nucleus
ST	solitary tract
stn	spinal trigeminal nucleus
STT	spinal trigeminal tract
tn	trochlear nucleus
TST	tectospinal tract
vmn	vagal motor nucleus

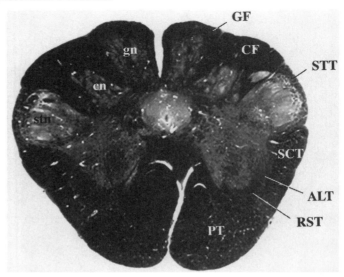

Figure B1   The closed medulla. The pyramidal tracts are just beginning their decussation at this level. The fiber fascicles streaming out of the dorsal column nuclei (gracile and cuneate) to curve ventromedially under the central gray core are called the internal arcuate fibers—they'll form the medial lemniscus seen in the more rostral sections.

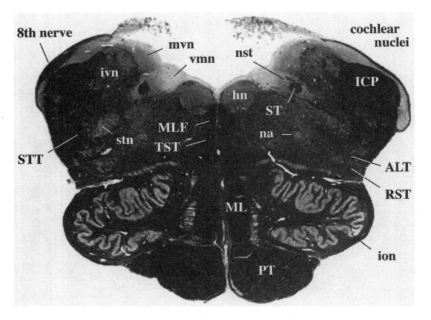

Figure B2   The open medulla.

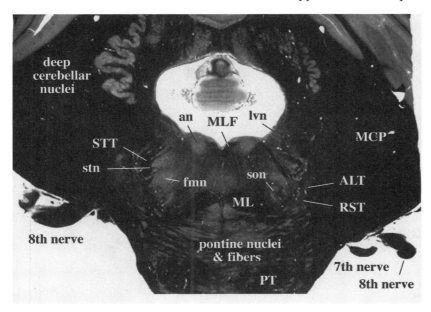

Figure B3 The lower pons. Most of the cerebellum lies beyond the dorsal and lateral edges of the image. The ventralmost part of the ventral pons has been cropped off as well.

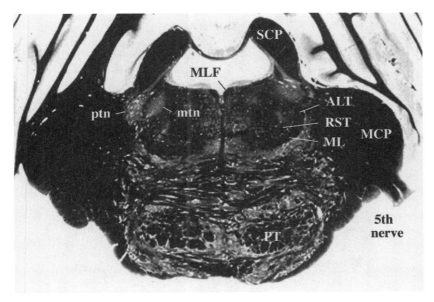

Figure B4 The middle pons. Again, only a small part of the cerebellar cortex is visible at the upper corners of the photograph.

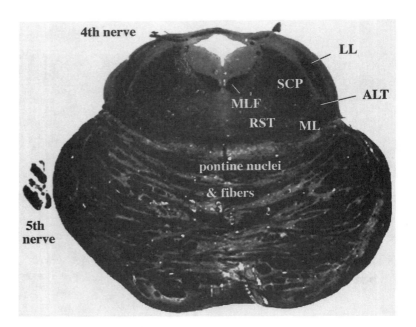

Figure B5 The juncture of the pons with the midbrain is called the isthmus. The gray matter surrounding the superior cerebellar peduncle is composed of the parabrachial nuclei, which are involved in the processing of viscerosensory information. The sliver of gray matter embedded in the lateral lemniscus is the nucleus of the lateral lemniscus, a nucleus involved with the processing of auditory information.

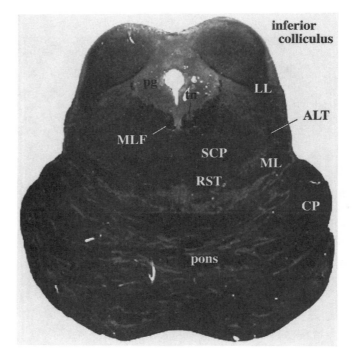

Figure B6 The caudal midbrain. The rostral end of the ventral pons is still present ventral to the midbrain tegmentum.

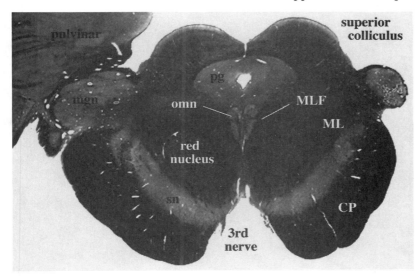

Figure B7 The rostral midbrain. The caudalmost end of the thalamus—the pulvinar and underlying medial geniculate nuclei—is visible in the upper left part of the image. Even a tiny scrap of the lateral geniculate nucleus is present just ventrolateral to the medial geniculate nucleus on the left.

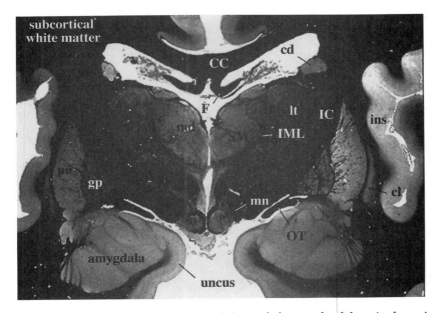

Figure B8 The caudal forebrain. Most of the cerebral hemisphere is outside the margins of the photograph. The black fiber bundle attached to the medial thalamus on either side of the third ventricle is the stria medullaris. The bundles attached to the medial sides of the mammillary nuclei are the original segments of the mammillothalamic tracts.

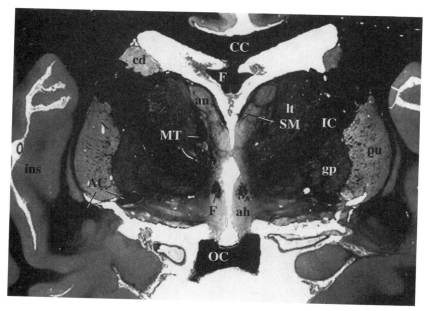

Figure B9 The middle forebrain. Again, most of the cerebral hemispheres are beyond the bounds of the image.

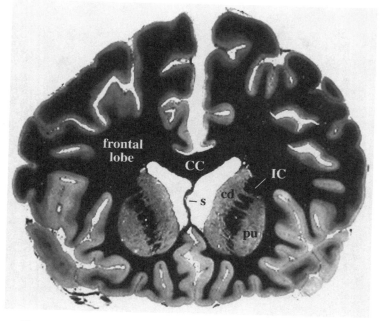

Figure B10 The rostral forebrain. Here, the magnification is low enough to include the entire cross section of the cerebral hemispheres. The clear area on either side of the septum is the lateral ventricle. The cortex at the base of the frontal lobe is called the orbital cortex, because it lies over the bony orbits of the skull.

Index

Abstract meaning, recognition of, 347–349
Accommodation reflex, 212
Accuracy in movement, 165–166
Acetylcholine (Ach), 32, 45, 64, 383
  agonists of, 160–161
  binding to nicotinic acetylcholine receptors, 50–51
  cotransmitters for, 34t
  deactivation of, 42, 43f
  motoneurons and, 166
  at neuromuscular junction, 170–171
  sympathetic ganglia and, 267–268
  synthesis of, 34–35, 36f
  visceromotor neurons and, 265–266
Acetylcholinesterase (Ache), 35b, 42, 43f, 170
Acoustic stria, 310
Actin, 171, 374, 385, 386
Actinomycin D, 382
Action potentials, 7, 45
  afterhyperpolarization and, 28, 29
  calcium and, 46
  conduction speed of, 30–31
  falling phase of, 28, 29
  inhibitory postsynaptic potentials and, 56
  mechanisms of, 28–29
  neurotransmitter release and, 58
  rising phase of, 28, 29
  triggering of, 29–30
Active zone, 8, 46, 47, 48f, 49
Adaptation, 129
Adenylyl cyclase, 61, 62, 63–65, 66, 255
Adrenalin. See Epinephrin
Adrenal medulla, chromaffin cells of. See Chromaffin
  cells
Afferent connections, 13
Afterhyperpolarization, 28, 29
Agnosia, 347–348

Agonists (compounds), 50b
Agonists (muscles), 163, 190
Aguayo, Albert, 399
Akinesia, 243–245
Alar plate, 90, 99
Allocortex, 117
Alpha adrenergic receptors, 268
Alpha motoneurons, 169, 172, 196, 201, 219–220
  breathing and, 194
  eye movement and, 202
  functions of, 166–167
  knee-jerk reflex and, 191
  visceromotor neurons compared with, 265–266
Alpha waves, 154–156
Alzheimer's disease, 71, 238, 245, 370
Amacrine cells, 318, 320–321
Amino acid neurotransmitters, 32–33, 39, 70, 133
  excitatory, 33, 150
α-Amino-3-hydroxy-5-methylisoxazole-4-propi onic
  acid (AMPA), 54
Ammon's horn. See Cornu ammonis
Amnesia, 296, 370
Amplitude of sound, 302, 308–309
Amygdala, 86, 110, 288
  central olfactory pathways and, 259–260
  emotion and, 294, 295
  structure and functions of, 291–292
Anencephaly, 88
Anosmia, 263
Antagonists (compounds), 50b, 51
Antagonists (muscles), 163, 190
Anterior cerebral artery, 411, 413
Anterior commissure, 111
Anterior communicating artery, 412
Anterior inferior cerebellar arteries (AICA), 411
Anterior perforated substance, 411

421

Anterior spinal artery, 411
Anterior white commissure, 144
Anterograde amnesia, 370
Anterograde transport, 40–41
Anterolateral system, 144–146
Anterolateral tract, 134, 144
Aphasia, 353
Appetitive behaviors, 262–264. *See also* Drinking;
    Eating
Arachnoid mater, 83
Archicortex, 117
Areal specialization, 341, 379–380
Arm movements, 165, 218–220
Arousal, 130, 145, 156–157
Arrestin, 319
Asanuma, Hiroshi, 215–216
Ascending reticular activating system, 156
Ascending somatosensory systems, 143–152
Aspartate (as a neurotransmitter), 33, 53, 242
Association cortex, 241, 353
    parietotemporal, 343–344
    prefrontal, 349–352
Association fibers, 111
Associative learning, 362, 368
Astrocytes, 8, 33f, 72–75
    cell adhesion molecules and, 387
    fibrous, 73, 383–384
    functions of, 78–79
    glutamate uptake in, 37
    injury and, 399
    markers of, 75
    phenotype determinants in, 383–384
    protoplasmic, 73, 383–384
Astrotactin, 376
Atonic bladder, 270
Automatic bladder, 270
Autonomic nervous system. *See* Peripheral
    visceromotor system
Autoreceptors, 58
Axons, 7–8, 13
    active zone of. *See* Active zones
    barrier molecules in extension prevention, 388–390
    cell adhesion molecules for extension of, 386–388
    GABA, 230–231, 271
    growth cone and, 385–386, 391f
    hillock, 7
    initial segment of, 7
    myelinated. *See* Myelinated axons
    neurotrophic factors and, 390–392
    propriospinal, 93
    regeneration failure in, 397–398, 399–403
    regeneration requirements for, 403–405
    retraction of, 394
    sensory. *See* Sensory axons
    somatosensory, 131–133
    sprouting of, 394

    staining of, 2–3b
    terminals, 32, 40
Axoplasmic transport, 7, 40–41, 398

Babinski sign, 195
Balance, 123, 197–201
Baroreflex, 269, 270–271, 272f
Barrel fields, 379–380, 393
Barrier molecules, 388–390
Basal eminence, 110, 111
Basal ganglia system, 238–252
    component nuclei of, 239–241
    movement and, 250–252
    outputs of, 248–249
    subthalamic nucleus and, 249–250
Basal plate, 90–91, 99
Basilar artery, 411
Basilar membrane, 304, 305, 306, 307, 308
Basket cells, 228, 230, 375
Behavioral strategies, 349–351
Békèsy, Georg von, 304
Bergmann glia, 375–376
Beta adrenergic receptors, 61, 64, 66, 268
Beta waves, 156
Binocularity, 328, 337
Biogenic amine neurotransmitters, 32–33, 34, 39, 70
Biological clock, 280–281
Biophysical equations, 20b
Bipolar cells, 318, 320, 321, 322f
Bitemporal hemianopsia, 328b
Björklund, Anders, 401
Black widow spider venom, 51
Bladder reflex, 269, 270
Blast cells, 99
Blindness, 314, 346
Blind sight, 346
Blind spots, 316, 346. *See also* Scotomas
Blobs, 331, 334–337, 338–339
Blood supply/drainage, 411–413
Body maps, 151
Bordetella pertussis toxin, 64–65
Brachial (pharyngeal) arches, 172
Brachium conjunctivum. *See* Superior cerebellar
    peduncles
Brachium pontis. *See* Middle cerebellar peduncles
Brain, 10, 13
Brain-derived neurotrophic factor (BDNF), 382, 404
Brain stem, 84
    motor nuclei of, 175–178
    ontogeny of, 95–107
    saccadic eye movements and, 208–209
    sensory nuclei in, 135–136, 137b
Breathing, 164–165, 194
Brightness, of color, 339
Broca, Paul, 3b, 288, 353
Broca's aphasia, 353

Broca's area, 353–354
Broca's limbic lobe, 291
Brodmann's area
   area 3,1,2, *See* Primary somatosensory cortex
   area 4, *See* Primary motor cortex
   areas 5 and 7, *See* Superior parietal lobule
   area 6, *See* Premotor cortex
   area 17, *See* Primary visual cortex
   area 18, *See* Secondary visual cortex
   area 19, *See* Third visual area
   area 41, *See* Primary auditory cortex
   area 8, *See* Frontal eye field
Bulbar reflexes, 193–194
α-Bungarotoxin, 51, 170, 388, 394
Burst cells, 208, 218
Bushy cells, 310

CA fields, 288, 290
Cajal, Santiago Ramón y, 2b, 4–5, 355
Calcium
   in *Aplysia*, 360
   exocytosis and, 8, 46–47, 48f
   hair cells and, 305
   nicotinic acetylcholine receptors and, 50, 51
   presynaptic inhibition and, 56
   resting membrane potential and, 18
   second messenger systems and, 65–66
   sodium exchange with, 46
   visual transduction and, 321–322
Calcium channels. *See also* Voltage-gated calcium
         channels
   hair cells and, 306
   memory and, 367–369
Calcium spike, 235
Calmodulin, 49, 65
Calmodulin-activated kinase type II (CAM kinase II),
         68, 368
Capacitative current, 24–26, 29
Cardiac muscle, 173
Card-sorting task, 349–350
Cataracts, 315, 327b, 358
Catecholamines, 33, 70, 268. *See also specific types*
   deactivation of, 42–44
   synthesis of, 36, 37f
Catechol-O-methyl transferase (COMT), 42–44
Cauda equina, 92
Caudal zone of hypothalamus, 277, 279
Cell adhesion molecules, 374–376, 385
   axon extension and, 386–388
   regeneration and, 404
Cell bodies, 7, 10, 29–30
Cell migration, 376–377
Central auditory pathways, 309–312
Central nervous system
   axon failure to regenerate in, 397–398, 399–
         403

basic plan of, 83–86
structure and functions of, 10–13
three main purposes for sensory information, 130
Central sulcus, 112–113
Central sensory pathways, 13
   auditory, 309–312
   olfactory, 258–260
   somatosensory, 143
   taste, 260–262
   viscerosensory, 260–262
   visual, 323–328
Cephalic flexure, 95
Cerebellar cortex
   cell adhesion molecules in, 375–376
   cellular organization of, 228–230
   functional organization of, 231–234
Cerebellum, 84, 86, 101f, 102, 226–237
   activity in circuit of, 230–231
   basal ganglia system compared with, 238
   gross structure of, 226–228
   movement precision and, 236–237
Cerebral aqueduct, 97
Cerebral cortex, 83–84
   areal specialization in, 341, 379–380
   columnar organization of, 118–119
   descending systems from, 178–179
   information flow across, 342–343
   layering, 377
   perception via serial processing in, 343–344
   pontocerebellum and, 234
   structures of, 116–119
Cerebral hemispheres, 83, 86
   left, 353, 354
   lobes of, 112–113
Cerebral peduncles, 102, 104f
Cerebrocerebellum. *See* Pontocerebellum
Cerebrospinal fluid, 83, 97
Cervical enlargements, 93
Cervical flexure, 95
Chemical driving forces, 19
Chemical synapses. *See* Synapses
Chemosensors, 255–258
Chloride, 17, 19–20, 21, 52
Chloride channels, 55
Cholecystokinin, 34t
Cholera toxin, 63
Choline, 34–35, 36f, 42, 43f
Choline acetyltransferase (ChAT), 34–35, 43f
Chondroitin sulfate proteoglycans, 389
Chopper cells, 310
Chorda tympani, 261
Choroid fissure, 111, 112f
Choroid membrane, 97, 112f
Choroid plexus, 80, 97, 108, 112f
Chromaffin cells, 36, 88, 266, 268
Chromatolysis, 397–398

Ciliary membrane, 255
Ciliary neurotrophic factor (CNTF), 384
Cingulate gyrus, 116, 288, 290–291, 293f
Cingulum bundle, 291
Circle of Willis, 411, 412
Climbing fibers, 229, 235
Cochlea, 303f, 304
Cochlear duct, 304, 305
Cognitive maps, 130
Collision-coupling, 61
Color-coding, 338–339
Color vision, 316, 335, 338–339
Coma, 157
Commissural fibers, 111
Commissural plate, 111
Complex cells, 331, 332–334, 335
Compound reflexes, 193
Computed tomography (CT), 223b
Conduction, 30–31
    decremental, 7
    nondecremental, 7
    saltatory, 31
Conductive deafness, 304
Cones, 316, 318, 319–320, 321, 327b
Conformation of ion channel proteins, 22
Conjugate eye movements, 203, 208–209
Consummatory behaviors, 262–264
Convergence, 8, 9
Cornea, 314–315, 327b
Cornu ammonis, 290. *See also* CA fields
Corollary discharge, 351
Corpus callosum, 86, 111, 152
Corpus cerebelli, 226
Corpus striatum, 86, 110, 239
    compartments of, 245–247
    inputs from, 241–245
    movement and, 251
    outputs of, 248–249
    projection to pallidum and substantia nigra, 247–248
Cortex, 13. *See also specific cortices*
    distinctions among, 342
    histogenesis of, 373–384
Corti, organ of. *See* Organ of Corti
Cortical columns, 151–152, 216
Cortical plate, 377, 378f
Corticospinal tract, 180–182, 186, 216–217, 219–220
Corticothalamic connections, 115
Cranial nerves, 137b, 177b
    3rd. *See* Oculomotor nerve
    5th, 131
    7th. *See* Facial nerve
    8th, 135, 199, 210, 304, 307–308, 309
    9th, 131, 260, 262
    10th, 260, 262
Critical periods of development, 358
Crossed-extensor reflex, 193

Cuneate fasciculus, 136
Cuneocerebellar tract, 140, 234
Curare, 381
Current flow, 24–28, 56
    capacitative, 24–26, 29
    resistive, 24
Cutaneous sensation, 123, 124
Cyclic adenosine monophosphate (CAMP), 64, 67, 360
Cyclic adenosine monophosphate-dependent protein kinase, 62–64
Cyclic adenosine monophosphate response element (CRE), 67–68
Cyclic adenosine monophosphate-response element-binding protein (CREB), 68, 69f
Cyclic guanosine monophosphate (cGMP), 318–319, 321f, 322, 369
Cyclic guanosine monophosphate (cGMP) phosphodiesterase, 65
Cycloheximide, 382
Cytochalasin, 386
Cytochrome oxidase, 35b
Cytoplasm, 17–18, 26, 27–28
Cytoplasmic dynein, 40
Cytoplasmic loop, 60, 61

Dale, Henry, 50b
Day-night cycle, 162
Deafness, 301, 303–304
Decerebrate rigidity, 196
Decremental conduction, 7
Decussation, 100
Deep layers of superior colliculi, 206–207, 249, 312
Degeneration, Wallerian, 398
    transneuronal, 398
DeLong, Mahlon, 242
Delta waves, 156
Dendrites, 7–8, 29–30, 105
Dendritic spine, 7, 58–59
Dentate gyrus, 288, 290
Depolarization
    action potentials and, 29, 30
    current flow and, 24–28
    defined, 21
    of hair cells, 199, 306, 308
    of light, 320
    membrane permeability and, 23
    voltage-gated potassium channels in, 27, 56
    voltage-gated sodium channels in, 24–27, 56
Depth perception. *See* Stereopsis
Dermatomes, 131, 132b
Descending systems, 178–179
    lower circuits controlled by, 195
    medial versus lateral, 186–189
Desensitization, 51
Diabetes insipidus, 286
Diacylglycerol, 65, 66, 68, 369

Diencephalon, 85f, 86, 108, 109
3,4-Dihydroxyphenylacetate (DOPAC), 44
L-Dihydroxyphenylalanine. *See* L-DOPA
Diplopia, 212
Discriminating touch
    lemniscal system and, 146–147
    primary somatosensory area and, 150–151
Disinhibition, 56
Disjunctive eye movements, 203
Divergence, 8, 9
L-DOPA, 36, 245, 352
Dopamine, 33, 70, 71
    cotransmitters for, 34t
    deactivation of, 42, 43f
    hypothalamus and, 282
    identification of in neurons, 35b
    Parkinson's disease and, 244–245
    schizophrenia and, 352
    synthesis of, 36
Dopamine β-hydroxylase, 36, 37f
Dopamine neurons, 35b
    striatum and, 242–245
    transplant of, 401
Dopamine receptors, 223b, 243, 245, 247
Dorsal funiculi, 134, 136
Dorsal horns, 93, 134, 135
Dorsal root ganglia, 88, 131
Dorsal spinocerebellar tract, 135, 233–234
Dorsolateral fasciculus (tract of Lissauer), 134
Dorsolateral sulcus, 133
Double vision, 212
Dreaming, 160–161
Drinking, 281. *See also* Appetitive behaviors
Dura mater, 83
Dynorphin, 34t

Eardrum (tympanic membrane), 302
Eating, 281. *See also* Appetitive behaviors
Efferent connections, 13
Electrochemical equilibrium, 19
Electroencephalograms (EEGs), 154–156, 157–159, 160
Electromagnetic driving forces, 19
Electrophysiology, 18b
Emotion, 285, 294–295
Encephalization, 10–13
End-feet, 74, 78, 387
Endocytosis, 2b
Endolymph, 198, 304, 305
Endoplasmic reticulum, 2b, 7
Endothelial cells, 74
Endplate potential, 171
Enkephalin, 70, 247
Enteric ganglia, 267
Enteric nervous system, 272–274
Entorhinal area, 289–290
Ependymal cells, 73, 80

Epilepsy, 295, 296, 347, 354
Epinephrin, 33, 36, 268
EPSPs. *See* postsynaptic potentials
Equilibrium potentials, 19t, 29
Evarts, Edward, 216–217
Excitation, 8, 9
Excitation-contraction coupling, 171
Excitatory amino acid neurotransmitters, 33, 150
Excitatory postsynaptic potentials (EPSPs), 53, 54, 55,
    70, 266, 360
Exocytosis, 7–8, 41, 45, 46–47, 48f
Extensor muscles, 175, 196
External cuneate nucleus, 135, 136, 140, 234
External granular layer, 117–118
External medullary lamina, 116
Extracellular fluid, 17–18, 26
Extraocular muscles, 176, 202–203
Eyeballs
    movement of, 202–203
    optics of, 314–315
Eye-head coordination
    semicircular canals and, 197
    superior colliculi and, 207
    tectospinal tract and, 183
    vestibulospinal tracts and, 185
Eye movements
    basic features of, 202–203
    conjugate, 203, 208–209
    disjunctive, 203
    saccadic. *See* Saccadic eye movements
    smooth pursuit, 204, 209
    vergence, 204, 212, 337
    vestibulo-oculomotor, 204, 210–212

Face, somatic sensations from, 140–142
Facial nerve, 131, 260, 303
Fascicles, 13
Fenestrated capillary beds, 80
Fetal tissue transplantation, 401–403
Fibronectin, 387
Fibrous astrocytes, 73, 383–384
Fight or flight responses, 269
Filopodia, 385
Fissures, 84
    choroid, 111, 112f
    lateral, 109
    posterolateral, 226
    primary, 226
Flexor motoneurons, 201, 217
Flexor muscles, 174–175
Flocculonodular lobe, 226, 235
Floor plate, of neural tube, 390
Folia, 84
Follicle-stimulating hormone (FSH), 280
Forebrain, 86
    caudal, 419f

Forebrain (*continued*)
  enlargements, 88
  middle, 420f
  ontogeny of, 108–119
  rostral, 420f
Forgetting, 357, 358
Fornix, 276, 277, 278, 286, 290
Fourth ventricle, 97, 99, 100
Fovea, 202, 316
Frequency codes, 128
Frequency of sound, 302, 308–309
Frontal eye field, 208–209
Frontal lobe, 113
  hypoactivity of, 350
Functional specialization, 342

G proteins, 64–66
  membrane excitability mediated by, 61–64
  metabotropic receptor association with, 60–61
  neurotransmission mediated by, 67
  subunits of, 61
GABA. *See* Gamma-aminobutyric acid
Galactocerebroside, 75
β-Galactosidase gene, 377
Gamma-aminobutyrate type A receptor (GABA$_A$-R), 49, 52
Gamma-aminobutyrate type B receptor (GABA$_B$-R), 60
Gamma-aminobutyric acid (GABA), 34
  astrocytes and, 79
  basal ganglia system and, 241f
  cotransmitters for, 34t
  deactivation of, 42
  genes influenced by, 70
  postsynaptic potentials and, 54
  primary visual cortex and, 330
  synthesis of, 37–39
Gamma-aminobutyric acid (GABA) cells, 35b, 39, 342
  baroreflex and, 271
  basal ganglia system and, 248, 249
  in cerebellum, 230–231
  in striatum, 247
Gamma-aminobutyric acid transaminase (GABA-T), 42
Gamma motoneurons, 169, 172, 196, 201, 219–220
  functions of, 166, 167–168
  stretch reflex and, 191–192
  visceromotor neurons compared with, 265–266
Ganglion cells, 9–10
  dorsal root, 88, 131
  enteric, 267
  parasympathetic, 88, 267–268
  spiral, 307
  sympathetic, 88, 267–268
  visceromotor neurons and, 266
GAP-43, 403
Gap junctions, 3
Gaze, 202–212. *See also* Vision

Gaze control areas
  horizontal, 205, 208–209
  reticular, 208–209
Gazzaniga, Michael, 354–355
Gene knockout, 368
Generator potential, 127
Geniculate cells, 261
  type X, 330–331, 334
  type Y, 331, 334
Geniculate channels, 330–334
Geniculostriate projection, 325, 326
G protein, 64–65
Glaucoma, 327b
Glial cells, 7, 72–80
  Bergmann, 375–376
  injury and, 399
  myelin and, 76–78
  neurons distinguished from, 74–75
  neurotransmitter transport and, 42
  phenotype determinants in, 383–384
  radial, 73, 374
  types of and selective markers, 72–75
Glial fibrillary acidic protein (GFAP), 75
Glial scars, 79, 399
Glioblasts, 72, 373–374
Globus pallidus, 110, 111, 239, 240f, 242, 247, 248, 249, 251
  external, 239
  internal, 239, 249–250
Glomerulus, 230, 258
Glutamate, 34, 242, 393
  Alzheimer's disease and, 71
  astrocytes and, 79
  cones and, 320
  cotransmitters for, 34t
  deactivation of, 42
  dorsal horn and, 135
  genes influenced by, 70
  memory and, 365, 367
  postsynaptic potentials and, 53, 55
  somatosensory axons and, 133
  subthalamic nucleus and, 250
  synthesis of, 37
L-Glutamate (as a neurotransmitter), 271
Glutamate decarboxylase, 35b
Glutamate-glutamine cycle, 37, 38f
Glutamate receptors, 52–53
Glutamatergic projection, 241
Glutamic acid decarboxylase, 39
Glutaminase, 37
Glutamine, 37, 38f
Glutamine synthase, 37
Glycine, 33, 34, 37–39, 54, 135
Glycine receptors, 49, 52
Goldman-Hodgkin-Katz equation, 20b, 21
G$_{olf}$ protein, 255, 257f

Golgi, Camillo, 2b, 4
Golgi apparatus, 40
Golgi cells, 228, 229, 230–231
Golgi stains, 2b, 3
Golgi tendon organs, 192
Gracile fasciculus, 136
Granule cells (of cerebellum), 228, 229, 230, 259, 376
Graybiel, Ann, 245
Gray matter, 11–12, 13, 84, 86, 93
Great cerebral vein, 413
Gross, Charles, 347
Growth cone, 385–386, 391f
Guanine nucleotide-binding proteins. *See* G proteins
Guanosine triphosphatase (GTPase), 62
Guanosine triphosphate (GTP), 61–62, 63
Guanylyl cyclase, 321f, 322
Gustatory nucleus, 261
Gut motility, 273
Gut peptides, 274
Gyrus, 84. *See also specific gyruses*
    cingulate, 116, 288, 290–291, 293f
    dentate, 288, 290
    parahippocampal, 288, 291
    postcentral, 116
    precentral, 214, 261

Habituation, 359–360
Hair cells, 197–201, 210, 304
    eighth cranial nerve axon innervation of, 307–308
    inner, 305, 307–308, 309
    mechanism of hearing transduction by, 305
    outer, 305, 308
    tuning of, 306–307
Head movements
    semicircular canals and, 197
    superior colliculi and, 207
    tectospinal tract and, 183
    vestibulospinal tracts and, 185
Hearing, 123, 301–313
    central auditory pathways for, 309–312
    coding of sound frequency and amplitude in, 308–309
    organs for, 302–304
    sound conduction in, 302–304
Hearing impairment, 301, 303–304
Helicotrema, 304
Hemiballismus, 249–250
Hemispheric (lateral) zones, 231
Hindbrain (rhombic) enlargements, 88
Hippocampal cortex, 241
Hippocampal formation, 290, 291
Hippocampus, 111, 117, 288
    memory and, 296, 364–369, 370
    sleep and, 160
    structure and functions of, 290
Histamines, 33
Histogenesis, 373–384

HIV. *See* Human immunodeficiency virus
Homocysteate, 33, 53
Homovanillic acid, 44
Horizontal cells, 318, 320
Horizontal gaze control center, 205, 208–209
5-HT. *See* Serotonin
Hubel, David, 119, 329, 335, 354, 393
Human immunodeficiency virus (HIV), 80
Huntington's disease, 251
Hybridoma cells, 35b, 405
Hydrocephalus, 88
6-Hydroxydopamine, 243
5-Hydroxytryptamine. *See* Serotonin
5-Hydroxytryptophan, 36
Hypercolumns, 336–337
Hyperpolarization
    current flow and, 27
    defined, 21
    of hair cells, 306, 308
    of light, 320
Hypothalamic sulcus, 108
Hypothalamus, 86, 108, 275–287
    biological clock in, 280–281
    central olfactory pathways and, 259
    clinicopathological conditions of, 286–287
    eating and drinking influenced by, 281
    emotion and, 285, 294–295
    lateral, 276–277
    major tracts and connections of, 277–279
    medial, 276
    pituitary relationship to, 281–285
    structural organization of, 275–277
    temperature regulation by, 279–280

Imaging techniques, 223b
Immediate early gene (IEG), 67, 68
Immunocytochemistry, 35b
Inferior cerebellar peduncles, 102, 228
Inferior colliculus, 311–312
Inferior temporal cortex, 345, 347–349
Infragranular layers (of cerebral cortex), 118
Infundibular stalk, 95
Inhibition, 8, 9
    tonic, 56, 208
Inhibitory postsynaptic potentials (IPSPs), 53, 55, 56, 64, 70
Initial segment, of axons, 7
Injury
    glia isolation of site, 399
    neuronal response to, 397–398
Inner hair cells, 305, 307–308, 309
Inner plexiform layer, 318
Inner segment of phototransducers, 318
Inositol triphosphate (IP3), 65, 66
In situ hybridization, 35b
Insomnia, 161

Insular cortex, 109
Intention tremor, 237
Interleukin-1, 280
Intermediate zone, spinal, 89, 93, 134
Internal arcuate axons, 147, 416f
Internal capsule, 86, 110f, 111
Internal carotid arteries, 411
Internal cerebral veins, 413
Internal granular layer, 117
Internal jugular veins, 413
Internal medullary lamina, 115–116
Interneurons, 9–10, 11, 12
    of *Aplysia*, 360
    arm movement and, 219, 220
    local circuit, 169–170
    Renshaw cells, 160
    type Ia. *See* Type Ia axons
Internodes, 76
Internuclear neurons, 204
Interstripe zones, 339
Intrafusal muscle fibers, 168
Inverse myotatic reflex (tendon reflex), 192
Ion channel proteins, 21–23, 64
Ionopores, 22, 23
Ionotropic receptors, 49, 52–53
Ions
    resting membrane potential and, 17
    voltage-gated channel selectivity for specific, 23–24
IPSPs. *See* Inhibitory postsynaptic potentials
Isthmus, 418f

Kainic acid, 54
Katz, Bernard, 45
Kinesin, 40
Kinocilium, 198, 199
Knee-jerk reflex, 190–191

Labeled line principle, 126
Lamellipodia, 385
Laminin, 387, 404
Laminin-heparan sulfate proteoglycan complex, 404
Language, 352–354
Lateral aperture, 97
Lateral descending systems, 186–189
Lateral fissure, 109
Lateral hypothalamus, 276–277
Lateralization of function, 342, 352–353, 354–355
Lateral lemniscus, 310, 311
Lateral ventricles, 97
Lateral vestibulospinal tract, 183–186
Learning
    associative, 362, 368
    in chicks, 362–364
    neurobiological view of, 357–358
    one-trial, 362–363, 364f

simple forms of, 358–362
    synaptic connection modification and, 393
Lemniscal system, 146–147, 156
Length constant, 26–27
Lens, 314–315, 327b
Lesions, of central visual pathway, 326, 327–328b
Leucine zipper, 68
Ligand-gated ion channels. *See* Ionotropic receptors
Light, 320
    complex cells and, 332
    ganglion cells and, 322
    optics of eyeball and, 314–315
    retinal adaptation to, 321–322
    simple cells and, 331
Limbic system, 286, 288–289
    functional role of, 292–297
Lipid bilayer, 21, 22, 24
Lobotomy, 351–352
Lobules (of cerebellum), 226–228
Local circuit interneurons, 169–170
Localization of function, 341, 342, 352–353
Locus coeruleus, 235
Long-loop reflexes, 164, 190, 196
Long-term memory, 361, 362f
Long-term potentiation, 365–369
Lorente de Nó, Rafael, 118–119
Lower motoneuron syndrome, 179
Lumbar enlargements, 93
Lund, Ray, 401
Luteinizing hormone (LH), 280

McLean, Paul, 289
Macrophages
    injury and, 399
    microglia as, 79–80
Macula, 197, 199, 200f
Magnetic resonance (MR), 223b, 350
Magnocellular zone (of reticular formation), 106, 196
Mammillotegmental tract, 278–279
Mammillothalamic tract, 278–279, 419f
Marginal zone, 89
Mechanoresponsive endings, 125
Medial descending systems, 186–189
Medial forebrain bundle, 276, 277, 278
Medial hypothalamus, 276
Medial lemniscus, 140, 147
Medial vestibular nuclei, 183, 210–211
Medial vestibulospinal tract, 183–186
Medial (vermal) zone, 231
Median aperture, 97
Medulla, 86, 95, 416f
    ontogeny of, 99–100
Membrane excitability, 61–64
Membrane permeability, 19–21
    ion channel proteins and, 21–23

Memory
in chicks, 362–364
clinicopathological insights into, 370
hippocampus and, 296, 364–369, 370
limbic system and, 295–297
long-term, 361, 362f
middle-term, 361, 362f
neurobiological view of, 357–358
short-term, 361, 362f
Mesencephalon, 85f, 86
caudal (midbrain), 418f
ontogeny of (midbrain), 102
rostral (midbrain), 419f
Metabotropic receptors, 49, 55, 60–61
Metencephalon, 85f, 86, 95
ontogeny of (pons), 100, 102
Methamphetamine, 352
Methionine-enkephalin, 34t
N-Methyl-D-aspartate (NMDA), 365–369, 393
1-Methyl-4-phenyl-1,2,3,6-tetrahydropyridine
(MPTP), 243–245
Meyer's loop, 328b
Microelectrodes, 18b
Microglia, 79–80
Microsaccades, 207
Microtubules, 40–41
Midbrain. See Mesencephalon
Midbrain enlargements, 88
Middle cerebellar peduncles, 102, 228
Middle cerebral artery, 411, 412
Middle-term memory, 361, 362f
Milk-ejection reflex, 283–284
Mitoses, 88–89
Mitotic ventricular zone, 101f, 104f, 110f
Mitral cells, 258
Molecular layer, 117, 228, 230, 290
Monoamine oxidase (MAO), 42–44
Monoamine receptors, 60
Monoaminergic neurons, 35b
Monoaminergic tracts, 188f
Monoclonal antibodies, 35b, 405
Mossy fibers, 229, 232, 234
Motoneurons, 3, 10, 11, 12, 13, 164
alpha. See Alpha motoneurons
of Aplysia, 360
eye movements and, 204
flexor, 201, 217
gamma. See Gamma motoneurons
local circuit interneurons and, 169–170
synapses of, 8–9
types and functions of, 166–168
Motor nuclei
brain stem, 175–178
column arrangement of, 172–173
distal, 173
ocular, 203–204, 207

spinal cord, 173–175
Motor pools, 93, 174–175
Motor units, 167
Mountcastle, Vernon, 119, 145
Movement, 130, 163–171
accuracy in, 165–166
basal ganglia system and, 250–252
cerebellum and, 236–237
classes of and levels of control in, 163–165
of neuroblasts and glioblasts, 373–374
precision in, 165–166, 236–237
volitional, 214–225, 236
MPTP. See 1-Methyl-4-phenyl-1,2,3,6-tetrahydropyridine
Multiform layer, 117–118
Multiple sclerosis, 77, 209
Muscarinic acetylcholine receptors (mACh-R), 60, 64, 268
imaging of, 223b
pharmacology of, 50b
visceromotor neurons and, 266, 271
Muscle spindle organ, 168–169
Muscle tone, 195–196, 197f
Myelencephalon, 85f, 86, 95
ontogeny of (medulla), 99–100
Myelinated axons, 7
conduction in, 30–31
glial cells and, 76–78
regeneration and, 404–405
sensory information and, 129
Myelin basic protein (MBP), 77
Myelomeningocele, 88
Myenteric plexus, 272–274
Myesthenia gravis, 42
Myosin, 171, 374, 385

Narcolepsy, 161
Nauta, Walle J., 2b
Nauta technique, 2b
Negative feedback inhibition, 36
Neglect syndromes, 225, 346
Nernst equation, 19, 20b
Nernst equilibrium potential, 21
Nerve growth factor (NGF), 381–382, 383
regeneration and, 403–404
synapse formation and, 390–392
Nervous system phylogeny, 1–4
Neural cell adhesion molecule (NCAM), 377, 387–388, 389, 404
Neural crest cells, 87–88, 266, 268
Neural-hemal organs, 282
Neural tube, 87–88, 389–390
Neuroanatomical techniques, 2–3b
Neuroblasts, 72, 373–374
Neuroendocrine transduction, 284
Neurogenesis, 88–89

Neurohormones, 283
Neurokinin, 34t
Neuromodulators, 34
Neuromuscular junction, 45
   neural cell adhesion molecule at, 388
   nicotinic acetylcholine receptor at, 49–51, 170
   role of in movement, 170–171
Neuronal membranes, 17–31
   current flow and, 24–28
   permeability of. *See* Membrane permeability
   restlessness of, 21
Neuron doctrine, 5
Neuron-glial cell adhesion molecule, 376
Neurons
   of *Aplysia*, 359–360
   archetypal, 6–8
   bipolar, 318, 320, 321, 322f
   burst, 208, 218, 310
   corticospinal, 180–182, 216–217, 219–220
   dopamine. *See* Dopamine neurons
   dynamic, 218
   GABAergic. *See* Gamma-aminobutyric acid cells
   glial cells distinguished from, 74–75
   injury and, 397–398
   inter-. *See* Interneurons
   internuclear, 204
   magnocellular neurosecretory, 283–284
   migration of, 376–377
   monoaminergic, 35b
   motor. *See* Motoneurons
   noradrenergic, 186
   of pallidum, 247
   pause, 208
   phenotypes of, 35b, 382–383
   phylogeny of, 1–4
   postsynaptic, 5–6, 8
   preganglionic parasympathetic, 175
   preganglionic sympathetic, 175
   presynaptic, 5
   of primary motor cortex, 220
   of primary somatosensory area, 151
   Purkinje. *See* Purkinje cells
   pyramidal. *See* Pyramidal cells
   response repertoire of, 60–71
   of retina, 316–318
   sensory. *See* Sensory neurons
   serotonergic, 35b, 106, 161, 186
   static, 218
   striatal, 246–247
   synaptic, 5
   tonic, 218
   total number in central nervous system, 10–11
   visceromotor, 166, 175, 265–266, 271
Neuropeptide tyrosine, 34t
Neurotensin, 34t

Neurotransmission
   far-ranging effects on cell function, 67
   genes encoding products of, 70
Neurotransmitter receptors
   categorization of, 49
   pharmacology of, 50b
Neurotransmitter release, 41, 45–46
   action potentials and, 58
   calcium in, 46–47, 48f
Neurotransmitters, 5–6, 7–8, 31, 32–44
   amino acid. *See* Amino acid neurotransmitters
   astrocytes and, 79
   biogenic amine, 32–33, 34, 39, 70
   classification of, 32–34
   criteria for, 32
   deactivation of, 42–44
   fast-acting, 34
   packaging and transport of, 40–41
   peptide. *See* Peptide cotransmitters
   production and replenishment of, 34–40
   slow-acting, 34
Neurotrophic factors, 390–392. *See also* Trophic
     molecules
Neurotrophin-3 (NT3), 382
Neurotrophin-4 (NT4), 382
NgCAM, 387, 404
NGF. *See* Nerve growth factor
Nicotinic acetylcholine receptors (nACh-R)
   behavior at neuromuscular junction, 49–51, 170
   ionotropic receptors compared with, 52
   pharmacology of, 50b
   subunits of, 49
   visceromotor neurons and, 266
Night blindness, 320, 327b
Night vision, 316
Nissl stains, 2b
Nissl substance, 2b, 7, 397
Nitric oxide synthase, 369
NMDA. *See* N-Methyl-D-aspartate
Nocisensory modality, 126, 143
Node of Ranvier, 76
Nondecremental conduction, 7
Noradrenergic neurons, 186
Norepinephrin, 33, 61, 107, 383
   cerebellum and, 235
   cotransmitters for, 34t
   deactivation of, 42
   identification of in neurons, 35b
   sympathetic ganglia and, 267–268
   synthesis of, 36
Nuclear regulatory proteins, 67
Nucleus, 12–13
   abducens, 176, 203(205
   accumbens, 239f, 240, 241
   ambiguus, 177
   anterior olfactory, 259

anterior thalamic, 116, 290–291, 293f
arcuate, 282
axial motor, 176
basolateral (of amygdala), 292, 294f
caudate, 110, 111, 239, 241
central (of amygdala), 292, 294f
centromedian, 116
cochlear, 135, 309–311
corticomedial (of amygdala), 292, 294f
cuneate, 136, 138–140, 147, 416f
deep cerebellar, 230, 231, 232, 235–236
dentate, 228
distal motor, 173
dorsal (of Clarke), 134–135, 233–234
dorsal column, 100, 136–140
Edinger-Westphal, 177, 212, 401
external cuneate, 135, 136, 140, 234
facial motor, 177
fastigial, 228, 235
gigantocellular reticular, 186
gracile, 136, 147, 416f
habenular, 279
histogenesis of, 373–384
hypoglossal, 176
inferior olivary, 104, 233f, 235
inferior salivary, 178
inferior vestibular, 183
intermediolateral, 93, 175, 268
interposed, 228
intralaminar, 145, 156 157
lateral geniculate, 160, 323–325, 326, 338, 393, 419f
lateral reticular, 235
lateral vestibular, 183
laterodorsal, 293f
lateroposterior, 116
lentiform, 239
locus coeruleus, 107, 186
magnocellular secretory, 281–284
mammillary (bodies), 276
medial geniculate, 116, 311–312
medial vestibular, 183, 210–211
mediodorsal, 116
mesencephalic trigeminal, 140, 142
ocular motor, 203–204, 207
oculomotor, 176, 203–204
of the lateral lemniscus, 311, 418f
of the solitary tract, 135 260f, 261, 295
parabrachial, 262, 418f
paraventricular, 281
parvicellular secretory, 282
pedunculopontine, 248–249
pontine, 101f, 232, 233f, 234
posterior, 145
posteromarginal, 134
precerebellar, 235
pretectal, 212, 324

principal trieminal, 140–142
proprius, 134
pulvinar, 116
putamen, 110, 111, 239, 241, 242
raphé, 106–107, 161, 186, 235
red, 102, 104–105, 182–183, 214, 228, 236
reticular (of thalamus), 116
sensory, 135–136, 137f
somatosensory, 133–135, 136
spinal trigeminal, 100, 140–142, 148–149
subthalamic, 240f, 249–250
superior olivary, 308, 310–311
superior salivary, 178
superior vestibular, 183, 210–211
suprachiasmatic, 162, 280–281, 324
supraoptic, 281
thalamic, 113,196116
trigeminal, 140–142
trigeminal motor, 142, 177
trochlear, 176, 203–204
tuberal, 282
vagal motor, 177,196178
ventral anterior, 116, 248, 250
ventral forebrain, 86, 110
ventral lateral 116, 248, 250
ventral posterolateral, 116, 145, 147, 148, 149, 149f
    oral part of, 235–236
ventral posteromedial, 148, 149, 260f, 261
vestibular, 135, 183, 189, 201
visceromotor, 173
viscerosensory, 135
Nystagmus
    optokinetic, 209
    postrotatory, 212
    vestibular, 211

O2A progenitor, 383–384
Occipital cortex, 160, 379
Occipital eye field, 209
Occipital lobe, 113, 329, 346
Ocular dominance columns, 335–336, 392–393
Oculomotor nerve, 212
Ohm's law, 20b
Olfactory bulbs, 110
Olfactory cortex, 241
Olfactory tract, 259
Olfactory transducer cells, 255–256
Olfactory tubercle, 239f, 240, 241
Oligodendrocyte-derived proteins, 405
Oligodendrocytes, 72–73, 76, 77–78
    markers of, 75
    phenotype determinants in, 383–384
Oligodeoxynucleotide probes, 35b
One-trial learning, 362–363, 364f
Ontogeny
    of the brain stem, 95–107

Ontogeny (*continued*)
  of the forebrain, 108–119
  of the medulla, 99–100
  of the mesencephalon (midbrain), 102
  of the pons, 100–102
  of the reticular formation, 102–106
  of the spinal cord, 89–92
  of the telencephalon, 108–111
Opsin, 60, 318–319
Optic chiasm, 324
Optic cups, 95
Optic disc, 316
Optic nerve, 314, 316, 324, 403–404
Optic tracts, 95
Optic vesicles, 95
Organ of Corti, 302, 304, 305, 309
Otitis externa (swimmer's ear), 304
Otitis media, 304
Outer hair cells, 305, 308
Outer plexiform layer, 318
Outer segment of phototransducers, 318
Oxytocin, 281–282, 284

Pacinian corpuscles, 124–125
Pain, 152
  anterolateral system and, 144–146
  spinal trigeminal nucleus and, 148
Papez, James, 288, 291
Papez circuit, 290–291
Papilledemas, 327b
Parahippocampal gyrus, 288, 291
Parallel fibers, 229, 236
Parasympathetic ganglia, 88, 267–268
Parietal cortex, 224–225, 351
Parietal lobe, 113
Parieto-occipital sulcus, 113
Parietotemporal association cortex, 343–344
Parkinson's disease, 71, 194, 238, 245, 352, 401
Parvicellular reticular formation, 193–194
Patch clamp method, 18b
Pattern codes, 128
Pavlovian conditioning, 368
Penfield, Wilder, 215
Peptide cotransmitters, 32–33, 71
  common types of, 34t
  genes influenced by, 70
  release of, 47–48
  somatosensory axons and, 133
  synthesis of, 39–40
  visceromotor neurons and, 265
Perception, 123, 130
  serial cortical processing and, 343–344
  of space, 344–346
Perforant path, 290
Periaqueductal gray substance, 102
Periglomerular cells, 258

Perilymph, 199, 306
Peripheral nerve grafts, 399–400
Peripheral neuropathy, 126
Peripheral visceromotor system, 266–268
Perseveration, 350
Personality, 351–352
PGO spikes, 160–161
Phasic transducers, 129
Phenotypes
  glial, 383–384
  neuronal, 35b, 382–383
Phenylethanolamine-*N*-methyltransferase (PNMT),
    36, 37f
Pheromones, 264
Phospholipase C, 65
Phototransducers, 315, 316, 318–320, 321–322
Phylogeny, of neurons and nervous systems, 1–4
Pia mater, 83
Pineal gland, 97
Piriform cortex, 117, 259–260
Pituitary gland, 95
  hypothalamus relationship to, 281–285
Platelet-derived growth factor (PDGF), 384
Poliomyelitis, 167
Pons, 95, 99, 103f, 417f
  ontogeny of, 100–102
  sleep and, 160
  ventral, 86
Pontine arteries, 411
Pontine flexure, 95, 99
Pontine tegmentum, 86, 101f, 102, 196
Pontocerebellum, 232, 234
Population codes, 128, 309
Positron emission tomography (PET), 223b, 341, 345, 353
Postcentral gyrus, 116
Posterior cerebral artery, 412, 413
Posterior communicating arteries, 412
Posterior inferior cerebellar arteries (PICA), 411
Postsynaptic neurons, 5–6, 8
Postsynaptic potentials, 53–55, 154
Posture, 195–196, 197f
Potassium
  astrocytes and, 78
  ATPase of. *See* Sodium-potassium adenosine
    triphosphatase
  nicotinic acetylcholine receptors and, 50
  postsynaptic potentials and, 54–55
  resting membrane potential and, 18, 19–21
  sodium exchange with, 78
Potassium channels. *See also* Voltage-gated potassium
    channels
  in *Aplysia*, 360
  hair cells and, 305, 306
  presynaptic excitation and, 58
  presynaptic inhibition and, 56
  visual transduction and, 318, 320

Potency, of synapses, 9
Precentral gyrus, 214, 261
Precision in movement, 165–166, 236–237
Prediction, 349–351
Prefrontal cortex
    in behavioral strategy planning, 349–351
    personality and, 351–352
    schizophrenia and, 352
Prefrontal lobotomy, 351–352
Premigratory zone, 375
Premotor area, 214, 220–224
Premotor cortex, 179, 214, 236, 251, 351
Preoptic area, 162, 276, 277, 279–280
Presynaptic excitation, 56–58
Presynaptic inhibition, 56–58
Presynaptic membranes, 8, 58
Presynaptic neurons, 5
Primary auditory cortex, 113, 311, 312–313
Primary gustatory cortex, 261–262
Primary motor cortex, 113, 215–218, 221f, 341, 350–351
    red nucleus and, 218–220
Primary somatosensory cortex, 113, 116, 149f, 150–152, 379–380
    representational maps in, 394–395
    volitional movement and, 220
Primary somatosensory processing, 131–142
Primary visual cortex, 325, 328, 339, 344, 379, 392
    basic circuit of, 329–330
    blobs in, 331, 334–337
    geniculate channel segregation in, 330–334
    slabs in, 334–337
Programmed cell death, 380–382
Projection fibers, 111
Prolactin inhibitory factor, 282
Propeptides, 39–40
Propriosensory modality, 126, 143
Propriospinal axons, 93
Protein kinase A (PKA), 68, 360
Protein kinase C (PKC), 68
Protein zero (P0), 77
Proteolipid protein (PLP), 77
Protoplasmic cells, 73, 383–384
Pseudopodium, 373–374
Psychic blindness, 347
Psychomotor epilepsy, 295
Pupillary reflex, 401–402
Purkinje cells, 228–229, 230–231, 232, 235, 236
    cell adhesion molecules and, 375, 376
    phenotype determinants in, 382
Putamen, 110, 111, 239, 241, 242
Pyramidal cells, 117, 179
    electroencephalogram recordings of, 154
    phenotype determinants in, 382
    in primary visual cortex, 329–330
    volitional movement and, 214, 216
Pyramidal fibers, 179–180

Pyramidal tract, 100, 111, 416f
    structure and functions of, 179–182
    volitional movement and, 214

Radial glia, 73, 374
Rapid eye movement (REM) sleep, 160–162
Reactive gliosis, 399
Receptive fields, 127
Receptor proteins, 32
Reciprocal inhibition, 191
Recurrent inhibition, 169
Reflexes, 163–164, 166
    accommodation, 212
    baro, 269, 270–271, 272f
    bladder, 269, 270
    bulbar, 193–194
    compound, 193
    crossed-extensor, 193
    knee-jerk, 190–191
    long-loop, 164, 190, 196
    milk-ejection, 283–284
    tendon (inverse myotatic), 192
    vestibulo-oculomotor, 204, 210–212
    visceromotor, 265–274
Refractory period, 29
Regeneration
    failure of, 397–398, 399–403
    molecular requirements for, 403–405
Release phenomena, 195
Renshaw cells, 169
Representational maps, 126
    alteration by experience, 394–395
Reproduction
    preoptic area and, 280
    smell, taste, and viscerosensation in, 262–264
Restiform body. See Inferior cerebellar peduncles
Resting membrane potential, 17–21, 54–55
    in *Aplysia*, 360
    biophysical equations for, 20b
    current flow and, 25, 27
    defined, 17
Reticular formation, 94, 99, 102, 149, 189
    arousal and, 145, 156–157
    baroreflex and, 271
    cerebellum and, 236
    descending systems from, 178–179
    dreaming and, 160–161
    gaze control and, 208–209
    lateral hypothalamus compared with, 277
    movement and, 164
    ocular motor nuclei and, 204
    ontogeny of, 102–106
    parvicellular, 193–194
    in posture and muscle tone, 196, 197f
    sleep and, 157–160
    zones of, 106–107

Reticular gaze control areas, 208–209
Reticulospinal tracts, 187, 188f
  posture and muscle tone controlled by, 195–196
  structure and functions of, 186
Retina, 95, 315, 329
  adaptation to light, 321–322
  axons of, 392–393
  processing circuits of, 320–321
  pupillary reflex and, 401–402
  structure of, 316–318
Retinal (vitamin), 319, 320
Retinal ganglion cells, 318, 320, 321, 382, 387, 404–405
  as contrast or change detectors in vision, 322–323
Retinitis pigmentosa, 327b
Retinogeniculostriate pathway, 324
Retraction, axonal, 394
Retrograde amnesia, 370
Retrograde transport, 40–41, 398
Rexed, Bror, 93
Rhinencephalon, 288
Rhodopsin, 317f, 318–319
Rhombic (hindbrain) enlargements, 88
Rhomboid fossa. *See* Fourth ventricle
Rhombomeres, 88
Rhythmic movements, 163–165, 166, 194–195
Robinson, David Lee, 345
Rods, 316, 317f, 318–319, 321, 327b
Roof plate, of neural tube, 389–390
Rose, Steven P.R., 363
Rubrospinal tract, 186, 219–220
  structure and functions of, 182–183

Saccadic eye movements, 204–205, 218, 249
  blindness and, 346
  brain stem and, 208–209
  superior colliculi and, 205–207
Saccule, 197, 199, 200f, 210
Saltatory conduction, 31
Satellite cells, 72
Schizophrenia, 350, 352
Schwann cells, 72–73, 76, 77–78, 404
Scotomas, 328b, 346. *See also* Blind spots
Secondary sensory fibers, 143
Secondary somatosensory cortex, 152
Secondary visual cortex, 329, 338–339, 344
Second messenger systems, 65–66, 70, 361, 368
Segmental innervation, 132b
Seizures, 289
Semicircular canals, 197, 210
Sensitization, 360–361
Sensory axons, 10, 129. *See also* Sensory neurons
  type Ia, 168–169, 191
  type Ib, 192
  type II, 168–169
Sensory coding, 128–129
Sensory cranial nerves, 137b

Sensory evoked potentials, 156
Sensory neurons, 3, 10, 13. *See also* Sensory axons
  of *Aplysia*, 360
  synapses of, 8–9
Sensory transduction
  mechanisms of, 127
  three main purposes of, 130
  transducer location and, 126–127
  transducer specializations and, 123–126
Septal veins, 413
Septohypothalamomesencephalic continuum, 277, 292–294
Septum, 111
Serial cortical processing, 343–344
Serotonergic neurons, 35b, 106, 161, 186
Serotonin (5-HT), 33, 67
  in *Aplysia*, 360
  astrocytes and, 79
  cerebellum and, 235
  cotransmitters for, 34t
  deactivation of, 42
  dorsal horn and, 135
  genes influenced by, 70
  identification of in neurons, 35b
  striatum and, 243
  synthesis of, 36, 38f
Sherrington, Charles, 5, 18b, 169–170, 214
Short-term memory, 361, 362f
Shunting, 54, 55
Signal transduction, 49
  G proteins in, 61–64
  multiple interactive pathways of, 64–66
Simple cells, 331–333, 335
Single photon emission computed tomography (SPECT), 223b
Slabs, 334–337, 339
Sleep
  dreaming in, 160–161
  electroencephalogram recordings of, 154–156, 157–159, 160
  importance of, 161–162
  rapid eye movement, 160–162
  reticular formation and, 157–160
Sleep apnea, 161
Sleep-waking cycle, 162
Slow-wave sleep, 157–159, 160
Smell, 123
  behaviors guided by, 262–264
  central olfactory pathways for, 258–260
  chemosensors and, 255–258
Smooth pursuit eye movements, 204, 209
Socialization, 351–352
Sodium
  hair cells and, 198
  movement and, 171
  neurotransmitter deactivation and, 42

nicotinic acetylcholine receptors and, 50, 51
postsynaptic potentials and, 54, 55
resting membrane potential and, 17, 19–21
sensory transduction and, 127
Sodium-calcium exchange, 46
Sodium channels. *See also* Voltage-gated sodium
    channels
smell and, 255
taste and, 258
visual transduction and, 318, 319, 320, 322
Sodium-potassium adenosine triphosphatase (ATPase)
astrocytes and, 78
in myelinated axons, 31
resting membrane potential and, 18–19, 21
visual transduction and, 318
Sodium-potassium exchange, 78
Soma. *See* Cell bodies
Somatic sensations, 140–142
Somatosensory axons, 131–133
Somatosensory discrimination, 136–140
Somatosensory systems
ascending, 143–152
primary processing in, 131–142
Somatosensory thalamocortical system, 149–152
Somatostatin, 34t, 247, 282
Sound
amplitude of, 302, 308–309
conduction to transducer organ, 302–304
defined, 301–302
frequency of, 302, 308–309
Space, perception of, 344–346
Spatial summation, 55
Specific hungers, 264
Speech, 301
Sperry, Roger, 354
Spikes. *See* Action potentials
Spina bifida, 88
Spinal border cells, 234
Spinal cord, 10, 13, 83–94
cerebellum and, 233f
mature, 93–94
motor nuclei of, 173–175
neurogenesis in, 88–89
ontogeny of, 89–92
overview of, 11–12
rhythmic movements and, 165
somatosensory nuclei in, 133–135
Spinal trigeminal tract, 140
Spinocerebellum, 231–234
Spinoreticulothalamic pathways, 156
Spiral ganglion, 307
Split-brain patients, 354–355
Sprouting, axonal, 394
Staining techniques, 2–3b
Stellate cells, 117, 228, 230–231, 310, 326, 329–330, 331,
    375

Stem cells, 72
Stereocilia, 197–201, 305, 306
Stereopsis, 212, 337
Stimuli, 126
duration, 129
intensity, 128
Stratum opticum, 205
Stretch reflex, 190–192
Stria medullaris, 278–279, 419f
Striate arteries, 411
Stria terminalis, 277, 278
Striatum. *See* Corpus striatum
Stria vascularis, 304
Striola, 200f, 201
Stroke, 150–151, 195, 209, 342
neglect syndromes and, 225, 346
perception deficits and, 345–346
Subiculum, 277, 290
Submucosal plexus, 272–274
Substance P, 34t, 70, 134, 247
Substantia gelatinosa, 134
Substantia nigra, 102, 247–248
Sulcus, 84. *See also specific sulcuses*
Sulcus limitans, 90, 99, 100, 108, 175
Superior cerebellar arteries, 411
Superior cerebellar peduncles, 102, 228, 418f
Superior colliculi, 102, 189, 323–324
deep layers of, 206–207, 249, 312
descending systems from, 178–179
saccidic eye movements and, 205–207
superficial layers of, 206
Superior laryngeal nerve, 261
Superior parietal lobule, 224–225, 344–346
Superior sagittal sinus, 413
Supplementary motor area, 214, 220–224, 248
Sympathetic ganglia, 88, 267–268
Sympathetic nervous system, 269
Synapses, 4–6. *See also* Synaptic transmission
connections of, 8–10
discovery of, 4–5
maintenance and modification of connections in,
    393–394
neurotrophic factors in formation of, 390–392
potency of, 9
selection of during development, 392–393
significance of differences in, 55–56
of somatosensory axons, 131–133
Synapsin I, 49
Synaptic cleft, 79
Synaptic transmission, 8, 31, 45–58. *See also*
    Synapses
calcium dependence in, 46–47
dendritic spine in, 58–59
genes influenced by, 70
gene transcription induced by, 67
postsynaptic potentials in, 53–55, 154

Synaptic transmission (*continued*)
  presynaptic phase of, 45–46
  synapse differences and, 55–56
Synaptoporin, 47
Synaptotagmin, 46–47

Tabes dorsalis, 134, 152
Taste, 123, 124
  behaviors guided by, 262–264
  central pathways for, 260–262
  chemosensors and, 255–258
Taste transducer cells, 256–258
Tectospinal tract, 183, 187
Tectum, 102
Tegmentum, 102
Telencephalic vesicles, 95, 108, 109f, 111
Telencephalon, 85f, 86
  ontogeny of, 108–111
Temperature sensation, 152
  anterolateral system and, 144–146
  hypothalamus regulation of, 279–280
  spinal trigeminal nucleus and, 149
Temporal lobe, 113, 344
Temporal summation, 55
Tenascin, 376
Tendon reflex (inverse myotatic reflex), 192
Teuber, Hans-Lukas, 285, 351
Thalamic radiations, 149
Thalamocortical projections, 113–115
Thalamostriate vein, 413
Thalamus, 86, 97, 108, 110f, 228, 419f
  arousal and, 156
  central olfactory pathways and, 260
  nuclear organization of, 113–116
  striatal input and, 242
  volitional movement and, 220, 221f, 222
Third ventricle, 97, 108
Threshold, 30, 127
Thrombospondin, 376, 387
Thyrotropin-releasing hormone, 282
Tinnitus, 304
Tract of Lissauer (dorsolateral fasciculus), 134
Tracts, 13
Transducers
  hearing, 302–304
  olfactory, 255–256
  phasic, 129
  photo. *See* Phototransducers
  sensory. *See* Sensory transduction
  specificity of, 125
  taste, 256–258
  tonic, 129
  visual. *See* Visual transduction
Transducin, 319
Transporter proteins, 42
Trapezoid body, 310

Tricyclic antidepressants, 42
Trigeminal motor nerve, 303
Trigeminocerebellar tract, 234
Trigeminothalamic system, 148–149
Trophic molecules, 79, 387, 390. *See also* Neurotrophic
  factors
Troponin, 171
Tryptophan hydroxylase, 36
Tuberoinfundibular tract, 282
Twitch, 171
Tympanic membrane (eardrum), 302
Tyrosine hydroxylase, 35b, 36, 70
Tyrosine kinase, 247

Upper motoneuron syndrome, 179
Utricle, 197, 199, 200f, 210

Vasopressin, 34t, 281–282, 286
Ventral amygdalofugal pathway, 277
Ventral anterior-ventral lateral (VA-VL) complex, 116,
  248, 250
  deep cerebellar nuclei output to, 235–236
  volitional movement and, 222
Ventral horns, 93
Ventral pallidum, 240
Ventral pons, 86
Ventral spinocerebellar tract, 234
Ventral striatum, 240, 241
Ventricular system, 83, 97
Ventricular zone, 89
Vergence eye movements, 204, 212, 337
Vermis, 226, 235
Vertebral arteries, 411
Vesicles, 8, 45–46
  calcium-dependent exocytosis in, 46–47
  peptide cotransmitters in, 47–48
  preparation of for release, 48–49
Vestibular nuclei, 135, 183, 189, 201
  cerebellum and, 231, 232, 233f, 236
  descending systems from, 178–179
  ocular motor nuclei and, 204
Vestibulocerebellum, 231–232
Vestibulocochlear nerve. *See* Eighth cranial nerve
Vestibulo-oculomotor reflex, 204, 210–212
Vestibulospinal tracts, 187
  reflexes and, 197–201
  structure and functions of, 183–186
Visceral sensation, 123
Visceromotor neurons, 166, 175, 265–266, 271
Visceromotor reflexes, 265–274
Viscerosensations, 126
  behaviors guided by, 262–264
  central pathway for, 260–262
Vision, 123. *See also* Gaze; Visual processing; Visual
  transduction perception of space and, 344–346

Visual cortex, 113, 160, 346. *See also* Primary visual
    cortex; Secondary visual cortex
Visual processing, 329–340
  color-coding in, 338–339
  cortical areas involved in, 339–340
Visual transduction, 314–328
  central pathways for, 323–328
  lesions and, 326, 327–328b
  optics of eyeball in, 314–315
Vitamin A deficiency, 320
Volitional movement, 214–225, 236
Voltage clamp method, 18b
Voltage-gated calcium channels, 41, 46, 48f, 171, 360
Voltage-gated channels, 22, 23–24. *See also specific types*
Voltage-gated potassium channels, 55–56
  action potentials and, 29, 30
  astrocytes and, 78
  current flow and, 27
  depolarization and, 27, 56
  hair cells and, 198–199

  membrane permeability and, 22, 23
  in myelinated axons, 31
  selectivity for specific ions, 23–24
Voltage-gated sodium channels, 55–56
  action potentials and, 29, 30
  depolarization and, 24–27, 56
  membrane permeability and, 22–23
  in myelinated axons, 31
  selectivity for specific ions, 23–24

Weisel, Torsten, 119
Wernicke, Carl, 353
Wernicke-Korsakoff syndrome, 296
Wernicke's aphasia, 353
Wernicke's area, 353–354
White matter, 12, 13, 84, 86, 93, 100
Wiesel, Torsten, 329, 335, 354, 393
Withdrawal reflex, 166, 193, 359, 361f
Wurtz, Robert, 249